Functional Lubricating Materials

Functional Lubricating Materials

Editor

Jiusheng Li

Basel • Beijing • Wuhan • Barcelona • Belgrade • Novi Sad • Cluj • Manchester

Editor
Jiusheng Li
Laboratory for Advanced
Lubricating Materials
Shanghai Advanced Research
Institute, Chinese Academy
of Sciences
Shanghai
China

Editorial Office
MDPI AG
Grosspeteranlage 5
4052 Basel, Switzerland

This is a reprint of articles from the Special Issue published online in the open access journal *Lubricants* (ISSN 2075-4442) (available at: www.mdpi.com/journal/lubricants/special_issues/F111W3MY3E).

For citation purposes, cite each article independently as indicated on the article page online and as indicated below:

Lastname, A.A.; Lastname, B.B. Article Title. *Journal Name* **Year**, *Volume Number*, Page Range.

ISBN 978-3-7258-1402-2 (Hbk)
ISBN 978-3-7258-1401-5 (PDF)
doi.org/10.3390/books978-3-7258-1401-5

© 2024 by the authors. Articles in this book are Open Access and distributed under the Creative Commons Attribution (CC BY) license. The book as a whole is distributed by MDPI under the terms and conditions of the Creative Commons Attribution-NonCommercial-NoDerivs (CC BY-NC-ND) license.

Contents

About the Editor . . . . **vii**

Preface . . . . **ix**

Xiaomei Xu, Fan Yang, Hongmei Yang, Yanan Zhao, Xiuli Sun and Yong Tang
Preparation and Tribological Behaviors of Sulfur- and Phosphorus-Free Organic Friction Modifier of Amide–Ester Type
Reprinted from: *Lubricants* **2024**, *12*, 196, doi:10.3390/lubricants12060196 . . . . **1**

Sung-Jun Lee, Yoonchul Sohn, Dawit Zenebe Segu and Chang-Lae Kim
An Evaluation of the Tribological Characteristics of Diaphragm Plates for High-Pressure Hydrogen Gas Compressor Applications
Reprinted from: *Lubricants* **2023**, *11*, 411, doi:10.3390/lubricants11090411 . . . . **13**

Xufei Wang, Shuguang Fan, Ningning Song, Laigui Yu, Yujuan Zhang and Shengmao Zhang
Effect of Copper Nanoparticles Surface-Capped by Dialkyl Dithiophosphate on Different Base Oil Viscosity
Reprinted from: *Lubricants* **2024**, *12*, 137, doi:10.3390/ubricants12040137 . . . . **30**

Jose Jaime Taha-Tijerina, Karla Aviña, Victoria Padilla-Gainza and Aditya Akundi
Halloysite Reinforced Natural Esters for Energy Applications
Reprinted from: *Lubricants* **2023**, *11*, 65, doi:10.3390/lubricants11020065 . . . . **44**

Yaping Xing, Ebo Liu, Bailin Ren, Lisha Liu, Zhiguo Liu, Bocheng Zhu, Xiaotian Wang, Zhengfeng Jia, Weifang Han and Yungang Bai
Preparation and Tribological Behavior of Nitrogen-Doped Willow Catkins/MoS_2 Nanocomposites as Lubricant Additives in Liquid Paraffin
Reprinted from: *Lubricants* **2023**, *11*, 524, doi:10.3390/lubricants11120524 . . . . **56**

Can Zhu, Zhongyi He, Liping Xiong, Jiusheng Li, Yinglei Wu and Lili Li
Study on the Influence of the MoS_2 Addition Method on the Tribological and Corrosion Properties of Greases
Reprinted from: *Lubricants* **2023**, *11*, 517, doi:10.3390/lubricants11120517 . . . . **71**

Shaokun Jia, Jiahuan Zhao, Guangzhen Hao, Jifeng Feng, Chuanbo Zhang, Zhihui Wang, et al.
Preparation and Tribological Behavior of Nitrogen-Doped Carbon Nanotube/Ag Nanocomposites as Lubricant Additives
Reprinted from: *Lubricants* **2023**, *11*, 443, doi:10.3390/lubricants11100443 . . . . **87**

Ge Du, Hongmei Yang, Xiuli Sun and Yong Tang
Tribological Behavior and Wear Protection Ability of Graphene Additives in Synthetic Hydrocarbon Base Stocks
Reprinted from: *Lubricants* **2023**, *11*, 200, doi:10.3390/lubricants11050200 . . . . **101**

Yumeng Wang, Ning Shi, Min Liu, Sheng Han and Jincan Yan
Enhanced Thermally Conductive Silicone Grease by Modified Boron Nitride
Reprinted from: *Lubricants* **2023**, *11*, 198, doi:10.3390/lubricants11050198 . . . . **113**

Jinjun Lu, Rong Qu, Fuyan Liu, Tao Wang, Qinglun Che, Yanan Qiao and Ruiqing Yao
Tribological Property of Al_3BC_3 Ceramic: A Lightweight Material
Reprinted from: *Lubricants* **2023**, *11*, 492, doi:10.3390/lubricants11110492 . . . . **127**

Meirong Yi, Taoping Wang, Zizheng Liu, Jin Lei, Jiaxun Qiu and Wenhu Xu
Tribological Performance of Steel/W-DLC and W-DLC/W-DLC in a Solid–Liquid Lubrication System Additivated with Ultrathin MoS_2 Nanosheets
Reprinted from: *Lubricants* **2023**, *11*, 433, doi:10.3390/lubricants11100433 . . . . **134**

Guodong Huang, Tao Zhang, Yi Chen, Fei Yang, Huadong Huang and Yongwu Zhao
Graphite Fluoride as a Novel Solider Lubricant Additive for Ultra-High-Molecular-Weight Polyethylene Composites with Excellent Tribological Properties
Reprinted from: *Lubricants* **2023**, *11*, 403, doi:10.3390/lubricants11090403 . . . . **150**

About the Editor

Jiusheng Li

Professor Li, graduated from Shanghai JiaoTong University in 2002 and received his doctorate in Materials Science. Since 2012, he has been working as a full Professor and Doctoral supervisor at Shanghai Advanced Research Institute, China Academy of Science. Professor Li is mainly engages in the research and development of green synthesis processes for fine chemicals and high-end lubricating materials such as synthetic base oils, high-performance additives and formulations.

As the project leader, he has undertaken and completed more than 30 projects commissioned by the Chinese Academy of Sciences, the Ministry of Science and Technology, the Shanxi Provincial Department of Science and Technology and enterprises. He has published over 100 SCI papers, applied for more than 50 patents and authorized 28 projects.

Preface

Over the past decade, both education and economic development have reached a new level, and a series of breakthroughs have been made in lubricating materials. The current Special Issue will present the state of the art of functional lubricating materials. It aims to display a number of recent representative advances in basic and application research on lubricating base oil, lubricant additives, liquid lubricating materials, solid lubricating materials, polymer-based lubricating materials, intelligent lubricating materials, nanostructured functional polymers and functional metal matrix composites.

In this Special Issue, the authors have dedicated their time, energy and expertise to exploring and shedding light on various aspects of this subject, presenting insightful research, innovative perspectives and thought-provoking analyses. We are grateful for the dedication and hard work of each and every author who has contributed to this Special Issue, as well as the reviewers and editors who have helped bring these papers to fruition. Their commitment to excellence and their passion for their respective fields is truly commendable.

As we delve into the pages of this Special Issue, we invite readers to embark on a journey of discovery, reflection and inspiration. May the papers within this reprint spark new ideas, ignite curiosity and fuel further exploration and understanding of functional lubricating materials. It is our hope that this Special Issue will serve as a catalyst for continued dialogue, collaboration and progress in this vital area of study.

We are excited to present this collection of papers and eagerly look forward to the discussions and discoveries that they will inspire. Thank you for joining us on this journey of exploration and learning.

Jiusheng Li
Editor

Article

Preparation and Tribological Behaviors of Sulfur- and Phosphorus-Free Organic Friction Modifier of Amide–Ester Type

Xiaomei Xu [1,2], Fan Yang [1,2], Hongmei Yang [2,*], Yanan Zhao [2], Xiuli Sun [2] and Yong Tang [2]

1. School of Materials and Chemistry, University of Shanghai for Science and Technology, Shanghai 200093, China
2. State Key Laboratory of Organometallic Chemistry, Shanghai Institute of Organic Chemistry, Chinese Academy of Sciences, Shanghai 200032, China
* Correspondence: yanghm@sioc.ac.cn

Abstract: With the increasingly demanding engine conditions and the implementation of "double carbon" policies, the demand for high-quality lubricants that are cost-effective and environmentally friendly is increasing. Additives, especially high-performance friction modifiers, play an important role in boosting lubricant efficiency and fuel economy, so their developments are at the forefront of lubrication technologies. In this study, 1,3-dioleoamide-2-propyloleate (DOAPO), which incorporates polar amide, ester, and nonpolar alkyl chains, was synthesized from 1,3-diamino-2-propanol to give an eco-friendly organic friction modifier. Nuclear magnetic resonance (NMR), high-resolution mass spectrometry (HR-MS), Fourier-transform infrared spectroscopy (FT-IR), and thermogravimetric analysis (TGA) were used to characterize the structure and thermal stability of DOAPO. Meanwhile, the storage stability and tribological behaviors of DOAPO in synthetic base oil were studied and compared with a commercial oleamide. The results show that DOAPO has better thermal stability and better storage stability in synthetic base oil. Additionally, 0.5 wt.% of DOAPO could shorten the running-in period and reduce the average friction coefficient (ave. COF) and wear scar diameter (ave. WSD) by 8.2% and 16.2%, respectively. The worn surface analysis and theoretical calculation results show that the ester bond in DOAPO breaks preferentially during friction, which can reduce the interfacial shear force and easily react with metal surfaces to form iron oxide films, thus demonstrating a better friction-reducing and anti-wear performance.

Keywords: sulfur- and phosphorus-free; amide–ester; tribological behavior; synthetic base oil

Citation: Xu, X.; Yang, F.; Yang, H.; Zhao, Y.; Sun, X.; Tang, Y. Preparation and Tribological Behaviors of Sulfur- and Phosphorus-Free Organic Friction Modifier of Amide–Ester Type. *Lubricants* **2024**, *12*, 196. https://doi.org/10.3390/lubricants12060196

Received: 31 March 2024
Revised: 27 May 2024
Accepted: 28 May 2024
Published: 30 May 2024

Copyright: © 2024 by the authors. Licensee MDPI, Basel, Switzerland. This article is an open access article distributed under the terms and conditions of the Creative Commons Attribution (CC BY) license (https://creativecommons.org/licenses/by/4.0/).

1. Introduction

In recent years, energy conservation and emission reduction have become one of the most urgent challenges for the automobile industry. The pursuit of improved fuel efficiency and "dual carbon" goals emphasizes the growing trend toward the use of low-viscosity oils [1], which could minimize the shear resistance between friction counterparts [2–4]. However, the shift to low-viscosity lubricants carries a certain risk of wear resistance as the lubrication regime changes from a favorable hydrodynamic lubrication to a less favorable mixed or boundary lubrication for engines with more stringent operating conditions. Under the boundary lubrication state, the lubricating films of low-viscosity oils are thin and lack of strength, resulting in direct contact and making the films break during high-strength engine operations, which would increase friction and wear [5]. Therefore, friction modifiers, which can form thick boundary films under mixed or boundary lubrication regimes, were applied to reduce or prevent direct friction solid–solid contact on friction pairs [6–8].

In general, the friction modifications used in engine oil are metal friction modifications (such as didithiophosphate zinc (ZDDP), organic molybdenum [9,10], etc.) and non-metallic friction modifications (such as oleamide [11], glycerol monooleate [3,12,13], etc.). Metal friction improvement agents are mostly metal or metal compounds containing sulfur and phosphorus. Although they show excellent performance and are most widely used [14,15], the metal they contain will increase the ash content of lubricating oil, and the metal compounds containing sulfur and phosphorus can poison automobile catalysts used for emission control, causing adverse effects on the engine and the environment. Therefore, green non-metallic organic friction modifiers (OFMs) composed only of carbon, hydrogen, oxygen, and nitrogen atoms are attracting increasing attention [5,16,17].

OFMs tend to have amphiphilic structures, in which polar groups could adsorb onto metal surfaces, while nonpolar hydrocarbons arrange outward within the lubricant [18,19]. This arrangement establishes a hydrocarbon surface with low shear strength on metal surfaces. At present, the developed OFMs incorporate various polar functional groups, such as carboxyl [20–22], alcohol [23–25], amine [26,27], amide [5,28–30], and ester [31] functionalities. Biresaw [31] synthesized seven lipoic acid esters using various alcohols, and the study showed that the performance of thioic acid multifunctional additives in base oils is related to its structure. When the addition is 5 wt.%, 2-ethylhexyl thioctiocate and dodecyl thioctiocate with straight chains increased the kinematic viscosity at 40 °C from 40.8 mm^2/s to 78.7 mm^2/s and 69.5 mm^2/s, kinematic viscosity at 100 °C from 8.7 mm^2/s to 18.2 mm^2/s and 15.0 mm^2/s, the viscosity index from 200 to 253 and 229, showing a good viscosity improvement performance. Compared with the base oil, the addition of 20 wt.% lipoic acid ester makes the onset oxidation temperature and extreme pressure load increase from 187.2 °C to 218.4–221.5 °C, and 120 kgf to 420–480 kgf, respectively, showing a good anti-oxidation and extreme pressure performance.

Hou [5] prepared a novel organic friction modifier N-(2,2,6,6-tetramethyl-1-oxyl-4-piperidyl) dodecenamide (C_{12}Amide-TEMPO) and found it can form a unique double-layer boundary film on the iron oxide surface, i.e., the strong surface adsorption layer formed by chemical interactions between amide oxygen, free radicals, and iron oxide surfaces, as well as the interlayer hydrogen bond films formed by amide hydrogen and free radicals or oxygen. Meanwhile, the combination of intra-layer and inter-layer hydrogen bonds also increases the strength of the boundary film by enhancing cohesion, so C_{12}Amide-TEMPO is better than the traditional glyceryl monooleate (GMO) and stearic acid in terms of bearing capacity, friction reduction, and friction stability. Compared to GMO and stearic acid at an effective load of 5.0 N, C_{12}Amide-TEMPO demonstrates a more stable instantaneous friction coefficient (COF), with over 60% reduction in wear rate and surface roughness. The groove width and wear rate of wear scar lubricated with C_{12}Ester-TEMPO or C_{12}Amino-TEMPO is 569.0 μm, 544.0 μm and 461.2 $μm^3/(N·mm)$, 196.9 $μm^3/(N·mm)$, which is significantly higher than 365.0 μm and 42.2 $μm^3/(N·mm)$ that lubricated with C_{12}Amide-TEMPO. This indicates that C_{12}Amide-TEMPO with an amide-linked structure outperforms C_{12}Ester-TEMPO and C_{12}Amino-TEMPO in terms of friction-reducing and anti-wear properties. However, the long-term stability and durability study of these OFMs remains limited.

The reported studies show that ester- or amide-based compounds exhibit good performance in improving friction; however, the prepared additives are all individual esters or amides. Compared to a single-functional group, molecules with multiple functional groups would enhance adsorption strength through multi-site adsorption or chelation effects, improving the stability and durability of tribofilms and demonstrating excellent tribological performance. Additionally, most of the reported tribological properties were evaluated in PAO6, whose viscosity is relatively higher (kinematic viscosity of ~5.80 cSt at 100 °C). With increasingly stringent global emission regulations, low-viscosity lubricant technology has become a well-known trend in recent years [32], so the performance of additives should be conducted in lower-viscosity oils, such as PAO4 (kinematic viscosity of ~3.90 cSt at 100 °C). In this study, we designed and synthesized an eco-friendly OFM with a ternary structure based on amide, ester, and hydrocarbons. 1,3-diamino-2-propanol and oleic acid (OA) were used to prepare the 1,3-dioleoamide-2-propyloleate (DOAPO), which was characterized by NMR, HR-MS, FT-IR, and TGA. Meanwhile, the storage stability and tribological behaviors of DOAPO were investigated in a low-viscosity synthetic base oil and compared with a commercial oleamide. Additionally, micro-IR, XPS, and DFT calculations were applied to clarify its micro-lubrication mechanism.

2. Materials and Methods

2.1. Materials

1,3-diamino-2-propanol (98%) and 4-dimethylaminopyridine (DMAP, 99%) were obtained from Beijing Innochem Co., Ltd. (Beijing, China). Oxalyl chloride (98%), N, N-dimethylformamide (DMF, 99.5%), and triethylamine (TEA, 99%) were received from Energy Chemical. Dichloromethane (DCM, Shanghai Titan Scientific Co., Ltd. (Shanghai, China), 99.9%), oleic acid (OA, Alfa Aesar Chemical Co., Ltd. (Hangzhou, China), 99%), and all other reagents were commercially obtained and used as received for the synthesis of DOAPO.

Durasyn®164 (PAO4, INEOS, London, UK) and Priolube 3970 (3970, CRODA, Snaith, UK) were separately purchased from Shanghai Qicheng Industrial Co., Ltd. (Shanghai, China) and Hersbit Chemical Co., Ltd. (Shanghai, China), which were applied as base oils for the tribological evaluation of DOAPO, and oleamide (Tokyo Chemical Industry Co., Ltd. (Tokyo, Japan), 65%) was used as a commercial additive to compare with DOAPO.

2.2. Synthesis of 1,3-Dioleoamide-2-Propyloleate (DOAPO)

OA (12.55 g, 44.44 mmol), dry DCM (20 mL), and 2–3 drops of DMF were mixed in a round-bottom flask under an Ar atmosphere, and oxalyl chloride (11.28 g, 88.88 mmol) was slowly added dropwise into the mixture at 0 °C. After that, the reaction was stirred at room temperature for 4 h until the OA was transformed completely. The excess oxalyl chloride was removed by reduced pressure to yield the oily, colorless oleoyl chloride. Subsequently, the prepared oleoyl chloride was dissolved with dry DCM (20 mL) and added dropwise into a mixture of 1,3-diamino-2-propanol (1.00 g, 11.11 mmol), TEA (4.60 g, 45.55 mmol), DMAP (0.41 g, 3.33 mmol), and dry DCM (30 mL). The mixture was refluxed at 50 °C for 2 h and quenched with water when 1,3-diamino-2-propanol was completely consumed using TLC monitoring. The organic phase was extracted with DCM, washed with saturated $NaHCO_3$ and NaCl solutions, and dried with anhydrous Na_2SO_4. The crude product was purified by column chromatography (eluent: $V_{(DCM)}/V_{(MeOH)}$ = 30/1) to obtain 1,3-dioleoamide-2-propyloleate as a pale-yellow liquid (6.20 g, yield: 63%), which was recorded as DOAPO.

^{1}H NMR (400 MHz, $CDCl_3$) δ 6.22 (s, 2H), 5.34 (s, 6H), 4.83 (s, 1H), 3.49 (s, 2H), 3.30 (d, J = 14.1 Hz, 2H), 2.29 (t, J = 7.5 Hz, 2H), 2.21 (t, J = 7.5 Hz, 4H), 2.00 (d, J = 5.5 Hz, 12H), 1.63 (s, 6H), 1.28 (d, J = 14.2 Hz, 60H), 0.87 (t, J = 6.3 Hz, 9H) (Figure S1a). ^{13}C NMR (101 MHz, $CDCl_3$) δ 174.11, 173.18, 130.19, 130.15, 129.86, 129.82, 77.48, 77.16, 76.84, 71.21, 39.13, 36.94, 34.44, 33.95, 32.06, 29.92, 29.88, 29.68, 29.48, 29.42, 29.36, 29.31, 29.29, 27.39, 27.34, 25.88, 25.02, 22.83, 14.24 (Figure S1b). HR-MS (ESI) calcd. for $C_{57}H_{107}N_2O_4$ [M+H]$^+$: 883.82254, found: 883.82341 (Figure S1c). FT-IR (ATR): ν = 3292.0, 3079.6, 2922.7, 2853.3, 1739.1, 1651.7, 1548.5, 1465.0, 1377.5, 1246.1, 1172.6, 1083.8, 722.6 cm^{-1} (Figure S1d).

2.3. Characterization

The nuclear magnetic resonance (NMR) characterization, including ^{1}H NMR and ^{13}C NMR, was conducted on a 400-MR (Varian, Palo Alto, CA, USA) using $CDCl_3$ as the solution. High-resolution mass spectrometry (HR-MS) was carried out on JMS-T100LP AccuTOF LC-plus 4G (Nippon Electronics Corporation, Tokyo, Japan) using electrospray ionization. Nicolet iN10MX (Thermo Fisher, Waltham, MA, USA) was applied to record Fourier-transform infrared (FT-IR) and micro-infrared (micro-IR) spectroscopy by scanning from 400 to 4000 cm^{-1}. Thermogravimetric analysis (TGA) was performed on Q500 (TA, Milford, MA, USA) under a N_2 atmosphere with a flow rate of 60 mL/min and a heating rate of 10 °C/min from 25 to 600 °C. The morphology and elemental composition of metal surfaces are analyzed by scanning electron microscope with energy dispersive spectrometer (SEM-EDS) using QUANTAX (Bruker, San Jose, CA, USA). The chemical state of specific elements and potential tribochemical films formed on the frictional surface were analyzed using X-ray photoelectron spectroscopy (XPS, Thermo Scientific K-Alpha, Waltham, MA, USA) with an Al-Kα radiation source, and the obtained spectra were analyzed using the Avantage 5.9931 software.

2.4. Oil Preparation

In this study, the synthetic hydrocarbon PAO4 (90 wt.%) and saturated polyol ester 3970 (10 wt.%), which are both low viscosity, were blended after heating and stirring at 60 °C for 2 h to obtain the base oil. Oil samples containing additives were prepared as follows: 0.05~1.0 wt.% of self-prepared DOAPO or purchased oleamide, and the base oil was mixed at 60 °C for 2 h.

2.5. Tribological Test

The tribological behaviors of DOAPO in base oil were evaluated on a Tenkey MS-10A four-ball tester (Xiamen TenKey Automation Co., Ltd., Xiamen, China), which was compared with that of commercially available oleamide. A picture of the four-ball tester and its schematic are shown in Figure 1; all balls used are made of GCr15-bearing steel with a diameter of 12.7 mm. According to the standard NB/SH/T 0189-2017 [33], the tribological tests were operated at 75 °C for 1 h, where the rotational speed of the upper steel ball was 1200 rpm, and the load was 392 N. Each test was conducted at least three times to ensure the repeatability of the average friction coefficient (ave. COF) and average wear scar diameter (ave. WSD).

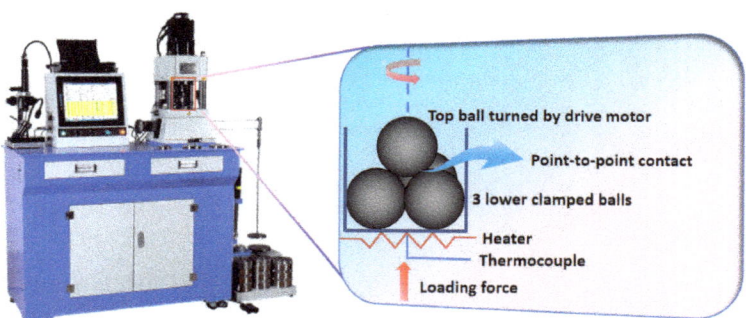

Figure 1. A picture of the MS-10A four-ball tester and its schematic.

3. Results and Discussion

3.1. Synthesis Route of DOAPO

At present, the synthesis of amides mainly includes direct amidation of carboxylic acids and amines; amidation of acyl halogens, anhydrides, or esters; hydrolysis of amides by oximes or nitriles; amidation of alcohol oxidation; and so on. Among them, the amidation of acyl halogens, namely the Schotten–Baumann reaction, is the most convenient and efficient method. Meanwhile, the reaction rate of carboxylic acid activated to acyl chloride is fast, even for substrates with large site resistance. Therefore, oxalyl chloride is used to activate OA to oleoyl chloride in this work, which can react simultaneously with the amine and hydroxyl groups of 1,3-diamino-2-propanol to obtain DOAPO; the synthesis route is shown in Scheme 1. NMR, HR-MS, and FT-IR were used to confirm the structure of DOAPO, and the results can be seen in Section 2.2.

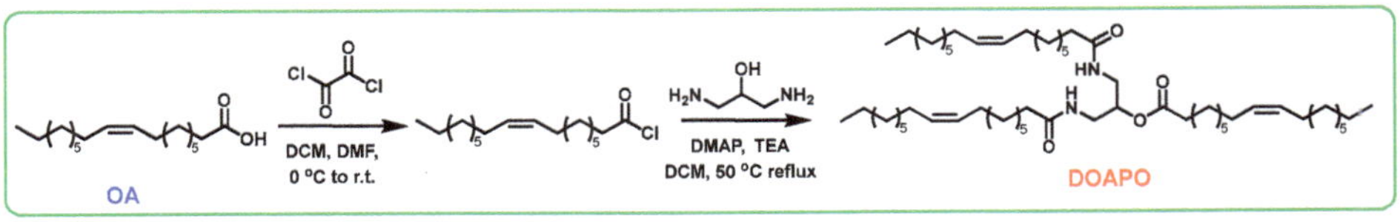

Scheme 1. The synthesis route of DOAPO.

3.2. Thermal Stability of DOAPO

In general, high-quality lubricants such as engine oils, anti-wear hydraulic oils, compressor oils, etc., all require a good high-temperature resistance. Although the thermal stability of lubricants mainly depends on base oils, it is worth noting that many additives with lower decomposition temperatures will adversely affect the overall stability of lubricants, thus reducing their comprehensive performance and service life. So, thermal stability is a key index to estimate the effectiveness of additives. The TG and DTG curves of 1,3-diamino-2-propanol, DOAPO, and commercial oleamide are shown in Figure 2. Combing the comparison data in Table S1, the initial and terminal decomposition temperatures of 1,3-diamino-2-propanol, DOAPO, and oleamide are 58.4 °C, 291.8 °C, and 273.4 °C and 188.9 °C, 495.0 °C, and 329.9 °C, respectively; their maximum decomposition temperatures are 178.1 °C, 395.5 °C, and 307.8 °C, respectively. Meanwhile, the residual masses of 1,3-diamino-2-propanol, DOAPO, and oleamide at 300 °C and 400 °C are 0.06%, 86.9%, and 33.9% and 0.02%, 20.1%, and 0.3%, respectively. The results show that the thermal stability of the three can be ranked as DOAPO > oleamide > 1,3-diamino-2-propanol. Due to the introduction of an oleacyl group with a long carbon chain, the thermal stability of DOAPO is significantly improved compared with the raw material 1,3-diamino-2-propanol, even better than that of commercial oleamide.

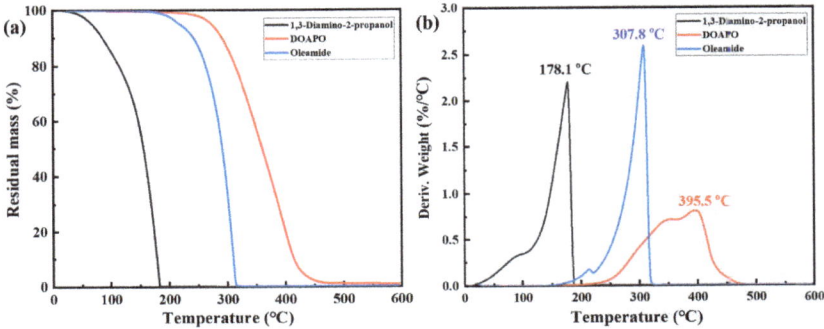

Figure 2. (a) TG and (b) DTG curves of 1,3-diamino-2-propanol, DOAPO, and oleamide.

3.3. Storage Stability of DOAPO in Synthetic Base Oil

Good storage stability is the basic requirement to ensure the performance of lubricants, which is primarily determined by the stability of additives in base oils. Therefore, we have investigated the storage stability of oils with different additions of DOAPO, as well as the same addition of DOAPO and commercial oleamide. As shown in Figure 3a and Table S1, after 30 days of storage at room temperature, the oil samples with 0.05~1.0 wt.% DOAPO remained clear and bright without any precipitation (Figure 3a), while the bottom of 0.5 wt.% oleamide appeared obvious precipitation (Figure 3b), indicating that the storage stability of DOAPO in synthetic base oil is better than that of oleamide.

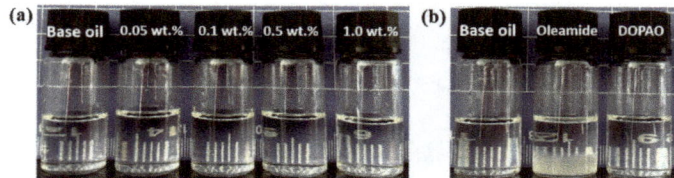

Figure 3. Appearance of oil samples after 30 d: (**a**) different additions of DOAPO; (**b**) the same addition (0.5 wt.%) of DOAPO or oleamide.

3.4. Tribological Properties of DOAPO

3.4.1. Different Additions of DOAPO

The performance of additives in base oils usually varies with different additions, exhibiting better comprehensive properties within an optimal addition range [34]. Figure 4 displays the tribological performance of oil samples with different additions of DOAPO. The friction profiles (Figure 4a) show that the running-in period of oil with a low DOAPO addition, i.e., 0.05 wt.%, is much longer than that with 0.1~1.0 wt.% (600 s vs. 120 s), which is similar to base oil. During the relative-stability period (600~3600 s), oil samples with 0.05~1.0 wt.% DOAPO are much more stable compared to base oil, even the COF of 0.05 wt.% DOAPO oil is high. It is worth noting that the COF of 0.1 wt% DOAPO oil increases at the end of friction, which is maintained stable for 0.5 wt% DOAPO oil. Nonetheless, the COF of 1.0 wt% DOAPO oil fluctuates slightly in the initial phase of the relative-stability period. Overall, oils with 0.1~1.0 wt% DOAPO exhibit better tribological properties, i.e., lower COF and smaller WSD, when compared to the base oil (Figure 4b). However, the tribological performance of 0.05 wt.% DOAPO oil is slightly worse than base oil, which may be related to the higher friction during the running-in period (see the insertion in Figure 4a). Overall, the oil with 0.5 wt.% DOAPO shows the best friction-reducing and anti-wear performance, namely reducing COF and WSD by 8.2% and 16.2% compared to base oil, respectively.

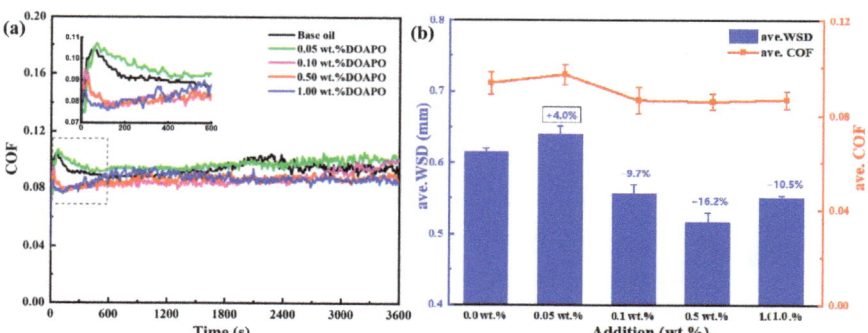

Figure 4. (**a**) Friction profiles and (**b**) ave. COF and ave. WSD of oil samples with different additions of DOAPO.

3.4.2. Comparison with the Commercial Oleamide

The tribological performance of DOAPO was compared with the commercial oleamide that has a similar structure, applying 0.5 wt% as the optimal addition. As demonstrated in Figure 5a, both DOAPO and oleamide could shorten the running-in period to some extent when compared to the base oil, but the friction of the oil containing DOAPO is more stable. The ave. COF values during the running-in period of base oil, 0.5 wt.% DOAPO oil, and 0.5 wt.% oleamide oil are 0.094, 0.080, and 0.082, respectively. The results in Figure 5b show that oils with DOAPO and oleamide exhibit lower COF and smaller WSD than base oil, i.e., both of them have effectiveness in friction reduction and anti-wear. Specifically, 0.5 wt.% DOAPO decreases the COF and WSD by 8.2% and 16.2%, while 0.5 wt.% oleamide decreases the COF and WSD by 2.6% and 12.0%, which indicates that the friction-reducing and anti-wear properties of DOAPO are better.

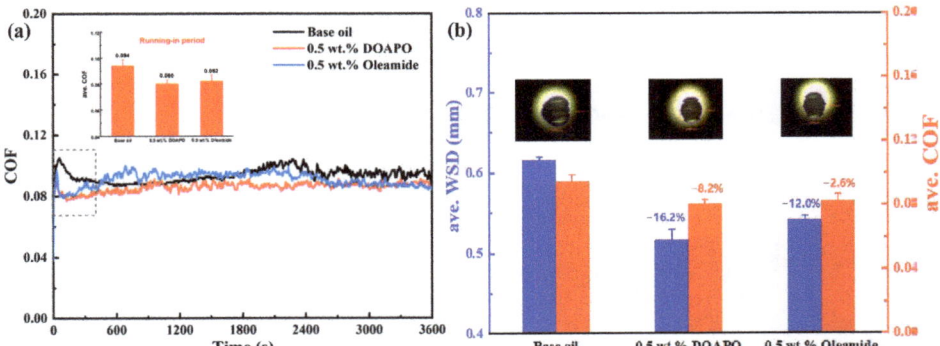

Figure 5. (**a**) Friction profiles and (**b**) ave. WSD and COF of oils with 0.5 wt.% of additives.

3.5. Micro-Lubrication Mechanism

3.5.1. Worn Surface Analysis

To investigate the micro-lubrication mechanism of the as-prepared amide–ester in synthetic base oil, micro-IR, SEM-EDS, and XPS were applied to analyze the composition of tribofilms on worn and non-worn surfaces lubricated with 0.5 wt.% DOAPO oil before and after tribological tests (marked as DOAPO_Non-wear and DOAPO_Wear, respectively), which were also compared to that with base oil (marked as Base oil_Wear). In Figure 6a, DOAPO_Non-wear not only has the stretching vibrations at 3357 cm^{-1} and 3177 cm^{-1} (υ_{N-H}), bending vibration at 1632 cm^{-1} (δ_{N-H}), and δ_{C-H} at 722 cm^{-1} of long alkyl chains but also has υ_{C-O-C} at 1058 cm^{-1} and 1021 cm^{-1} and $\delta_{C=C-H}$ at 892 cm^{-1}, which is characteristic for OA-based amide–ester. However, the υ_{C-O-C} at 1058 cm^{-1} and 1021 cm^{-1} and $\delta_{C=C-H}$ at 892 cm^{-1} that are characteristic of OA-based ester disappeared for DOAPO_wear, which is most likely caused by the breaking of long alkyl chain for ester in DOAPO during friction. When compared to the Base oil_Wear (Figure 6b), it has a characteristic $\upsilon_{(CO)O-H}$ at 1696 cm^{-1}, indicating an ester chain broken in ester 3970, which composed the base oil. According to the micro-IR results, it can be speculated that the ester group in DOAPO is more prone to be broken than the amide group when friction occurs.

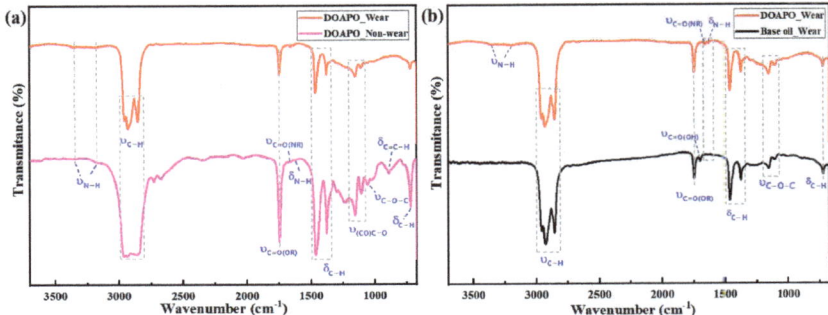

Figure 6. Micro-IR analysis of (**a**) worn and non-worn surfaces lubricated with 0.5 wt.% DOAPO oil and (**b**) worn surfaces lubricated with 0.5 wt.% DOAPO oil and base oil.

The morphology and elemental composition were analyzed by SEM-EDS, as shown in Table S3. The SEM images show that the surface wear is significantly improved when lubricated with 0.5 wt.% DOAPO. Compared with base oil, the surface lubricated with 0.5 wt.% DOAPO had lower C and slightly higher O and Fe before the tribological test, but it had higher C and Fe and lower O after the tribological test, indicating that DOAPO is involved in the formation of friction films. By further comparing the element composition of wear and non-wear surfaces lubricated with 0.5 wt.% DOAPO and base oil, it can be seen that the content of C and O for the wear surface is significantly higher than that of non-wear, and C content of surface lubricated with 0.5 wt.% DOAPO increases more, indicating that C is a key component of friction films.

XPS is mainly used to determine the binding energy of electrons. By comparing the chemical composition, bond state, and surface state before and after friction, XPS is beneficial for obtaining the chemical change information of the material surface during friction [35]. The bonding states of C, O, Fe, and N elements on worn surfaces lubricated with base oil and 0.5% DOAPO oil were further analyzed by XPS. After deconvolution (in Figure 7 and Table S4), there are three major peaks in the C1s spectra for DOAPO_Wear, i.e., C-C/C=C (284.80 eV, ~69.0%), C-O/C-N (285.94 eV, ~10.1%), and C=O (288.58 eV, ~20.90%) [36,37], whose C=O content is less than that of DOAPO_Wear. while there are only C-C/C=C (~91.6%) and C=O (~8.4%) for Base oil_Wear. In the O1s spectra, both of them have peaks at 530.33 eV, 531.93 eV, and 532.86 eV, which are ascribed to Fe-O, C=O, and C-O bonds, respectively [37,38]. In the Fe2p spectrum, the peaks at 707.32 eV (2p3/2) and 719.87 eV (2p1/2) are attributed to iron atoms arising from the steel ball. Peaks at 724.21 eV (2p1/2), 713.17 eV (2p3/2), and 710.81 eV (2p3/2) correspond to Fe^{2+} (2p1/2), Fe^{3+} (2p3/2), and Fe^{2+} (2p3/2), respectively, signifying that local high-temperature and high-pressure during friction lead to chemical reactions between iron in the steel balls and oxygen in the air [38,39]. Combining the O1s spectra, iron oxide films are formed for the DOAPO_Wear during friction, which are potentially composed of Fe_2O_3, FeOOH, FeO, and Fe_3O_4 [40]. In addition, the N1s spectrum of DOAPO_Wear exhibits peaks at 399.50 eV and 402.63 eV, corresponding to C-N and N-O bonds, respectively, suggesting that there are some amides turn into nitrogen oxides [41,42]. The results support that the tribofilm formed by DOAPO is composed of organic oxides and iron oxides, which would improve friction-reducing and anti-wear performance.

3.5.2. DFT Calculation

In order to reveal the influence of amide-only and amide–ester structure on the tribological properties as lubricating additives, DFT theoretical calculations of electrostatic potential (ESP) were conducted using the Gaussian16 software. Geometric optimizations were performed for both DOAPO and oleamide, applying the B3LYP hybrid exchange-correlation function. The optimized structures were characterized by harmonic vibration frequency with the minimum (Nimag = 0) or transition state (Nimag = 1) to analyze the atomic ESPs of C, H, O, and N with a 6-31G (d) basis set. The ESPs of compound molecules were calculated using Multiwfn based on the efficient algorithm, with reference to the van der Waals surface, while the molecular surface was defined as an isosurface with electron density r = 0.001 a.u.). Figure 8 reveals that the minimum and maximum ESP values for oleamide and DOAPO are -0.0689, 0.0706 and -0.0656, 0.0710, respectively, which suggests that oleamide has stronger adsorption to metal surfaces compared to DOAPO. However, it is worth noting that DOAPO is superior to oleamide in friction-reducing and anti-wear performance, which indicates that strong adsorption does not necessarily demonstrate better tribological performance. According to the micro-IR analysis, the ester group in DOAPO is more prone to be broken than the amide group when friction occurs. It means that DOAPO could produce ester chain fractures during friction, which is more convenient to react with metal surfaces to form metal-oxide films and achieve the anti-wear effect. Meanwhile, the interfacial shear force is reduced when the ester bond is broken, which improves friction-reducing performance. In general, DOAPO can not only form a strong adsorption film with metal surfaces through amide and ester groups but also produce chain fractures during friction to reduce interfacial shear force, improving tribological performance.

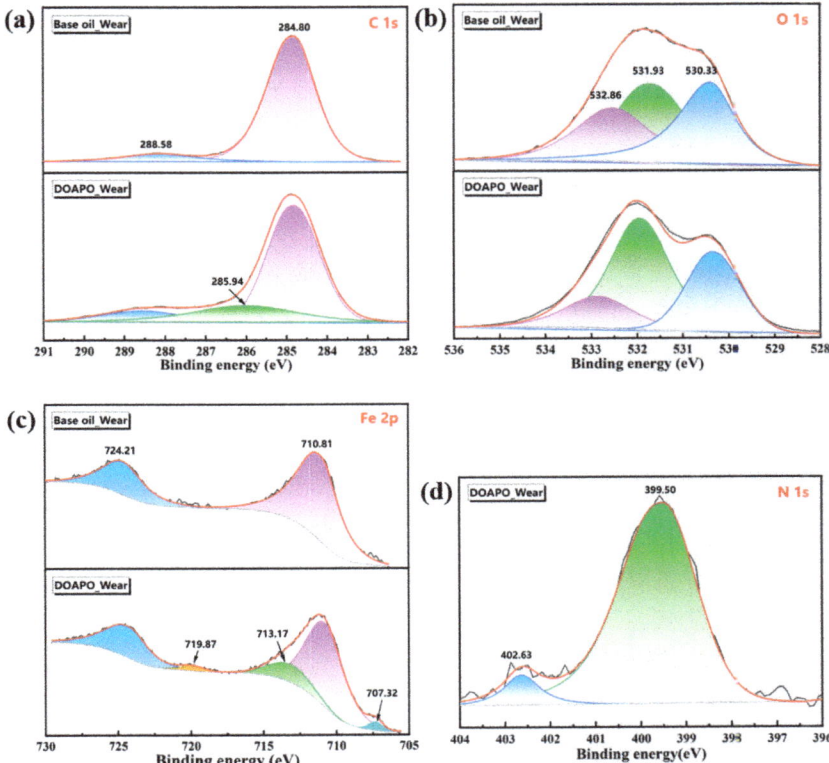

Figure 7. Worn surface analysis of base oil and oil with 0.5 wt.% DOAPO by XPS spectra: (**a**) C1s; (**b**) Fe2p; (**c**) O1s; (**d**) N1s.

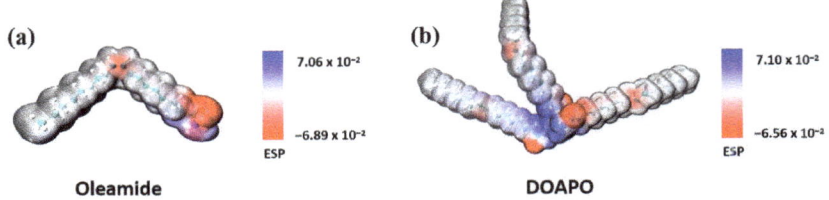

Figure 8. Theoretical electrostatic potential (ESP) calculation of (**a**) oleamide and (**b**) DOAPO.

Combining the DFT calculation and worn surface analysis, although the adsorption of DOAPO on the metal surface is slightly weaker than that of commercial oleamide (ESP: −0.0656 vs. −0.0689), ester-/amide-bonds in DOAPO are easier broken to produce polar carboxyl groups and alkyl chains during friction, as illustrated in Figure 9. While the broken chains reduce the interface shear force, and the carboxyl groups react with metal surfaces to form iron oxide protective films, so DOAPO shows good friction-reducing and anti-wear performance.

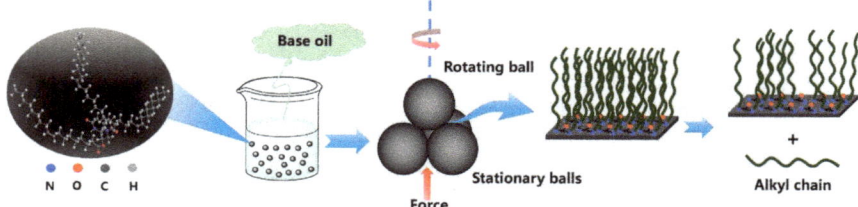

Figure 9. Schematic diagram of the lubrication mechanism with DOAPO.

4. Conclusions

In this work, a new sulfur- and phosphorus-free amide–ester, DOAPO, containing polar amide, ester, and nonpolar alkyl chains was synthesized from simple 1,3-diamino-2-propanol and OA. Its structure and thermal stability were characterized by NMR, HR-MS, FT-IR, and TGA. The tribological properties of DOAPO in synthetic base oil were studied and compared with commercial oleamide, and the micro-lubrication mechanism was disclosed by combining worn surface analysis and theoretical calculations. The following conclusions were drawn from this study:

(1) The introduction and multi-structure of long alkyl chains make DOAPO exhibit better storage stability in synthetic base oil and better thermal stability than that of commercial oleamide, whose residual mass at 300 °C is 86.9% vs. 33.9%.

(2) The optimal addition of DOAPO in the selected synthetic base oil is 0.5 wt.%, which can not only effectively shorten the running-in period compared to base oil (120 s vs. 600 s) but also reduce ave. COF and ave. WSD by 8.2% and 16.2%, respectively, which is better than that of commercial oleamide.

(3) The worn surface analysis and DFT calculation show that although the adsorption of DOAPO on metal surfaces is slightly weaker than oleamide (ESP: −0.0656 vs. 0.0689), its ester bond breaks preferentially during friction, which could reduce the interfacial shear force and easily react with metal surfaces to form iron oxide films, thus demonstrating better friction-reducing and anti-wear performance.

Supplementary Materials: The following supporting information can be downloaded at https://www.mdpi.com/article/10.3390/lubricants12060196/s1, Figure S1: (a) 1H NMR, (b) 13C NMR, (c) HR-MS and (d) FT-IR spectra of DOAPO. Table S1: Thermal stability comparison of 1,3-diamino-2-propanol, DOAPO and oleamide. Table S2: Storage stability of oil samples with different additions of DOAPO and the same addition (0.5 wt.%) of DOAPO or oleamide. Table S3: SEM-EDS analysis of surfaces lubricated with oils before and after tribological tests. Table S4: Surface XPS analysis after friction testing of base oils and oil samples supplemented with 0.5 wt.% DOAPO.

Author Contributions: Conceptualization, H.Y., X.S. and Y.T.; methodology, H.Y. and X.S.; validation, H.Y.; formal analysis, H.Y. and X.X.; investigation, X.X. and F.Y.; calculation, Y.Z.; data curation, X.X. and F.Y.; writing—original draft preparation, H.Y. and X.X.; writing—review and editing, H.Y. and X.X.; supervision, H.Y. All authors have read and agreed to the published version of the manuscript.

Funding: This work was funded by the National Natural Science Foundation of China Joint Fund (Shanghai) project (Grant No. U23A2084) and the Ling Chuang Research Project of China National Nuclear Corporation.

Data Availability Statement: Data is contained within the article.

Conflicts of Interest: The authors declare that they have no known competing financial interests or personal relationships that could influence the work reported in this paper.

References

1. Wong, V.W.; Tung, S.C. Overview of Automotive Engine Friction and Reduction Trends–Effects of Surface, Material, and Lubricant-Additive Technologies. *Friction* **2016**, *4*, 1–28. [CrossRef]

2. McQueen, J.S.; Gao, H.; Black, E.D.; Gangopadhyay, A.K.; Jensen, R.K. Friction and Wear of Tribofilms Formed by Zinc Dialkyl Dithiophosphate Antiwear Additive in Low Viscosity Engine Oils. *Tribol. Int.* **2005**, *38*, 289–297. [CrossRef]
3. Spikes, H. Friction Modifier Additives. *Tribol. Lett.* **2015**, *60*, 5. [CrossRef]
4. Vaitkunaite, G.; Espejo, C.; Wang, C.; Thiébaut, B.; Charrin, C.; Neville, A.; Morina, A. MoS2 Tribofilm Distribution from Low Viscosity Lubricants and Its Effect on Friction. *Tribol. Int.* **2020**, *151*, 106531. [CrossRef]
5. Hou, J.; Tsukamoto, M.; Hor, S.; Chen, X.; Yang, J.; Zhang, H.; Koga, N.; Yasuda, K.; Fukuzawa, K.; Itoh, S.; et al. Molecules with a TEMPO-Based Head Group as High-Performance Organic Friction Modifiers. *Friction* **2023**, *11*, 316–332. [CrossRef]
6. Zhou, Y.; Qu, J. Ionic Liquids as Lubricant Additives: A Review. *ACS Appl. Mater. Interfaces* **2017**, *9*, 3209–3222. [CrossRef] [PubMed]
7. Tang, Z.; Li, S. A Review of Recent Developments of Friction Modifiers for Liquid Lubricants (2007–Present). *Curr. Opin. Solid State Mater. Sci.* **2014**, *18*, 119–139. [CrossRef]
8. Meng, Y.; Xu, J.; Jin, Z.; Prakash, B.; Hu, Y. A Review of Recent Advances in Tribology. *Friction* **2020**, *8*, 221–300. [CrossRef]
9. Peeters, S.; Restuccia, P.; Loehlé, S.; Thiebaut, B.; Righi, M.C. Tribochemical Reactions of MoDTC Lubricant Additives with Iron by Quantum Mechanics/Molecular Mechanics Simulations. *J. Phys. Chem. C* **2020**, *124*, 13688–13694. [CrossRef]
10. Sarin, R.; Tuli, D.K.; Sureshbabu, A.V.; Misra, A.K.; Rai, M.M.; Bhatnagar, A.K. Molybdenum Dialkylphosphorodithioates: Synthesis and Performance Evaluation as Multifunctional Additives for Lubricants. *Tribol. Int.* **1994**, *27*, 379–386. [CrossRef]
11. Casford, M.T.L.; Puhan, D.; Davies, P.B.; Bracchi, G.L.; Smith, T.D. Thermal Behaviour of Synovene and Oleamide in Oil Adsorbed on Steel. *Tribol. Lett.* **2020**, *68*, 52. [CrossRef]
12. Shu, J.; Harris, K.; Munavirov, B.; Westbroek, R.; Leckner, J.; Glavatskih, S. Tribology of Polypropylene and Li-Complex Greases with ZDDP and MoDTC Additives. *Tribol. Int.* **2018**, *118*, 189–195. [CrossRef]
13. Tsagkaropoulou, G.; Warrens, C.P.; Camp, P.J. Interactions between Friction Modifiers and Dispersants in Lubricants: The Case of Glycerol Monooleate and Polyisobutylsuccinimide-Polyamine. *ACS Appl. Mater. Interfaces* **2019**, *11*, 28359–28369. [CrossRef] [PubMed]
14. Zhang, Z.; Yamaguchi, E.S.; Kasrai, M.; Bancroft, G.M.; Liu, X.; Fleet, M.E. Tribofilms Generated from ZDDP and DDP on Steel Surfaces: Part 2, Chemistry. *Tribol. Lett.* **2005**, *19*, 221–229. [CrossRef]
15. Zhang, Z.; Yamaguchi, E.S.; Kasrai, M.; Bancroft, G.M. Tribofilms Generated from ZDDP and DDP on Steel Surfaces: Part 1, Growth, Wear and Morphology. *Tribol. Lett.* **2005**, *19*, 211–220. [CrossRef]
16. Li, W.; Kumara, C.; Meyer, H.M.; Luo, H.; Qu, J. Compatibility between Various Ionic Liquids and an Organic Friction Modifier as Lubricant Additives. *Langmuir* **2018**, *34*, 10711–10720. [CrossRef] [PubMed]
17. Ewen, J.P.; Gattinoni, C.; Morgan, N.; Spikes, H.A.; Dini, D. Nonequilibrium Molecular Dynamics Simulations of Organic Friction Modifiers Adsorbed on Iron Oxide Surfaces. *Langmuir* **2016**, *32*, 4450–4463. [CrossRef]
18. Lundgren, S.M.; Persson, K.; Mueller, G.; Kronberg, B.; Clarke, J.; Chtaib, M.; Claesson, P.M. Unsaturated Fatty Acids in Alkane Solution: Adsorption to Steel Surfaces. *Langmuir* **2007**, *23*, 10598–10602. [CrossRef] [PubMed]
19. Ratoi, M.; Niste, V.B.; Alghawel, H.; Suen, Y.F.; Nelson, K. The Impact of Organic Friction Modifiers on Engine Oil Tribofilms. *RSC Adv.* **2014**, *4*, 4278–4285. [CrossRef]
20. Khatri, P.K.; Sadanandan, A.M.; Thakre, G.D.; Jain, S.L.; Singh, R.; Gupta, P. Tribo-Performance of the Ionic Liquids Derived from Dicarboxylic Acids as Lubricant Additives for Reducing Wear and Friction. *J. Mol. Liq.* **2022**, *364*, 119941. [CrossRef]
21. Campen, S.; Green, J.H.; Lamb, G.D.; Spikes, H.A. In Situ Study of Model Organic Friction Modifiers Using Liquid Cell AFM; Saturated and Mono-Unsaturated Carboxylic Acids. *Tribol. Lett.* **2015**, *57*, 18. [CrossRef]
22. Cyriac, F.; Yamashita, N.; Hirayama, T.; Yi, T.X.; Poornachary, S.K.; Chow, P.S. Mechanistic Insights into the Effect of Structural Factors on Film Formation and Tribological Performance of Organic Friction Modifiers. *Tribol. Int.* **2021**, *164*, 107243. [CrossRef]
23. Zhang, T.; Liu, S.; Zhang, X.; Gao, J.; Yu, H.; Ye, Q.; Liu, S.; Liu, W. Fabrication of Two-Dimensional Functional Covalent Organic Frameworks via the Thiol-Ene "Click" Reaction as Lubricant Additives for Antiwear and Friction Reduction. *ACS Appl. Mater. Interfaces* **2021**, *13*, 36213–36220. [CrossRef] [PubMed]
24. Kuwahara, T.; Romero, P.A.; Makowski, S.; Weihnacht, V.; Moras, G.; Moseler, M. Mechano-Chemical Decomposition of Organic Friction Modifiers with Multiple Reactive Centres Induces Superlubricity of Ta-C. *Nat. Commun.* **2019**, *10*, 151. [CrossRef] [PubMed]
25. Fry, B.M.; Chui, M.Y.; Moody, G.; Wong, J.S.S. Interactions between Organic Friction Modifier Additives. *Tribol. Int.* **2020**, *151*, 106438. [CrossRef]
26. Hu, W.; Xu, Y.; Zeng, X.; Li, J. Alkyl-Ethylene Amines as Effective Organic Friction Modifiers for the Boundary Lubrication Regime. *Langmuir* **2020**, *36*, 6716–6727. [CrossRef] [PubMed]
27. Fry, B.M.; Moody, G.; Spikes, H.A.; Wong, J.S.S. Adsorption of Organic Friction Modifier Additives. *Langmuir* **2020**, *36*, 1147–1155. [CrossRef] [PubMed]
28. Faujdar, E.; Singh, R.K. Amide Polymers Based on N-phenyl-p-phenylenediamine with α-olefins-co-maleic Anhydride as Multifunctional Additives for Lubricant Application. *Polym. Adv. Techs* **2022**, *33*, 2820–2834. [CrossRef]
29. Lee, J.; Kim, B.; Lee, J.W.; Hong, C.Y.; Kim, G.H.; Lee, S.J. Bioinspired Fatty Acid Amide-Based Slippery Oleogels for Shear-Stable Lubrication. *Adv. Sci.* **2022**, *9*, 2105528. [CrossRef]
30. Faujdar, E.; Negi, P.; Bhonsle, A.; Atray, N.; Singh, R.K. Efficiency of Dodecenylsuccinic Amide of n-Phenyl-p-Phenylenediammine as Novel Multifunctional Lubricant Additive for Deposit Control and Lubricity. *J. Surfactants Deterg.* **2021**, *24*, 173–184. [CrossRef]

31. Biresaw, G.; Compton, D.; Evans, K.; Bantchev, G.B. Lipoate Ester Multifunctional Lubricant Additives. *Ind. Eng. Chem. Res.* **2016**, *55*, 373–383. [CrossRef]
32. Tanaka, H.; Nagashima, T.; Sato, T.; Kawauchi, S. *The Effect of 0W-20 Low Viscosity Engine Oil on Fuel Economy*; No. 1999-01-3468; SAE International: Warrendale, PA, USA, 1999. [CrossRef]
33. *NB/SH/T 0189-2017*; Standard Test Method for Wear Preventive Characteristics of Lubricating Fluid—Four-Ball Method. National Energy Board: Beijing, China, 2017.
34. Wang, Y.; Lu, Q.; Xie, H.; Liu, S.; Ye, Q.; Zhou, F.; Liu, W. In-Situ Formation of Nitrogen Doped Microporous Carbon Nanospheres Derived from Polystyrene as Lubricant Additives for Anti-Wear and Friction Reduction. *Friction* **2023**, *12*, 439–451. [CrossRef]
35. De Barros, M.I.; Bouchet, J.; Raoult, I.; Le Mogne, T.; Martin, J.M.; Kasrai, M.; Yamada, Y. Friction Reduction by Metal Sulfides in Boundary Lubrication Studied by XPS and XANES Analyses. *Wear* **2003**, *254*, 863–870.
36. Fan, X.; Wang, L. High-Performance Lubricant Additives Based on Modified Graphene Oxide by Ionic Liquids. *J. Colloid Interface Sci.* **2015**, *452*, 98–108. [CrossRef] [PubMed]
37. Zhang, R.; Liu, X.; Guo, Z.; Cai, M.; Shi, L. Effective Sugar-Derived Organic Gelator for Three Different Types of Lubricant Oils to Improve Tribological Performance. *Friction* **2020**, *8*, 1025–1038. [CrossRef]
38. Huang, J.; Li, Y.; Jia, X.; Song, H. Preparation and Tribological Properties of Core-Shell Fe_3O_4@C Microspheres. *Tribol. Int.* **2019**, *129*, 427–435. [CrossRef]
39. Hu, J.; Zhang, Y.; Yang, G.; Gao, C.; Song, N.; Zhang, S.; Zhang, P. In-Situ Formed Carbon Based Composite Tribo-Film with Ultra-High Load Bearing Capacity. *Tribol. Int.* **2020**, *152*, 106577. [CrossRef]
40. Lee, A.Y.; Blakeslee, D.M.; Powell, C.J.; John, R.; Rumble, J. Development of the Web-Based NIST X-ray Photoelectron Spectroscopy (XPS) Database. *Data Sci. J.* **2006**, *1*, 1. [CrossRef]
41. Otero, I.; López, E.R.; Reichelt, M.; Fernández, J. Tribo-Chemical Reactions of Anion in Pyrrolidinium Salts for Steel–Steel Contact. *Tribol. Int.* **2014**, *77*, 160–170. [CrossRef]
42. Lu, Q.; Wang, H.; Ye, C.; Liu, W.; Xue, Q. Room Temperature Ionic Liquid 1-Ethyl-3-Hexylimidazolium-Bis(Trifluoromethylsulfonyl)-Imide as Lubricant for Steel–Steel Contact. *Tribol. Int.* **2004**, *37*, 547–552. [CrossRef]

Disclaimer/Publisher's Note: The statements, opinions and data contained in all publications are solely those of the individual author(s) and contributor(s) and not of MDPI and/or the editor(s). MDPI and/or the editor(s) disclaim responsibility for any injury to people or property resulting from any ideas, methods, instructions or products referred to in the content.

Article

An Evaluation of the Tribological Characteristics of Diaphragm Plates for High-Pressure Hydrogen Gas Compressor Applications

Sung-Jun Lee [1,†], Yoonchul Sohn [2,†], Dawit Zenebe Segu [1] and Chang-Lae Kim [1,*]

1. Department of Mechanical Engineering, Chosun University, Gwangju 61452, Republic of Korea
2. Department of Welding and Joining Science Engineering, Chosun University, Gwangju 61452, Republic of Korea
* Correspondence: kimcl@chosun.ac.kr
† These authors contributed equally to this work.

Abstract: Diaphragm plates, a key part of high-pressure hydrogen gas compressors, are easily cracked or broken due to repeated shape deformations caused by pressure, resulting in increasing difficulties in maintenance. This study aimed to improve the durability of diaphragm plates. This investigation focuses on the potential for friction and wear reduction through the application of surface polishing and Teflon coating on two diaphragm plate materials, namely stainless steel 301 and Inconel 718. To achieve this, various metal substrates with diverse surface morphologies were prepared and subjected to comprehensive assessments of their surface, mechanical, and tribological properties. Research findings revealed that the surface hardness and tensile strength of stainless steel 301 surpassed those of Inconel 718. Through friction and wear analysis, it was observed that Teflon-coated diaphragm plate material with a microstructure demonstrated superior friction performance. Furthermore, finite element analysis was employed to investigate the stress behavior of stainless steel 301 under different applied loads and conditions, offering valuable insights into the diaphragm's performance. From the results of this study, the excellence of the Teflon coating applied to the surface of stainless steel 301—the material of the hydrogen compressor diaphragm plate—was confirmed.

Keywords: friction; wear; solid lubricant; diaphragm; hydrogen gas compressor

1. Introduction

The rapid growth of global industries has presented significant challenges such as the energy depletion crisis and environmental pollution [1]. Among pressing concerns, the detrimental impact of greenhouse gas emissions on industrial development has become a critical issue [2]. In response to industrial evolution witnessed since the twentieth century, there is an urgent need for a transition toward greenhouse gas reduction and the adoption of eco-friendly energy systems to sustain the continuous growth of national economies [3].

Recognizing the potential of water and hydrogen as renewable and sustainable resources for eco-friendly energy conversion, hydrogen energy utilization has emerged as a viable alternative to address environmental pollution and energy scarcity [4,5]. Hydrogen, being a common element, constitutes over 70% of the universe's mass and is a component of water, which accounts for approximately 70% of the Earth's mass [6]. In addition to its abundance, hydrogen boasts low harmful gas emissions and renewability, rendering it indispensable for the development of eco-friendly energy systems.

Hydrogen refueling stations, which form the core of hydrogen infrastructure, are responsible for supplying fuel to vehicles and storage bases [7,8]. Hydrogen gas is compressed at high pressure, reaching approximately 700 bar, and is stored in gas tanks at these refueling stations [9]. Among the compression methods employed, diaphragm compressors have garnered attention for their ability to produce high-purity gas without leakage under high pressure, making them advantageous for hydrogen gas production compared to other compressor types [10].

The diaphragm compressor comprises essential components, including a driving part for power generation and transmission, a head part responsible for gas compression, and a diaphragm plate that isolates the gas from the actuating oil and facilitates gas compression [11]. The diaphragm plate is a critical element consisting of three layers: an oil diaphragm in contact with the actuating oil, a gas diaphragm in contact with the gas, and a middle diaphragm positioned between these two layers. This unique triple structure effectively isolates gas from the actuating oil, ensuring the production of high-purity gas, thereby making it the preferred method for high-pressure hydrogen gas compression. Despite its advantages, diaphragm compressors are subject to certain limitations, particularly with regard to flow capacity. To enhance gas production, it becomes necessary to expand the area, leading to stress caused by plate sagging when it contacts the cavity surface, ultimately causing fatigue failure.

In the field of diaphragm compressors for hydrogen gas production, several studies have explored stress and deformation in the cylinder head under high-temperature and high-pressure conditions, predicting the component's lifespan based on thermal-structural combined analysis [12]. Additionally, a work of research proposed a model to investigate the fatigue characteristics of diaphragm plates under various load conditions, aiming to verify their reliability [13]. Another research work revealed the significance of actuating oil pressure and outlet hole diameters as factors affecting the fatigue life of the diaphragm [14].

While previous studies have primarily focused on the radius stress and cavity shape, limited research has been dedicated to diaphragm plates, which play a pivotal role in the fatigue life of diaphragm compressors. During compressor operation, diaphragm plates experience elastic deformation within their limits. However, repeated deformation processes may lead to micro-slipping scratches between diaphragm plates and plastic deformation damage, potentially resulting in severe repercussions for gas production. External forces from the piston deform the diaphragm plate, causing bending and tensile stress. The type of stress experienced by the diaphragm plate determines the conversion of bending stress into radial tensile stress from the plate center, resulting in impacts and micro-friction between diaphragm plates. As such, the diaphragm plate emerges as the most critical factor influencing gas compressibility and the lifespan performance of diaphragm compressors. Therefore, it is imperative to investigate the characteristics of friction and wear to prevent damage caused by micro-impacts and friction on the diaphragm plate and to enhance its lifespan.

Improving the friction and wear characteristics of diaphragm plates necessitates accompanying surface treatments. Techniques such as coating and polishing enhance the wear resistance of materials by reducing their adhesive properties or improving their mechanical strength. Surface polishing can reduce unnecessary contact and friction during interactions by eliminating irregularities or protrusions on the material's surface [15]. Notably, a smooth surface has the advantage of dispersing wear and reducing concentrated stress, thereby mitigating temperature increases and wear due to friction. However, polishing demands precise and consistent performance, typically involving complex processes.

Among numerous coating materials, Teflon is well known for its non-stick properties and high-temperature resistance [16,17]. This suggests that when applied as a diaphragm surface coating, Teflon exhibits low-friction characteristics that could potentially enhance its lifespan. However, given that Teflon has lower strength compared to metal, it is essential to investigate its friction and wear characteristics during contact sliding processes with metal.

This study aims to assess the durability of diaphragm plates, focusing on the impact of surface treatment and material types. Specifically, we investigated the effects of surface polishing and coating on the mechanical properties, friction, and wear characteristics of diaphragm plates. Through a comprehensive evaluation, we analyzed the surface, mechanical, and tribological properties of the diaphragm plate, further utilizing finite element analysis simulation to examine stress distribution and deformation characteristics.

By conducting a thorough investigation into diaphragm plate durability, this research seeks to enhance the performance and reliability of hydrogen diaphragm compressors. The outcomes hold the potential to drive advancements in hydrogen energy utilization, significantly contributing to the addressing of environmental challenges and promoting sustainable energy practices in industrial applications.

2. Materials and Methods

2.1. Materials

In this study, diaphragm materials, specifically stainless steel (SUS) 301 and Inconel 718, were prepared to investigate the friction and wear characteristics of diaphragm plates, with a focus on the influence of surface treatment. Changes in friction and wear characteristics between thin film-coated and polished surfaces were thoroughly evaluated and compared.

For the thin film coating, Teflon, known for its excellent low friction properties, was chosen and deposited onto the surface of stainless steel 301. The Teflon coating had an approximate thickness of 50 μm. In contrast, the surface of the Inconel 718 substrate was subjected to a smooth polishing process using lapping. To serve as reference points, untreated bare Inconel and bare stainless steel substrates were also prepared. All specimens had a uniform thickness of approximately 0.4 mm and were precisely cut into 20 mm × 20 mm sizes to ensure consistency during testing.

As indicated in Table 1, the designations used for the specimens are as follows: the untreated bare Inconel and bare stainless steel substrates are denoted as BI and BS, respectively, while the polished Inconel and Teflon-coated stainless steel substrates are denoted as PI and TCS, respectively.

Table 1. Specimen names of bare Inconel 718, polished Inconel 718, stainless steel 301, and Teflon coated stainless steel 301.

Specimen Name	Material	Surface Treatment	Thickness [mm]	Size [mm × mm]
BI	Inconel 718	Non	0.4	
PI		Polishing	0.4	20 × 20
BS	Stainless steel 301	Non	0.4	
TCS		Coating	0.45	

2.2. Experiments

To determine the mechanical properties of the specimens, both hardness tests and tensile tests were conducted. Hardness measurements were performed using a micro hardness tester (TH712, Beijing TIME High Technology Ltd., Beijing, China) with a constant load of 1 N. A load was applied for 10 s, and upon removing the indenter, the surface area of the resulting indentation mark was calculated to determine the hardness.

Tensile tests were carried out employing a universal testing machine (SGA-E-20AD, Shin Gang Precision Ind Co. Ltd., Seoul, Republic of Korea), equipped with mechanical wedge grips and an extensometer capable of measuring a maximum axial force of 200 kN. For the tests, two sets of metal specimens were prepared, each consisting of four identical samples, which were precisely cut using a laser. The gauge length of the tensile specimens was set at 50 mm, and the diameter was 12.5 mm. Tensile testing was performed at a crosshead speed of 0.75 mm/min. The strain rate used for the tensile test was set at $2.5 \times 10^{-4}\ \text{s}^{-1}$.

The surface characteristics of the fabricated specimens were analyzed by measuring surface roughness and water droplet contact angles. Surface roughness of each specimen was assessed using a 2D profiler (SV-2100M4, Mitutoyo Korea Corporation, Gunpo, Republic of Korea) with a load of 0.75 mN and a scanning speed of 1 mm/s. The profiler scanned over a length of 4 mm to obtain the centerline average roughness value (Ra). The contact angle was determined by carefully dropping 10 μL of deionized water onto the substrate surface and measuring the angle formed by the water droplet on the substrate surface.

To evaluate the friction and wear characteristics of diaphragm plate materials based on different surface treatment conditions, a reciprocating sliding tribotester (RFW 160, NEOPLUS, Co., Ltd., Daejeon, Republic of Korea) was employed. The tribotester operates by converting the rotational motion of a motor into linear motion, enabling a repetitive sliding of the stage. Experimental conditions were selected considering the operating environment of the diaphragm compressor. The specific tribotest conditions are provided in Table 2. For the counter tip material, stainless steel, commonly used for diaphragm plate materials, was selected. The experiment was conducted by fixing the specimen on the stage and horizontally leveling the counter tip. A normal load was then applied by raising the weights, and the experiment was performed under these conditions. The test was conducted for more than 5000 cycles with conditions of a load greater than 100 mN, a sliding stroke of 2 mm, and sliding speeds of 4 mm/s and 16 mm/s. Following the experiment, the wear morphology of the specimen surface was examined using a microscope and a 2D profiler, and the wear area was quantified.

Table 2. Tribotest conditions.

Tribotest (Reciprocating Type)	
Tip material (diameter)	Stainless steel ball (1 mm)
Normal load	100 mN, 200 mN
Sliding speed	4 mm/s, 16 mm/s
Sliding stroke	2 mm
Sliding cycle	5000 cycles, 10,000 cycles

3. Results and Discussion

Figure 1a presents the Vickers hardness values of stainless steel 301 and Inconel 718, measuring 457 HV and 303 HV, respectively. Notably, stainless steel 301 exhibits a higher surface hardness value, approximately 50% greater than that of Inconel 718. Moving on to Figure 1b, the stress–strain curve of Inconel 718 showcases a tensile strength of 916 MPa and a yield strength of 536 MPa. The Inconel 718 demonstrated a maximum displacement of 36.7 mm during the tensile test, with an elongation of 73.5% and an elastic modulus of 65 GPa. In contrast, stainless steel 301 exhibits higher tensile strength (1168 MPa) and yield strength (716 MPa) compared to Inconel 718, with a smaller maximum displacement of 27 mm. The deformation of the stainless steel 301 specimen resulted in an elongation of 54.1%, and its elastic modulus slightly surpassed that of Inconel 718, measuring 67 GPa. Upon analyzing the stress–strain curve, it becomes evident that stainless steel 301 possesses higher tensile and yield strength, indicating superior resistance to deformation under applied loads [18]. However, Inconel 718 showcases remarkable ductility and plastic deformation properties, evident from its higher elongation and maximum displacement values [19]. Moreover, the elastic modulus values indicate that the two materials have a similar degree of stiffness.

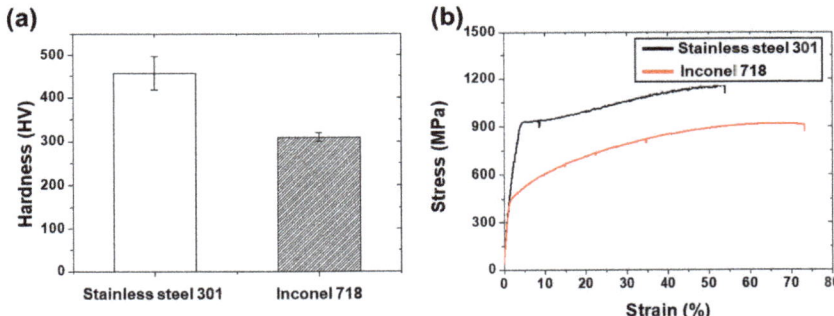

Figure 1. The mechanical properties of stainless steel 301 and Inconel 718: (**a**) Vickers hardness and (**b**) tensile stress–strain curve.

The surface characteristics of the diaphragm materials were analyzed based on their surface treatments using measurements of surface roughness and water droplet contact angles. Figure 2a presents the surface roughness values of all specimens (BI, PI, BS, TCS), which are 0.32 µm, 0.1 µm, 0.13 µm, and 1.33 µm, respectively. The surface roughness of PI is the smallest due to the surface polishing process, whereas TCS exhibits the highest value owing to the presence of Teflon particles coated on its surface. Figure 2b displays the water droplet contact angles of each specimen, with contact angles of 77°, 76°, 74°, and 96° for BI, PI, BS, and TCS, respectively. Inconel 718 and stainless steel 301 are considered hydrophilic materials, resulting in relatively small contact angles on the substrate surface due to their high affinity with water at the interface when in contact [20,21]. However, Teflon exhibits a higher contact angle compared to the other metal substrates due to its inherent hydrophobic nature [22]. Although there is no significant difference in the water droplet contact angle between the two metals, the contact angle increased with the application of the Teflon coating. Figure 3 illustrates the surface morphologies of the specimens. BI exhibits a very rough surface morphology, while the surface of PI appears very smooth. On the other hand, the surface of TCS displays an uneven distribution of rough particles, and BS exhibits a surface morphology similar to that of PI.

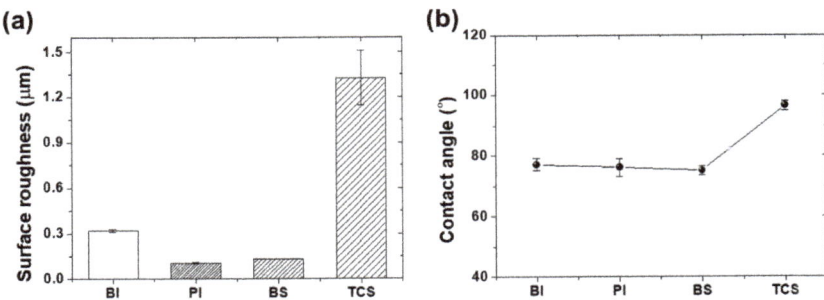

Figure 2. The (**a**) surface roughness and (**b**) water droplet contact angle of each specimen.

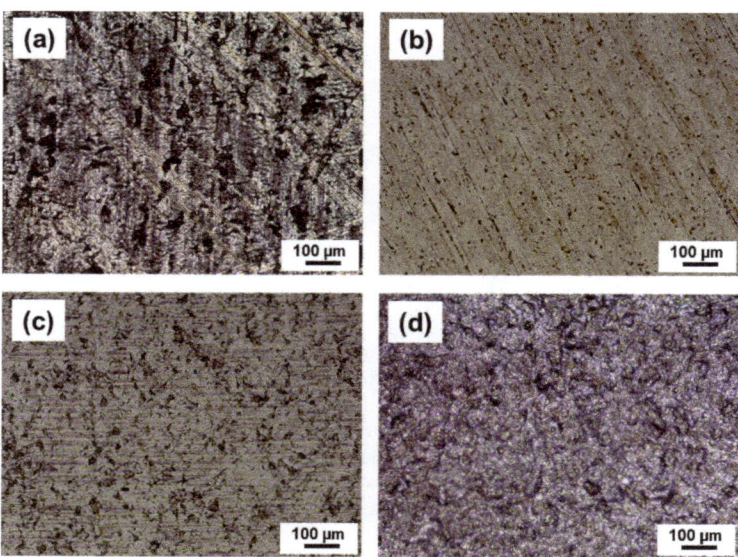

Figure 3. Optical microscope images of the specimen: (**a**) BI, (**b**) PI, (**c**) BS and (**d**) TCS.

In this study, the friction and wear characteristics of the specimens were evaluated through tribotests conducted over 5000 cycles at a speed of 4 mm/s under a normal load of 100 mN. Based on Hertzian contact theory, the contact pressures for Inconel 718 and stainless steel 301 were calculated as 612 MPa and 613 MPa, respectively, while that of Teflon was 34 MPa [23]. Figure 4a illustrates the friction coefficient history of each specimen during the sliding cycles. For bare Inconel (BI), the friction coefficient started below 0.5 initially, rapidly increased to over 0.9 within 500 cycles, and then remained relatively constant with slight fluctuations throughout the entire sliding cycle. Bare stainless steel (BS) began with a higher initial friction coefficient compared to BI, but after gradually increasing to approximately 0.9, the value was sustained with severe fluctuations. Polished Inconel (PI) began with a low friction coefficient of approximately 0.1, rapidly increased to 0.6 within the first 100 cycles, then decreased temporarily, followed by a gradual increase to 0.7 over hundreds of cycles. Subsequently, the friction coefficient remained constant, with a sudden decrease occurring around 2300 cycles, after which it returned to the original value and remained stable for the remainder of the test. Teflon-coated stainless steel (TCS) maintained a constant friction coefficient of 0.08 throughout the entire test duration. Figure 4b depicts the average friction coefficient values of each specimen. The average friction coefficients for BI and BS, which are bare Inconel and stainless steel specimens without surface treatment, were as high as 0.92 and 0.95, respectively. In contrast, the average friction coefficient of polished Inconel (PI) decreased slightly to 0.73, while that of TCS coated with Teflon on stainless steel was 0.08, representing the lowest average friction coefficient with a 92% decrease compared to BS. Following the tribotests, the wear tracks formed on the specimen surfaces were analyzed through a 2D profiler, and the wear rate was calculated, as shown in Figure 4c. The wear rate ($mm^3/N \cdot mm$) was calculated by dividing the wear volume obtained from measuring the wear width and depth of the wear track by the normal load and total sliding distance [24,25]. The wear rates for BI, PI, BS, and TCS were 0.92×10^{-7} $mm^3/N \cdot mm$, 0.34×10^{-7} $mm^3/N \cdot mm$, 0.40×10^{-7} $mm^3/N \cdot mm$, and 0.46×10^{-7} $mm^3/N \cdot mm$, respectively. The wear rate of BI was the highest, while PI exhibited the smallest wear rate. The surface of BI, being bare Inconel without surface treatment, displayed numerous patterns and protrusions, such as processing traces, leading to a large surface roughness, thus resulting in a high friction coefficient and increased wear.

In contrast, the smooth polished surface of PI significantly reduced surface roughness, leading to a notable decrease in the wear rate, along with the decrease in the friction coefficient [26]. The wear rate of BS, which is bare stainless steel without surface polishing, is 52% smaller than that of BI, which is bare Inconel. This is considered to be because the mechanical properties of BS are superior to those of BI. Compared to BI, BS had a slightly higher friction coefficient despite its smaller surface roughness, but significantly less wear due to its excellent mechanical properties. As such, in the case of a bare surface without any surface treatment, it was confirmed that the excellent mechanical properties of the material can reduce the wear rate. Compared to BI and BS, in the case of polished Inconel PI, the wear rate was smaller because the surface roughness and friction coefficient were smaller. In particular, from the fact that the wear rate of PI is lower than that of BS despite the inferior mechanical properties of Inconel compared to stainless steel, it can be confirmed that surface durability can be further improved through the control of surface roughness rather than mechanical properties. However, although the surface roughness of TCS coated with Teflon on stainless steel was much higher than that of other specimens, the friction coefficient was the lowest, and the wear rate was smaller than BI but larger than PI and BS. Since the water droplet contact angle of TCS is much larger than that of other specimens, it is expected that the surface energy is relatively small. The smaller surface energy of TCS explains why the friction coefficient is lower than that of other specimens. That is, Teflon is a typical low-friction material, and even if the surface roughness of TCS is large, the friction coefficient is low because it has low surface energy. However, since the mechanical properties of the Teflon coating are weaker than those of the two metal materials, Inconel and stainless steel, it can be easily damaged by stainless steel, the material of the counter tip. In addition, the rough surface formed by the Teflon particles is easily peeled off by repeated contact and sliding movements with the counter tip, and it is thought that the wear of the Teflon coating was intensified due to worn-out wear particles [27].

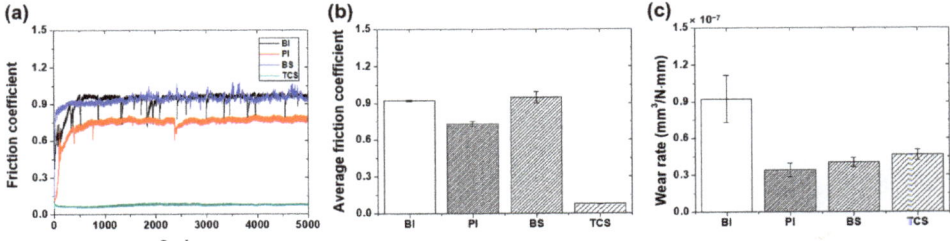

Figure 4. The friction and wear characteristics of the specimen under 100 mN at 4 mm/s during 5000 cycles: (**a**) friction coefficient history, (**b**) average friction coefficient, and (**c**) wear rate.

Figure 5 depicts optical microscope images illustrating the wear tracks that formed on the surfaces of the specimens following the tribotesting procedure. The wear morphology exhibited distinctive variations based on the specific contact conditions. With the exception of TCS, the wear tracks observed on the specimens appeared broad and featured numerous scratch marks. In contrast, the wear tracks on TCS demonstrated a relatively narrower width yet exhibited the greatest wear depth. This phenomenon is attributed to the likelihood of coating particles detaching from the surface [28,29].

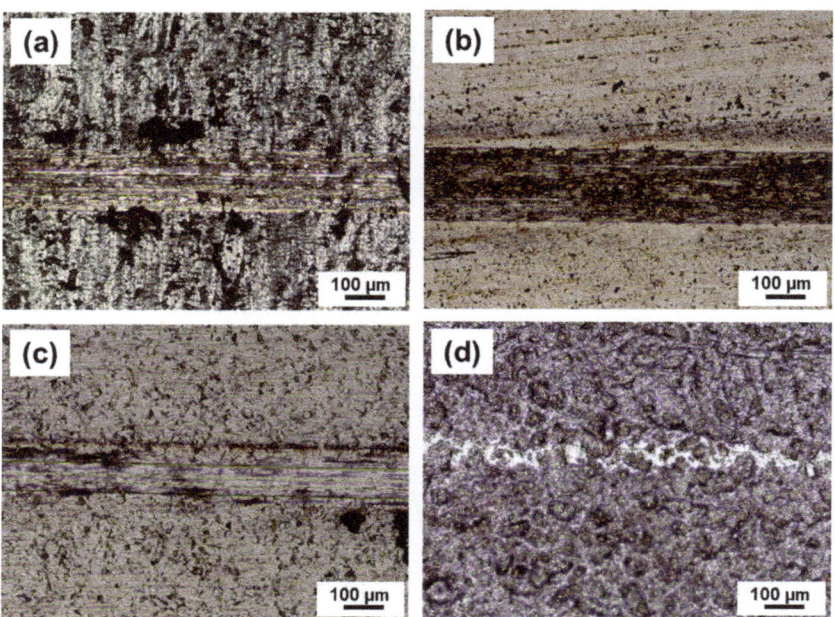

Figure 5. Optical microscope images of wear track of the specimen under 100 mN at 4 mm/s during 5000 cycles: (**a**) BI, (**b**) PI and (**c**) BS and (**d**) TCS.

To investigate the effect of sliding speed on friction and wear characteristics, an experiment was conducted at the sliding speed of 16 mm/s, which was four times the original sliding speed, while keeping other conditions constant. As illustrated in Figure 6a, the friction coefficients for all specimens were tracked across sliding cycles. Specifically, PI and BI exhibited relatively elevated initial friction coefficients of 0.75 and 0.45, respectively, which then stabilized at 0.6 after 1000 cycles. In contrast, BS (Bare Stainless Steel) commenced with a low friction coefficient of less than 0.1 at the outset of sliding, gradually increased until 1000 cycles, and subsequently experienced a slight elevation, reaching 0.7. Remarkably, TCS (Teflon-Coated Stainless Steel) exhibited a constant friction coefficient of 0.08 at the sliding speed of 16 mm/s, demonstrating no fluctuation in friction coefficient throughout the entire sliding cycle. Figure 6b shows the average friction coefficient values for each specimen, with BS registering the highest average friction coefficient at 0.64, followed by BI, PI, and TCS with values of 0.61, 0.6, and 0.08, respectively. Notably, TCS exhibited the lowest average friction coefficient of 0.08, representing an 88% reduction compared to BS. Overall, the friction coefficients at the sliding speed of 16 mm/s were lower than those at the sliding speed of 4 mm/s. As the sliding speed increased, the friction attributed to adhesive bonds between contacting interfaces diminished, leading to a reduction in friction coefficient owing to the augmented sliding speed [30]. Additionally, it is recognized that friction coefficient declines as the contact area between the specimen and the counter tip decreases with increased sliding speed [31]. Conversely, TCS exhibited consistent friction coefficient values irrespective of sliding speed, attributed to its robust inherent low-friction characteristics. Depicted in Figure 6c are the wear rates across all specimens. BI exhibited the highest wear rate of 3.47×10^{-7} mm^3/N·mm, whereas BS displayed the lowest wear rate of 0.87×10^{-7} mm^3/N·mm. The wear rates for PI and TCS were 2.21×10^{-7} mm^3/N·mm and 1.04×10^{-7} mm^3/N·mm, respectively. Wear rates at the sliding speed of 16 mm/s were higher than those at 4 mm/s. The trend in wear rate variation with sliding speed exhibited an opposing aspect compared to the friction coefficient trend. Generally, multiple factors contribute to an increase in wear rate as sliding speed escalates.

Higher sliding speeds entail renewed contact sliding, generating greater frictional heat residue from previous sliding contacts. Enhanced frictional heat leads to surface material softening and accelerated wear [32]. Moreover, heightened sliding speed instantaneously disrupts adhesive bonding between the two contacting surfaces, hindering a rise in friction coefficient owing to adhesive forces. However, rapid collisions amid surface asperities can induce significant stress within the material, prompting deformation and fatigue-induced failure [33]. Repetitive loading cycles encountered by contact areas during successive relative sliding at elevated speeds can induce microcracks, leading to surface fatigue and consequent detachment of wear particles, thereby accelerating wear [34]. Consequently, as depicted in Figure 7, a coarser wear pattern emerged at the sliding speed of 16 mm/s compared to the 4 mm/s condition. All specimens exhibited broader wear widths with numerous surface scratches. Notably, as the sliding speed increased, the friction coefficient of PI experienced marginal change, while the wear rate surged approximately sevenfold; for other specimens, friction coefficients decreased, and wear rates escalated by three to four times. PI, polished Inconel, characterized by minimal surface roughness and smoothness, showcased lower friction coefficients and wear rates than unpolished BI and BS at a relatively gradual sliding speed of 4 mm/s. However, at the accelerated sliding speed of 16 mm/s, PI's friction coefficient remained comparable, while the wear rate increased significantly compared to BS. The smooth surface of PI is deemed significantly influenced by sliding speed. Additionally, TCS exhibited markedly low friction coefficients compared to other specimens, yet its wear rate was comparatively elevated. This discrepancy in TCS wear rate was attributed to the peeling of the Teflon coating due to repeated sliding friction, with the underlying stainless steel base material underneath the Teflon coating remaining protected. The friction coefficient and wear rate of TCS were comparatively less affected by the sliding speed change. During the tribotest spanning 5000 cycles, BI exhibited the highest friction coefficient and wear rate. Consequently, an extended experiment involving 10,000 cycles was conducted to assess the durability of the specimens. This extended experiment specifically included PI, BS, and TCS specimens, with the exclusion of BI for comparison purposes.

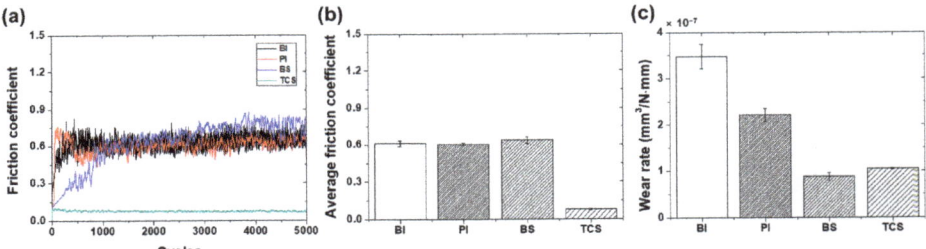

Figure 6. The friction and wear characteristics of the specimen under 100 mN at 16 mm/s during 5000 cycles: (**a**) friction coefficient history, (**b**) average friction coefficient, and (**c**) wear rate.

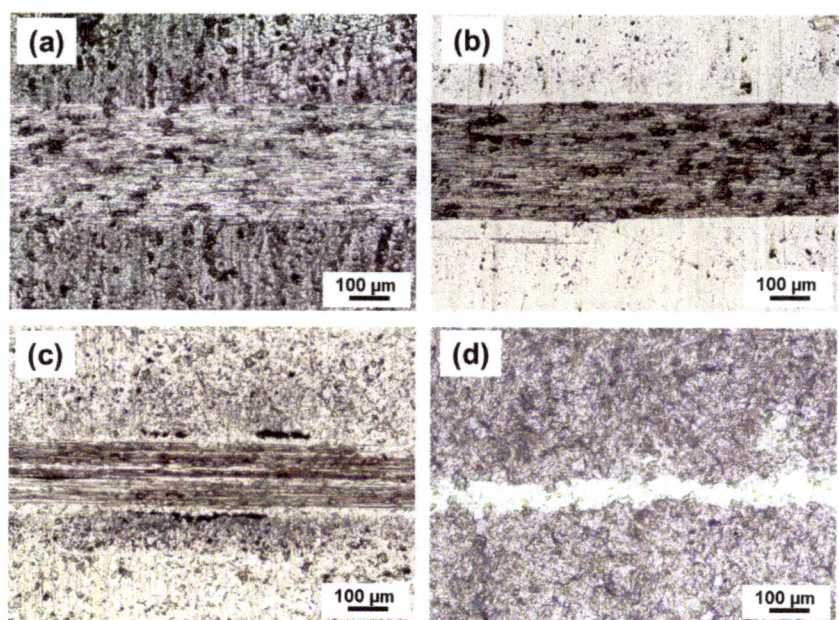

Figure 7. Optical microscope images of wear track of the specimen under 100 mN at 16 mm/s during 5000 cycles: (**a**) BI, (**b**) PI, (**c**) BS and (**d**) TCS.

Figure 8a illustrates the average friction coefficients of the PI, BS, and TCS specimens subjected to a tribotest over 10,000 cycles under a normal load of 100 mN at a sliding speed of 16 mm/s. The TCS specimen demonstrated the lowest average friction coefficient of 0.08, whereas PI and BS exhibited friction coefficients of 0.59 and 0.74, respectively. The friction coefficient of PI remained comparable to that observed in the 5000-cycle experiment, whereas that of BS escalated by roughly 15%. As depicted in Figure 6a, PI commenced with a low friction coefficient due to its meticulously polished surface but underwent rapid elevation during the initial cycles due to wear. Subsequently, it tapered off, stabilizing around 0.6 up to 5000 cycles. The experiments with extended sliding cycles displayed analogous trends, maintaining a friction coefficient of 0.6 up to 10,000 cycles. This signified that, following the initial rapid increase prompted by the impairment of PI's smooth surface and the emergence of scratches, no substantial fluctuations in the friction coefficient occurred beyond 10,000 cycles following the surface conditioning process. Conversely, BS, characterized by a relatively coarse surface as depicted in Figure 6a, commenced with a low friction coefficient, which gradually ascended. PI exhibited an average friction coefficient of 0.64, showing no notable deviation from that of PI. However, a pronounced fluctuation and a gradual upward trajectory were evident after 3000 cycles. The upward trajectory persisted in the experiment when extended to 10,000 cycles. In contrast, the augmented sliding cycles on the TCS specimen exhibited minimal influence on friction coefficients due to the inherent low-friction properties of Teflon. Figure 8b presents the wear rates of the three specimens. PI exhibited the highest wear rate at 2.96×10^{-7} mm^3/N·mm, whereas BS and TCS displayed wear rates of 0.88×10^{-7} and 0.65×10^{-7} mm^3/N·mm, respectively. In the 5000-cycle experiment, the relatively elevated wear rate was attributed to the inclusion of Teflon coating thickness into the wear rate calculation, following its detachment from TCS. However, a distinct trend emerged when the sliding cycle experiment was extended to 10,000 cycles. With increasing sliding cycles, the Teflon particles that peeled off adhered to the counter tip, leading to a shift in contact characteristics from hard contact mode to soft contact mode [35]. Teflon-to-Teflon contact experiences lower contact pressure in comparison to metals, and the stress dispersion due to the cushioning effect of Teflon particles maintains low-friction/wear attributes [36]. Consequently, it is plausible that TCS exhibited the lowest friction coefficient and wear rate. Figure 9 shows optical microscope images of wear tracks formed on the specimens through tribotests under 100 mN at 16 mm/s during 10,000 cycles. In line with the wear rate findings, PI exhibited the widest wear width, replete with numerous scratches at the center, and worn particles adhered around the wear track. BS displayed a narrower wear width compared to PI, yet numerous scratch-shaped morphologies were evident, and wear debris was adhered across the wear track. Conversely, TCS exhibited the narrowest wear width, coupled with the peeling of Teflon particles, with scratches only manifesting on select substrate parts.

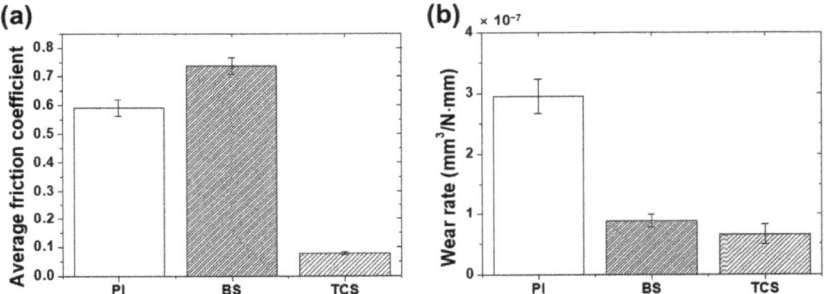

Figure 8. The friction and wear characteristics of PI, BS, and TCS under 100 mN at 16 mm/s during 10,000 cycles: (**a**) average friction coefficient and (**b**) wear rate.

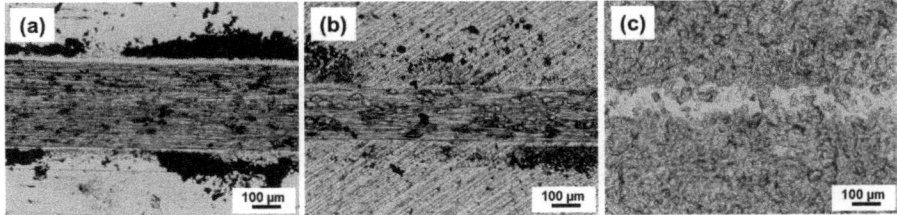

Figure 9. Microscope images of wear tracks under 100 mN at 16 mm/s during 10,000 cycles: (**a**) PI, (**b**) BS, and (**c**) TCS.

Figure 10a depicts the average friction coefficient observed during the tribotest conducted under a normal load of 200 mN at a sliding speed of 16 mm/s over 10,000 cycles. Under the normal load of 200 mN, the contact pressures applied to PI, BS, and TCS were 772 MPa, 773 MPa, and 43.2 MPa, respectively. The average friction coefficients of PI and BS were 0.61 and 0.66, respectively, significantly surpassing that of TCS (0.07). The mechanism generating friction forces on each specimen in the high-pressure contact regime remained consistent with that observed under low loads. The lack of a substantial divergence in the friction coefficient in relation to the normal load suggests a direct proportionality between friction force and normal load. Consequently, it can be anticipated that the friction force resulting from a diaphragm micro-slip will escalate commensurate with applied pressure. Notably, surface roughness significantly influenced the friction coefficient, with smoother surfaces demonstrating lower friction coefficients. However, in the case of TCS, a consistently low-friction characteristic was sustained irrespective of the applied normal load. Furthermore, as depicted in Figure 10b, TCS demonstrated the most minimal wear rate of 0.41×10^{-7} mm^3/N·mm. This underscores the potential of applying a Teflon coating to the diaphragm plate surface to yield consistent frictional attributes. In contrast, PI (with a wear rate of 1.64×10^{-7} mm^3/N·mm) and BS (with a wear rate of 0.63×10^{-7} mm^3/N·mm) displayed reduced wear rates under high-load conditions (200 mN) compared to low-load conditions (100 mN), signifying their superior wear resistance characteristics in elevated load conditions. It is, however, worth noting that considering the lower wear rate of BS in comparison to PI, and the minimal wear rate exhibited by TCS, the superior durability of stainless steel 301 is evident. Consequently, stainless steel 301 and Teflon coating are deemed more appropriate materials for diaphragm plates. As shown in Figure 11, the wear morphology formed in each specimen under a load of 200 mN was found to be analogous to that exhibited when only applying a normal load of 100 mN, while keeping the other experimental conditions constant. While the size of the wear track slightly increased, there were no significant differences in the damage morphology of the scratch marks caused by the relative sliding motion.

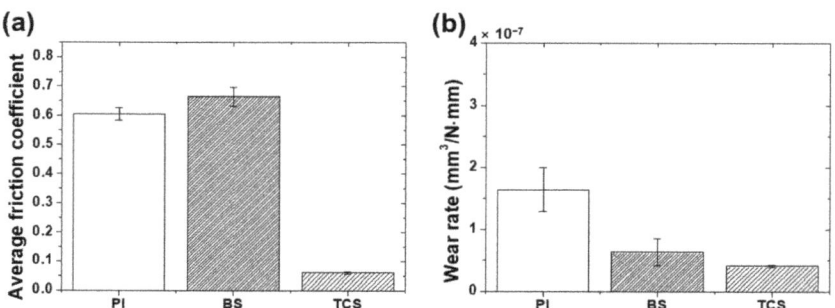

Figure 10. The friction and wear characteristics of PI, BS, and TCS under 200 mN at 16 mm/s during 10,000 cycles: (**a**) average friction coefficient and (**b**) wear rate.

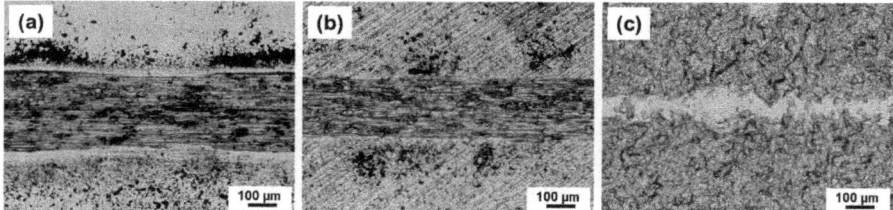

Figure 11. Microscope images of wear tracks under 200 mN at 16 mm/s during 10,000 cycles: (**a**) PI, (**b**) BS, and (**c**) TCS.

Finite element analysis (FEA) was conducted to assess the stress distribution within the diaphragm in response to the internal pressure variations of the compressor. As depicted in Figure 12, the model encompassed the head section of the diaphragm compressor, incorporating the triple diaphragm plates, diaphragm fixation section, and deformation-affected chamber segment. The head cover possessed an inwardly curved surface, resulting in a void between its inner surface and the diaphragm. The void space's height, representing the distance between the center point of the inner surface of the head cover and the diaphragm, was modeled as 3 mm, reflecting the maximum degree of diaphragm deformation. The diaphragm plates were simulated as circular plates with a diameter of 320 mm and a thickness of 0.4 mm. The configuration of the diaphragm incorporated three overlapping plates securely affixed through bolting. The triple diaphragm plates comprised the oil diaphragm, gas diaphragm, and middle diaphragm. The oil diaphragm resided on the side exposed to the actuating oil, the gas diaphragm was in contact with the gas, and the middle diaphragm was situated between the oil and gas diaphragm plates. Material properties derived from tensile testing were applied to the FEA simulation for both the head cover and diaphragm plates. To minimize the error in interpretation, the mesh of the diaphragm plate was densely constructed, and considering the circular shape of the diaphragm plate, the mesh was formed into a rectangular structure. Considering the joint between the head and diaphragm plate, it was fastened from the end of diaphragm plate towards its center up to 5 mm. The contact conditions between the inside of the head and diaphragm plate were determined using a surface-to-surface method, and pressure was applied to the entire surface of the oil diaphragm while in a state where diaphragm plate was fastened, inducing deformation in the diaphragm plate. The stress and deformation profiles of the head cover and diaphragm were analyzed as pressure was incrementally applied to reach 200 bar and 700 bar, respectively, followed by subsequent pressure release. Figures 13 and 14 showcase the stress responses of the diaphragm and head cover under pressure conditions of 200 and 700 bar. The maximum stress observed under both pressure conditions ranged from 139 MPa to 163 MPa, with no notable distinction in maximum stress contingent upon the diaphragm plate's location or pressure condition. However, for both pressure levels, the gas diaphragm experienced the highest stress, followed by the middle diaphragm and the oil diaphragm. This phenomenon is likely attributable to the bending of diaphragm plates in contact with each other under the pressure exerted on the oil diaphragm. Consequently, they convexly deformed toward the gas diaphragm and came into contact with the inner portion of the head cover of the diaphragm compressor [37]. Under the 200-bar pressure, stress gradually increased and subsequently decreased over the same period of deformation. In contrast, under the 700-bar pressure, stress exhibited rapid initial increase, followed by a period of constancy, and concluded with a decrease concurrent with the pressure reduction. This discrepancy underscores the impact of pressure and deformation time on the diaphragm's response. As pressure on the diaphragm escalated, the time required for stress increment shortened, leading to amplified contact pressure with the head cover. Additionally, the maximum stress applied to the head cover substantially rose with increasing pressure, reaching 70 MPa under 200 bar and 160 MPa under 700 bar. The dimensions and radius of the head cover, which accommodated space for diaphragm bending deformation, exerted notable influence on the extent of stress and deformation. As the void's size increased, stress and deformation levels augmented. Nevertheless, there remained no marked variance in the degree of stress and deformation among diaphragms within head covers sharing identical radii [12]. Consequently, the distribution of diaphragm stress was intrinsically influenced by gas pressure and head cover radius, necessitating post-treatment interventions to mitigate damage caused by contact with the head cover.

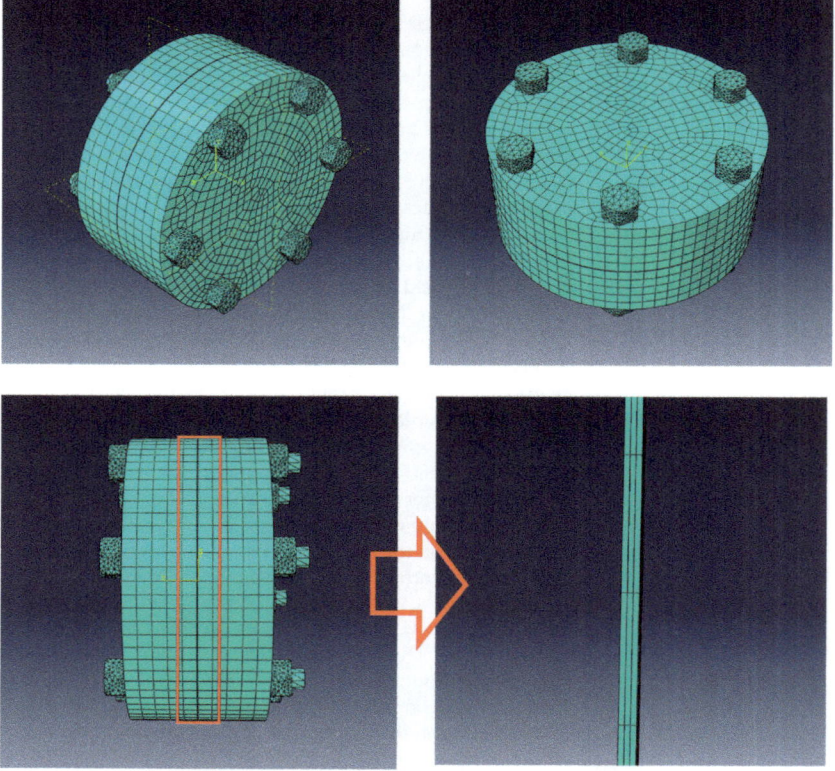

Figure 12. The finite element analysis model of the diaphragm compressor in this study.

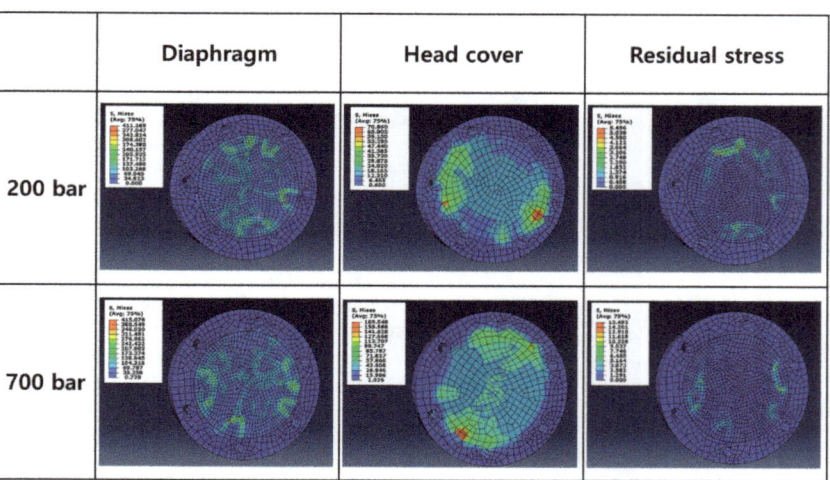

Figure 13. The FEA simulation results of the diaphragm and head cover at 200- and 700-bar pressure.

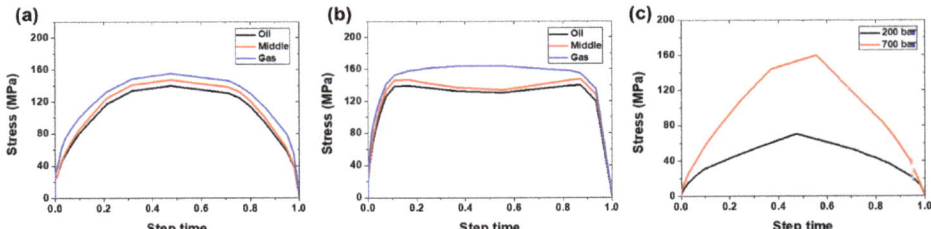

Figure 14. Stress variation curves: oil, middle and gas diaphragm plates (**a**) under 200 bar and (**b**) under 700 bar, and (**c**) cavity under the different pressure conditions (200 bar, 700 bar).

In this study, the friction and wear characteristics of Teflon-coated stainless steel 301 were observed to be superior. To determine the optimal coating conditions that exhibit the most favorable friction and wear characteristics, additional experiments considering various factors such as the temperature dependence of Teflon and P-V diagrams will be necessary. In particular, future research will likely require a comprehensive investigation into chemical, physical, and durability changes caused by environmental changes within a compressor during operation.

4. Conclusions

This paper presents a comprehensive assessment of friction characteristics pertinent to diaphragm materials employed in high-pressure hydrogen gas compressor applications. In this study, Inconel 718 and stainless steel 301 were identified as suitable diaphragm materials, with stainless steel 301 exhibiting superior mechanical strength compared to Inconel 718, as confirmed through hardness and tensile tests. Four distinctive conditions, namely BI, PI, BS, and TCS, were examined through specimen preparation to scrutinize their friction attributes. Notably, TCS demonstrated the most favorable friction and wear characteristics among the assessed materials. Particularly noteworthy was the fact that the friction coefficient of TCS was approximately 70% lower than that of the other specimens. The study further explored the influence of sliding speed and applied load on friction and wear characteristics. It was found that higher sliding speeds led to reduced friction coefficients but increased wear rates. Moreover, increasing the applied load resulted in higher friction but improved wear resistance for polished Inconel (PI) and bare stainless steel (BS). TCS maintained low-friction properties regardless of sliding speed or applied load. Finite element analysis (FEA) simulations were conducted to analyze stress distribution within the diaphragm under different pressure conditions. The results revealed that all three diaphragm plates experienced maximum stress near their contact areas with each other, with gas diaphragms experiencing the highest stress levels. Overall, this research contributes to enhancing the performance and reliability of hydrogen diaphragm compressors by providing insights into surface treatment techniques for improving friction and wear characteristics of diaphragm plates. These findings have implications for advancing hydrogen energy utilization, addressing environmental challenges, and promoting sustainable energy practices in industrial applications.

Author Contributions: S.-J.L.: Conceptualization, Methodology, Software, Validation, Formal analysis, Investigation, Data Curation, Writing—Original Draft, Writing—Review and Editing, Visualization. Y.S.: Conceptualization, Methodology, Software, Validation. D.Z.S.: Commented on manuscript, Resources, Validation. C.-L.K.: Conceptualization, Methodology, Resources, Writing—Review and Editing, Supervision, Project administration. All authors have read and agreed to the published version of the manuscript.

Funding: This research was supported by "Regional Innovation Strategy (RIS)" through the National Research Foundation of Korea (NRF) funded by the Ministry of Education (MOE) (2021RIS-002).

Data Availability Statement: Data is available on request from the corresponding author.

Conflicts of Interest: The authors declare no conflict of interest.

References

1. Siddiqua, A.; Hahladakis, J.N.; Al-Attiya, W.A.K. An overview of the environmental pollution and health effects associated with waste landfilling and open dumping. *Environ. Sci. Pollut. Res.* **2022**, *29*, 58514–58536. [CrossRef] [PubMed]
2. Dar, A.A.; Hameed, J.; Huo, C.; Sarfraz, M.; Albasher, G.; Wang, C.; Nawaz, A. Recent optimization and panelizing measures for green energy projects; insights into CO_2 emission influencing to circular economy. *Fuel* **2022**, *314*, 123094. [CrossRef]
3. Kuthiala, T.; Thakur, K.; Sharma, D.; Singh, G.; Khatri, M.; Arya, S.K. The eco-friendly approach of cocktail enzyme in agricultural waste treatment: A comprehensive review. *Int. J. Biol. Macromol.* **2022**, *209*, 1956–1974. [CrossRef]
4. Arsad, A.; Hannan, M.; Al-Shetwi, A.Q.; Mansur, M.; Muttaqi, K.; Dong, Z.; Blaabjerg, F. Hydrogen energy storage integrated hybrid renewable energy systems: A review analysis for future research directions. *Int. J. Hydrogen Energy* **2022**, *47*, 17285–17312. [CrossRef]
5. Xu, X.; Zhou, Q.; Yu, D. The future of hydrogen energy: Bio-hydrogen production technology. *Int. J. Hydrogen Energy* **2022**, *47*, 33677–33698. [CrossRef]
6. LeValley, T.L.; Richard, A.R.; Fan, M. The progress in water gas shift and steam reforming hydrogen production technologies–A review. *Int. J. Hydrogen Energy* **2014**, *39*, 16983–17000. [CrossRef]
7. Wang, L.; Jiao, S.; Xie, Y.; Xia, S.; Zhang, D.; Zhang, Y.; Li, M. Two-way dynamic pricing mechanism of hydrogen filling stations in electric-hydrogen coupling system enhanced by blockchain. *Energy* **2022**, *239*, 122194. [CrossRef]
8. Usman, M.R. Hydrogen storage methods: Review and current status. *Renew. Sustain. Energy Rev.* **2022**, *167*, 112743. [CrossRef]
9. Nguyen, D.H.; Kim, J.H.; Vo, T.T.N.; Kim, N.; Ahn, H.S. Design of portable hydrogen tank using adsorption material as storage media: An alternative to Type IV compressed tank. *Appl. Energy* **2022**, *310*, 118552. [CrossRef]
10. Zou, J.; Han, N.; Yan, J.; Feng, Q.; Wang, Y.; Zhao, Z.; Fan, Z.; Zeng, L.; Li, H.; Wang, H. Electrochemical compression technologies for high-pressure hydrogen: Current status, challenges and perspective. *Electrochem. Energy Rev.* **2020**, *3*, 690–729. [CrossRef]
11. Ren, S.; Jia, X.; Jiang, J.; Zhang, S.; Zhao, B.; Peng, X. Effect of hydraulic oil compressibility on the volumetric efficiency of a diaphragm compressor for hydrogen refueling stations. *Int. J. Hydrogen Energy* **2022**, *47*, 15224–15235. [CrossRef]
12. Wang, T.; Jia, X.; Li, X.; Ren, S.; Peng, X. Thermal-structural coupled analysis and improvement of the diaphragm compressor cylinder head for a hydrogen refueling station. *Int. J. Hydrogen Energy* **2020**, *45*, 809–821. [CrossRef]
13. Altukhov, S.; Zhukov, A. Determining the geometric parameters of the compression chamber of a diaphragm compressor with a hydraulic drive. *Chem. Pet. Eng.* **1987**, *22*, 309–312. [CrossRef]
14. Altukhov, S. Service life of compressor diaphragms. *Chem. Pet. Eng.* **1965**, *1*, 351–355. [CrossRef]
15. Voyer, J.; Klien, S.; Velkavrh, I.; Ausserer, F.; Diem, A. Static and Dynamic Friction of Pure and Friction-Modified PA6 Polymers in Contact with Steel Surfaces: Influence of Surface Roughness and Environmental Conditions. *Lubricants* **2019**, *7*, 17. [CrossRef]
16. Gürgen, S.; Celik, O.N.; Kushan, M.C. Tribological behavior of UHMWPE matrix composites reinforced with PTFE particles and aramid fibers. *Compos. Pt. B-Eng.* **2019**, *173*, 106949. [CrossRef]
17. Krick, B.A.; Ewin, J.J.; McCumiskey, E.J. Tribofilm formation and run-in behavior in ultra-low-wearing polytetrafluoroethylene (PTFE) and alumina nanocomposites. *Tribol. Trans.* **2014**, *57*, 1058–1065. [CrossRef]
18. Monteiro, S.N.; Nascimento, L.F.C.; Simonassi, N.T.; Lima, E.S.; de Paula, A.S.; de Oliveira Braga, F. High temperature work hardening stages, dynamic strain aging and related dislocation structure in tensile deformed AISI 301 stainless steel. *J. Mater. Res. Technol. -JMRT* **2018**, *7*, 571–577. [CrossRef]
19. McLouth, T.D.; Witkin, D.B.; Lohser, J.R.; Sitzman, S.D.; Adams, P.M.; Lingley, Z.R.; Bean, G.E.; Yang, J.-M.; Zaldivar, R.J. Temperature and strain-rate dependence of the elevated temperature ductility of Inconel 718 prepared by selective laser melting. *Mater. Sci. Eng. A-Struct. Mater. Prop. Microstruct. Process.* **2021**, *824*, 141814. [CrossRef]
20. Yang, Z.; Tian, Y.; Zhao, Y.; Yang, C. Study on the fabrication of super-hydrophobic surface on inconel alloy via nanosecond laser ablation. *Materials* **2019**, *12*, 278. [CrossRef] [PubMed]
21. Lee, S.-J.; Sohn, Y.-C.; Kim, C.-L. Tribological Effects of Water-Based Graphene Lubricants on Graphene Coatings. *Materials* **2023**, *16*, 197. [CrossRef]
22. Lee, S.-J.; Kim, G.-M.; Kim, C.-L. Self-lubrication and tribological properties of polymer composites containing lubricant. *RSC Adv.* **2023**, *13*, 3541–3551. [CrossRef] [PubMed]
23. Fischer-Cripps, A.C. The Hertzian contact surface. *J. Mater. Sci.* **1999**, *34*, 129–137. [CrossRef]
24. Lee, S.-J.; Kim, G.-M.; Kim, C.-L. Design and evaluation of micro-sized glass bubble embedded PDMS composite for application to haptic forceps. *Polym. Test.* **2023**, *117*, 107855. [CrossRef]
25. Lee, S.-J.; Kim, C.-L. Highly flexible, stretchable, durable conductive electrode for human-body-attachable wearable sensor application. *Polym. Test.* **2023**, *122*, 108018. [CrossRef]
26. Hanief, M.; Wani, M. Effect of surface roughness on wear rate during running-in of En31-steel: Model and experimental validation. *Mater. Lett.* **2016**, *176*, 91–93. [CrossRef]
27. Yu, C.; Wan, H.; Chen, L.; Li, H.; Cui, H.; Ju, P.; Zhou, H.; Chen, J. Marvelous abilities for polyhedral oligomeric silsesquioxane to improve tribological properties of polyamide-imide/polytetrafluoroethylene coatings. *J. Mater. Sci.* **2018**, *53*, 12616–12627. [CrossRef]

28. Peng, S.; Zhang, L.; Xie, G.; Guo, Y.; Si, L.; Luo, J. Friction and wear behavior of PTFE coatings modified with poly (methyl methacrylate). *Compos. Part B Eng.* **2019**, *172*, 316–322. [CrossRef]
29. Kameda, T.; Sato, H.; Oka, S.; Miyazaki, A.; Ohkuma, K.; Terada, K. Low temperature polytetrafluoroethylene (FTFE) coating improves the appearance of orthodontic wires without changing their mechanical properties. *Dent. Mater. J.* **2020**, *39*, 721–734. [CrossRef]
30. Xia, J.; Zhao, J.; Dou, S. Friction characteristics analysis of symmetric aluminum alloy parts in warm forming process. *Symmetry* **2022**, *14*, 166. [CrossRef]
31. Chey, S.K.; Tian, P.; Tian, Y. Estimation of real contact area during sliding friction from interface temperature. *Aip Adv.* **2016**, *6*, 065227. [CrossRef]
32. Reddyhoff, T.; Schmidt, A.; Spikes, H. Thermal conductivity and flash temperature. *Tribol. Lett.* **2019**, *67*, 1–9. [CrossRef]
33. Mao, L.; Cai, M.; Liu, Q.; He, Y. Effects of sliding speed on the tribological behavior of AA 7075 petroleum casing in simulated drilling environment. *Tribol. Int.* **2020**, *145*, 106194. [CrossRef]
34. Lee, H.-Y. Effect of changing sliding speed on wear behavior of mild carbon steel. *Met. Mater.-Int.* **2020**, *26*, 1749–1756. [CrossRef]
35. Schroeder, R.; Torres, F.; Binder, C.; Klein, A.; De Mello, J. Failure mode in sliding wear of PEEK based composites. *Wear* **2013**, *301*, 717–726. [CrossRef]
36. Kim, C.-L.; Jung, C.-W.; Oh, Y.-J.; Kim, D.-E. A highly flexible transparent conductive electrode based on nanomaterials. *NPG Asia Mater.* **2017**, *9*, e438. [CrossRef]
37. Li, X.; Chen, J.; Wang, Z.; Jia, X.; Peng, X. A non-destructive fault diagnosis method for a diaphragm compressor in the hydrogen refueling station. *Int. J. Hydrogen Energy* **2019**, *44*, 24301–24311. [CrossRef]

Disclaimer/Publisher's Note: The statements, opinions and data contained in all publications are solely those of the individual author(s) and contributor(s) and not of MDPI and/or the editor(s). MDPI and/or the editor(s) disclaim responsibility for any injury to people or property resulting from any ideas, methods, instructions or products referred to in the content.

Article

Effect of Copper Nanoparticles Surface-Capped by Dialkyl Dithiophosphate on Different Base Oil Viscosity

Xufei Wang, Shuguang Fan, Ningning Song, Laigui Yu, Yujuan Zhang * and Shengmao Zhang *

Institute of Nanoscience and Engineering, Henan University, Kaifeng 475004, China; wangxufei2016@163.com (X.W.); sgfan@vip.henu.edu.cn (S.F.); songning123456@163.com (N.S.); lgyu1963@sina.cn (L.Y.)
* Correspondence: cnzhangyujuan@henu.edu.cn (Y.Z.); zsm@henu.edu.cn (S.Z.)

Abstract: In order to more accurately characterize the effects of nanoparticles on lubricant viscosity, the effects of copper dialkyl dithiophosphate (HDDP)-modified (CuDDP) nanoparticles on the dynamic viscosity of mineral oils 150N, alkylated naphthalene (AN5), diisooctyl sebacate (DIOS), and polyalphaolefins (PAO4, PAO6, PAO10, PAO40, and PAO100) were investigated at an experimental temperature of 40 °C and additive mass fraction ranging from 0.5% to 2.5%. CuDDP exhibits a viscosity-reducing effect on higher-viscosity base oils, such as PAO40 and PAO100, and a viscosity-increasing effect on lower-viscosity base oils, namely, 150N, AN5, DIOS, PAO4, PAO6, and PAO10. These effects can be attributed to the interfacial slip effect and the shear resistance of the nanoparticles. The experimental dynamic viscosity of the eight base oils containing CuDDP was compared with that calculated by the three classical formulae of nanofluid viscosity, The predicted viscosity values of the formulae deviated greatly from the experimental viscosity values, with the maximum deviation being 7.9%. On this basis, the interface slip effect was introduced into Einstein's formula, the interface effect was quantified with the aniline point of the base oil, and a new equation was established to reflect the influence of CuDDP nanoparticles on lubricating oil viscosity. It can better reflect the influence of CuDDP on the viscosity of various base oils, and the deviation from the experimental data is less than 1.7%.

Keywords: copper nanoparticle; dialkyl dithiophosphate; nanofluid; viscosity

Citation: Wang, X.; Fan, S.; Song, N.; Yu, L.; Zhang, Y.; Zhang, S. Effect of Copper Nanoparticles Surface-Capped by Dialkyl Dithiophosphate on Different Base Oil Viscosity. *Lubricants* **2024**, *12*, 137. https://doi.org/10.3390/lubricants12040137

Received: 25 January 2024
Revised: 27 March 2024
Accepted: 31 March 2024
Published: 18 April 2024

Copyright: © 2024 by the authors. Licensee MDPI, Basel, Switzerland. This article is an open access article distributed under the terms and conditions of the Creative Commons Attribution (CC BY) license (https://creativecommons.org/licenses/by/4.0/).

1. Introduction

Lubricating oil has a significant impact on the working efficiency and service life of mechanical equipment, since lubricants can reduce the friction and wear between moving parts, as well as the energy loss in a mechanical movement. With the rapid development of nanomaterials, researchers have recently conducted extensive studies on the application of nanomaterials as lubricant additives. It has been found that nanomaterials with unique characteristics (e.g., small dimensions, large surface area, and high surface activity) can improve the friction-reducing and anti-wear abilities of lubricating oil [1–4]. Among various nanomaterials, nano-copper with low shear strength and grain boundary slip effect might be a promising multifunctional lubricant additive thanks to its synergistic friction reduction, anti-wear, and self-repairing abilities [5–8]. This new nanoadditive can perform excellently, providing more reliable support for the smooth operation of mechanical equipment. Considering energy and environmental perspectives, lubricating oil can effectively reduce energy loss and improve the energy efficiency of mechanical equipment, thereby relieving energy pressure to a certain extent. In addition, reducing friction and wear also helps to extend the service life of mechanical equipment and reduce resource consumption and waste generation. Therefore, studying nanomaterials as lubricant additives holds significance for the machinery industry. Nanoscale particles are dispersed in a conventional fluid medium (e.g., water, oil, or glycol) to form a homogeneous and stable fluid medium known as a nanofluid, in which the lubricating oil is used as a solvent known as a nanolubricant. In future developments, we can anticipate further optimization and broader application of nanolubricants to meet mechanical equipment's increasingly stringent performance and environmental requirements. This advancement promises new possibilities for industrial production. Consequently, in-depth investigation and application of nanolubricants possess significant scientific and practical value.

Viscosity is a pivotal property of lubricating oil, acting as a crucial indicator of lubricating oil fluidity and internal friction. Viscosity values that are either too high or too low can detrimentally affect the lubricant's performance: increased viscosity may lead to heightened frictional forces during fluid lubrication, while reduced viscosity can decrease the lubricant's load-bearing capacity [9,10]. Breki et al. [11] integrated Einstein's viscosity equation with dynamic lubrication theory, elucidating the relationship between the friction coefficient and the solid-phase volume fraction, underscoring the significance of viscosity in fluid lubrication. This underscores the importance of investigating the influence of nanoadditives on the viscosity of nanolubricants. However, the existing literature presents divergent results and conclusions concerning the impact of nanoadditives on lubricant viscosity [12–17]. Various experimental data show that the addition of nanomaterials may cause complex changes in lubricating oil viscosity, and the specific effect may depend on the type of nanoparticles, mass fraction, and operating conditions. For example, Ma et al. [18] demonstrated that introducing ZnO nanoparticles enhances the viscosity of SAE50 lubricant, likely due to augmented resistance to lubricant flow induced by nanoparticle agglomeration under van der Waals forces. Hemmat et al. [19] found that Al_2O_3 nanoadditives elevate the viscosity of 10W40 lubricant at 55 °C by 132%, noting that nanolubricant viscosity initially increases and then diminishes with rising temperature, attributed to the augmented shear thermal effect. They [20] further found that MgO nanoparticles reduce the viscosity of 5W30 lubricant, with the spherical nanoparticles functioning as roller balls between fluid layers. Mustafa et al. [21] observed that at low concentrations of TiO_2–CuO NPs, the mobility between nano-lubricating oil is facilitated, slightly reducing viscosity. However, at higher concentrations, due to agglomeration or increased particle size, the movement between oil layers is hindered, resulting in elevated viscosity. Sui et al. [22] examined the viscosification effects of four types of SiO_2 on PAO100, noting that while nanoparticles modified with different functional groups did not significantly impact PAO at 100 °C, variations in nanoparticle size did affect PAO viscosity at this temperature.

In the realm of nanofluid viscosity, comprehensive research has elucidated the influence of nanomaterials on the viscosity attributes of fluids. The viscosity of nanofluids is intricately linked to the nanomaterials' size, density, ultrasonic treatment time, and interfacial interactions. For example, Abdelhalim et al. [23] observed an increase in the viscosity of nanofluids with the incorporation of larger Au nanoparticles, reinforcing the notion that nanoparticle size is a critical determinant of nanofluid viscosity. Dehghani et al. [24] compared the viscosities of WO_3 and Al_2O_3 nanoparticles dispersed in deionized water and liquid paraffin, noting higher viscosities in the former, which may be attributed to WO_3's greater density and reduced Brownian motion velocity, underscoring density's role in influencing lubrication characteristics. Zhang et al. [25] conducted an experimental investigation into the viscosity of hydrophilic TiO_2–water and hydrophobic TiO_2–water nanofluids, discovering that hydrophilic nanoparticles form water-attracting layers more swiftly than hydrophobic ones, leading to higher viscosities in nanofluids with hydrophilic nanoparticles.

These investigations enhance our comprehension of how nanoparticles affect fluid viscosity and offer empirical data elucidating the mechanisms by which nanoparticles modulate viscosity. Nonetheless, a consolidated consensus on the impact of nanoparticles on viscosity and the associated mechanisms remains elusive. Therefore, to gain a deeper understanding of nanomaterials' effects on lubricant viscosity, further research is imperative. In this study, we examine the impact of dialkyl dithiophosphate (HDDP) and copper HDDP-modified (CuDDP) nanoparticles on the kinematic viscosity of various base oils. Diverging from prior research, we introduce the aniline point as a metric of lubricant polarity. By integrating the aniline point, we formulate a novel equation intended to characterize the viscosity of base oils infused with nanoparticles across different volume fractions. This innovative methodology offers a fresh lens through which to view the impact of nanoparticles on lubricant viscosity. Through rigorous analysis of the correlations between predicted dynamic viscosity and measured data, we aim to uncover regularity and determinants. This endeavor enhances our understanding of how nanoparticles influence lubricant viscosity. Finally, we undertake a thorough comparison of the experimental outcomes with the computational results to validate the accuracy of the newly developed formulae. This comparison is instrumental in ascertaining the practical applicability of our theoretical model and establishes a groundwork for future inquiries.

2. Materials and Methods

2.1. Material Characterization

Fourier-transform infrared (FTIR) spectroscopy (Tensor II, Bruker, Billerica, MS, USA) covering a wavelength range of 400 cm^{-1} to 4000 cm^{-1} was employed to ascertain the composition of the modifier. The thermal stability and modifier content of the sample were examined using a thermogravimetric analyzer (TGA/DSC3+, Mettler Toledo, Greifensee, Switzerland). The thermal analysis was conducted in a nitrogen atmosphere, with a heating rate of 10 °C per minute, spanning from 25 °C to 900 °C. To eliminate impurities, the sample was maintained at 100 °C for 5 min before the analysis. Additionally, the morphology and size of the CuDDP nanoparticles were characterized using transmission electron microscopy (TEM, JEM-F200, JEOL, Tokyo, Japan). This technique provided detailed insights into the nanostructure of CuDDP nanoparticles, which is crucial for understanding its interactions and performance in lubricant applications.

2.2. Sample Preparation

In our experiments, oil-soluble copper nanoparticles (CuNPs) and CuNPs surface-coated with oil-soluble dialkyl dithiophosphate (CuDDP) prepared by the Nanomaterials Engineering and Technology Research Center of Henan University (Kaifeng, China) were used as the nanoadditives. CuDDP nanoparticles were synthesized by means of a redox surface modification technique [26]. They were dispersed in base oils 150N, alkylated naphthalene (AN5), and diisooctyl sebacate (DIOS), as well as polyalphaolefins (PAO4, PAO6, PAO10, PAO40, and PAO100) at mass fractions of 0.5%, 1.0%, 1.5%, 2.0% and 2.5%. HDPP was dispersed in base oils 150N, AN5, DIOS and PAO6 at mass fractions of 0.5%, 1.0%, 1.5%, 2.0% and 2.5%, respectively. The eight base oils employed in the experiment were obtained from Qingdao Lubemater Group (Shandong, China). The typical physical properties of these oils are delineated in Table 1. The lubricant sample was mixed ultrasonically for 15 min to achieve uniform dispersion of the nanoadditives CuDDP and HDDP. CuDDP had good dispersion stability in the base oil used in the experiment, and no samples showed obvious precipitation after standing for 7 days.

Table 1. Physical properties of HDDP and various base oils at 40 °C.

Substances	Density (g/cm^3)	Kinematic Viscosity (mm^2/s)
HDDP	0.9829	12.88
150N	0.8288	34.09
AN5	0.8990	26.36
DIOS	0.9007	11.38
PAO4	0.8070	18.21
PAO6	0.8139	30.59
PAO10	0.8215	62.57
PAO40	0.8369	391.6
PAO100	0.8420	1010

2.3. Viscosity and Density Tests

A viscometer (SVM3001, Anton Paar, Styria, Austria) was employed to determine the kinematic viscosity of the lubricants. It uses the oscillating piston method to measure the density and dynamic viscosity of the sample and can adjust the temperature and calculate the kinematic viscosity automatically. At the end of the measurement, the measuring cell was fully washed with petroleum ether and anhydrous ethanol and dried by blowing. During the experimental procedure, the thermometer showed an expanded (k = 2) uncertainty of 0.03 °C. Relative expanded (k = 2) uncertainty of 0.35% was estimated for the kinematic viscosity.

The densities of CuDDP and PAO4 dispersions at various concentrations were quantified employing SVM3001. Subsequently, the densities of CuDDP were extrapolated utilizing the equations derived from curve fitting. The expanded (k = 2) uncertainty of density measurements performed with the SVM3001 was 0.0005 g·cm^{-3}.

2.4. Aniline Point Test

The aniline point of the eight base oils used in the experiment was tested with a petroleum product aniline point tester (DZY-013A, Dalian Instruments and Meters Co., Ltd., Dalian, China), and details about the test method are described in GB/T262 "Determination of Aniline Point of Petroleum Products".

3. Results and Discussion

3.1. Characterization of CuDDP

Figure 1a presents the thermogravimetric analysis (TGA) curve of CuDDP, where the weight loss observed below 100 °C is attributed to the removal of impurities. In the temperature range of 100 °C to 900 °C, CuDDP exhibits a weight loss of approximately 72% (mass fraction), suggesting that the organic modifier constitutes around 72% of its composition. Figure 1b illustrates the FTIR spectrum of CuDDP, with distinct C-H characteristic peaks (including CH_3 and CH_2) at 1380 cm^{-1}. Pertinent to this study are two pronounced absorption peaks near 1000 cm^{-1}, attributed to P-O-C at 636 cm^{-1} and P=S, indicative of the HDDP modifier's presence [27,28]. Figure 1c displays a TEM image and particle size distribution of CuDDP, revealing that the nanoparticles are spherical with an average diameter of approximately 5 nm, as shown in Figure 1d. These observations provide critical insights into the structural and compositional attributes of CuDDP, essential for understanding its behavior and efficacy as a lubricant additive.

Table 2 enumerates the densities of the PAO4 and CuDDP mixtures. Figure 1e illustrates the dependency of the mixture density on the CuDDP mass fraction, which has been subjected to a fitting procedure to derive Equation (1), where ρ_{nf} represents the density of the mixture, ρ_{bf} denotes the density of the base oil, ω symbolizes the mass fraction of CuDDP, and k is the fitting factor. This equation exhibits a correlation coefficient R^2 exceeding 0.99. Inserting a 100% mass fraction into Equation (1) results in a calculated density of CuDDP of 1.171 g/m^3, which is designated in this context the nanoparticle fitting density. It is pertinent to highlight that the nanoparticle fitting density (NFD) is a conceptualized density, formulated to precisely ascertain the volume fraction of nanoparticles within a dispersion, and its applicability is confined to scenarios of low concentration.

$$\rho_{nf} = \rho_{bf} + k\omega, \qquad (1)$$

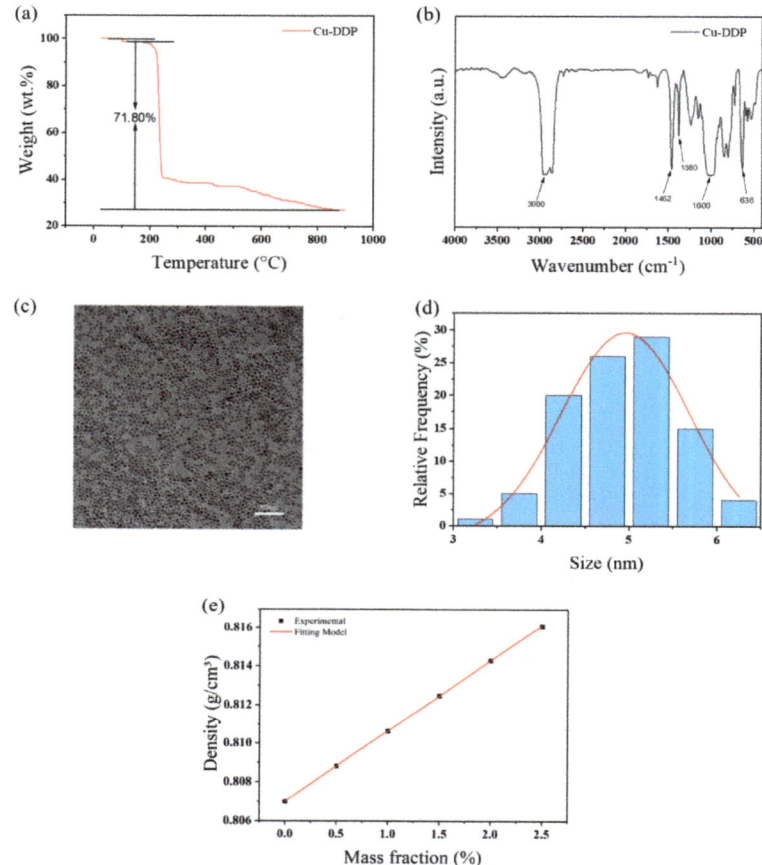

Figure 1. (**a**) TGA curves, (**b**) FTIR spectra of CuDDP, (**c**) TEM image of CuDDP, (**d**) and particle size distribution of CuDDP. (**e**) Density of CuDDP and PAO4 mixture changes with CuDDP mass fraction.

Table 2. Density of PAO4 and CuDDP mixtures at 40 °C.

Mass Fraction (%)	Density (g/cm³)
0	0.8070
0.5	0.8088
1	0.8106
1.5	0.8125
2	0.8143
2.5	0.8161

3.2. Kinematic Viscosity

In the study of lubricating oils, kinematic viscosity is a critical performance parameter that significantly influences the efficacy of lubricating oil in mechanical systems. Arrhenius proposed the following expression for calculating the viscosity of a mixed solution [29]:

$$\lg \nu_m = \omega_1 \lg \nu_1 + \omega_2 \lg \nu_2 \tag{2}$$

In the equation, ν_m represents the kinematic viscosity of the oil blend, while ν_1 and ν_2 denote the kinematic viscosities of component 1 and component 2 and ω_1 and ω_2 represent the mass fractions of component 1 and component 2. The kinematic viscosity of lubricating oils containing HDDP and CuDDP was calculated using Equation (2) and compared with experimental values. Table 3 shows the experimental kinematic viscosity of four base oils containing HDDP. The empirical outcomes indicate that the discrepancy between the measured kinematic viscosity of HDDP base oils and the theoretical values is minimal. This deviation is ascribed to the lipophilic nature of HDDP, which does not significantly affect the flow characteristics of the lubricating oil. Since HDDP and the base oil are merely physically mixed without substantial interactions, the kinematic viscosity of the samples with added HDDP closely align with the values predicted by Equation (2).

Table 3. Effect of HDDP on kinematic viscosity of four base oils at 40 °C: comparison of experimental values with computed ones (frac., Exp., Calc., and Devi. refer to fraction, experimental, calculated, and deviation).

Mass Frac. (%)	Kinematic Viscosity											
	150N + HDDP			AN5 + HDDP			DIOS + HDDP			PAO6 + HDDP		
	Exp. (m^2/s)	Calc. (m^2/s)	Devi. (%)	Exp. (m^2/s)	Calc. (m^2/s)	Devi. (%)	Exp. (m^2/s)	Calc. (m^2/s)	Devi. (%)	Exp. (m^2/s)	Calc. (m^2/s)	Devi. (%)
0	34.09	34.09	0	26.36	26.36	0.01	11.38	11.38	0	30.59	30.59	0
0.5	33.85	33.92	0.22	26.27	26.27	0.01	11.39	11.38	0.05	30.47	30.46	0.04
1	33.61	33.76	0.44	26.20	26.17	0.11	11.41	11.39	0.14	30.28	30.33	0.15
1.5	33.26	33.6	0.99	26.07	26.08	0.04	11.43	11.4	0.25	30.10	30.19	0.30
2	33.14	33.43	0.89	25.98	25.98	0.01	11.43	11.41	0.24	29.86	30.06	0.67
2.5	32.94	33.27	1.00	25.90	25.89	0.02	11.45	11.41	0.34	29.77	29.93	0.54

Furthermore, the data illustrate a declining trend in the kinematic viscosity of the blend comprising HDDP and PAO6, 150N, and AN5 as the mass fraction of HDDP escalates, as depicted in Figure 2. Conversely, under identical conditions, the kinematic viscosity of DIOS exhibits an increasing trend. Thus, the addition of HDDP predictably influences the kinematic viscosity of lubricating oils. This observation is logically consistent with the fact that HDDP's viscosity is lower than that of PAO6, 150N, and AN5, but higher than DIOS. In contrast, CuDDP is observed to increase the viscosity across all tested base oils, attributable to its stronger shear resistance compared to an equivalent amount of HDDP. This enhanced resistance significantly alters the lubricant's fluidity, resulting in notable viscosity increases in the base oils.

3.3. Dynamic Viscosity

In the investigation of nanofluid viscosity, the impact of nanoparticles on the fluid's viscosity is predominantly linked to the shear resistance associated with nanoparticle volume. Consequently, the correlation between the volume fraction of nanoparticles and the dynamic viscosity is frequently employed to illustrate how nanoparticles affect viscosity. The mathematical expression for converting the mass fraction of nanoparticles to their volume fraction is provided below:

$$\varphi = \frac{m_{np}/\rho_{np}}{m_{np}/\rho_{np} + m_{bf}/\rho_{bf}} \tag{3}$$

In 1905, Einstein focused on elucidating the dimensions of atoms and molecules, resulting in a seminal publication that delineated the relationship between the viscosity of dispersions (including suspensions and colloids) or dilute mixtures consisting of a liquid phase and small dispersed solid particles, and their volume concentration. This relationship was established under the premise of rigid spherical particles moving in an incompressible fluid, leading to the derivation of the subsequent equation [30]:

$$\mu_{nf} = \mu_{bf}(1 + 2.5\varphi) \tag{4}$$

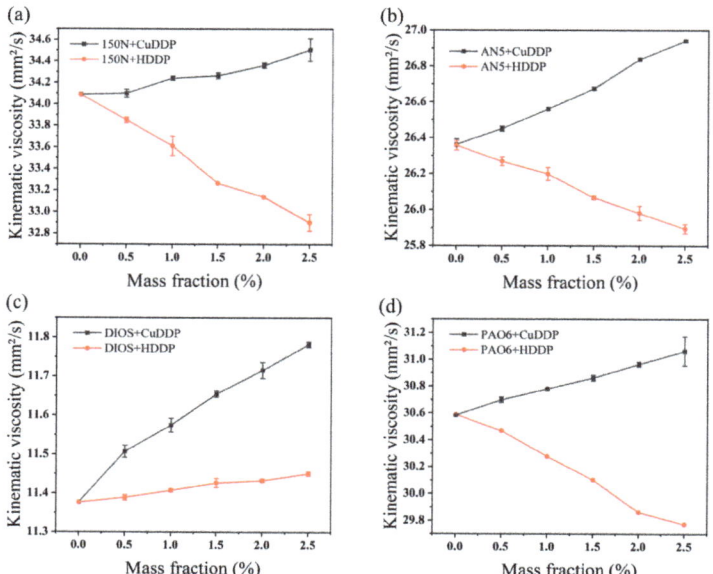

Figure 2. Effects of CuDDP and HDDP on kinematic viscosity of different base oils at 40 °C: (**a**) 150N, (**b**) AN5, (**c**) DIOS, (**d**) PAO6.

Batchelor [31] refined Einstein's equation integrating the influence of Brownian motion and addressing the rotational dynamics of nanoparticles. This augmentation resulted in the derivation of Equation (5):

$$\mu_{nf} = \mu_{bf}(1 + 2.5\varphi + 6.2\varphi^2) \qquad (5)$$

Brinkman [32] developed an expression to characterize the viscosity of solutions and suspensions at finite concentrations, considering the impact of solute molecule addition to the solution and treating the system as a continuum, as shown in Equation (6):

$$\mu_{nf} = \frac{\mu_{bf}}{(1-\varphi)^{2.5}} \qquad (6)$$

In Equations (3)–(6), the symbol ω represents the mass fraction, while φ denotes the volume fraction, which is derived from Equation (3). The variables m_{np} and m_{nf} correspond to the mass of nanoparticle and base oil, ρ_{np} is the density of nanoparticle, ρ_{nf} signifies the density of the nanolubricant, μ_{nf} refers to the viscosity of the nanofluid, and μ_{bf} denotes the viscosity of the base fluid.

The theoretical viscosity values of different base oils with CuDDP were calculated with Equations (4)–(6) and compared with the experimental values shown in Table 4. Pronounced discrepancies were observed between the predicted viscosity values and those obtained experimentally. Notably, the greatest deviations forecast by Einstein's formula, Batchelor's formula, and Brinkman's formula were 7.7%, 7.9%, and 7.9%, respectively. As can be seen in Figure 3, the predicted viscosity values of different base oils tend to increase with the increase in volume fraction of the nanomaterial. The experimental values are lower than predicted ones, which implies that—aside from the shear resistance of the nanoadditive—there are also other factors that cause decreased experimental viscosity values of the lubricants. In addition, the nanoadditive CuDDP leads to increased viscosity in low-viscosity base oils 150N, AN5, DIOS, PAO4, PAO6, and PAO10, as well as decreased viscosity in high-viscosity base oils PAO40 and PAO100. The reason might lie in that the nanoadditive mainly affects the viscosity of the base oil by an anti-shear effect [30], which is more significant in low-viscosity base oils. In high-viscosity base oils, the shear resistance of CuDDP is relatively small, which corresponds to its reduced anti-shear effect and the viscosity of the base oils.

Table 4. Effects of CuDDP on the dynamic viscosity of diverse base oils at 40 °C: experimental results.

Mass Fraction (%)	150N + CuDDP	AN5 + CuDDP	DIOS + CuDDP	PAO4 + CuDDP	PAO6 + CuDDP	PAO10 + CuDDP	PAO40 + CuDDP	PAO100 + CuDDP
0	28.22	23.69	10.34	18.21	24.89	51.42	326.9	1009.8
0.5	28.27	23.81	10.38	18.29	25.03	51.55	326.2	1002.9
1	28.43	23.94	10.45	18.37	25.13	51.69	325.5	994.0
1.5	28.49	24.07	10.53	18.45	25.24	51.80	324.6	989.8
2	28.61	24.24	10.60	18.53	25.36	51.99	323.9	987.5
2.5	28.78	24.37	10.67	18.61	25.45	52.13	323.2	979.7

3.4. Fitting Formula

To better understand the impact of CuDDP on the viscosity of base oils, we modified the relationship between the dynamic viscosity of nanomaterial additives and their volume fraction, based on Einstein's viscosity equation, as follows:

$$\mu_{nf} = \mu_{bf}(1 + \alpha\varphi) \tag{7}$$

In Equation (7), α is the fitting coefficient, and its value is presented in Table 5. The discrepancy between the dynamic viscosity value derived from Equation (6) and the empirically measured value is observed to be less than 0.5%, and the correlation coefficient R^2 is greater than 0.99, which indicates that the fitting equation can better describe the influence of CuDDP on the viscosity of the base oils. Furthermore, based on molecular dynamic simulations and experimental studies, we also established a mechanistic model to describe the solid–liquid interfacial slip behavior and shear viscosity of lubricants in relation to nanomodulation [33]. With the assumption that a fluid is a Newtonian fluid with constant viscosity, a rigid solid would not be deflected upon insertion into the liquid, nor would it be subjected to any inertial force in the flow field. In this case, the shear slip motion is only concentrated on the upper surface of the solid. As seen in Figure 4a, the velocity of the upper shear plate is v_{x1}, and the lower shear plate is static. In a pure liquid, the flow velocity distribution is $V_0(z)$, the shear stress is τ_0, and the viscosity is μ_0. When no solid is added, the shear stress in the Z_{h0} height range of the liquid would be $\tau_{h1} = \mu_0 v_1(Z_{h0})/Z_{h0} = \mu_0 v_{s1}/Z_{h0}$, and the shear stress in its Z_{h0} height range would be $\tau_{h0} = \mu_0 v_0(Z_{h0})/Z_{h0}$. When solids are added to the liquid in the absence of interface slip (Figure 4b), both the upper-surface velocity and the lower-surface velocity of the solids are V_{s1}. In this case, we have $V_{s1} > V_0(Z_{h0})$ and $\tau_{h1} > \tau_{h0}$, as well as a total stress of $\tau_1 > \tau_0$ and a total viscosity of $\mu_0 < \mu_1$, which corresponds to the Einstein model. When the interface slip effect is significant (Figure 4c), we would have $v_{s2} < v_0(Z_{h0})$ and a total viscosity of $\mu_2 < \mu_0$, which corresponds to the experimental value in Figure 3a–f. When the interface slip is less than the shear resistance caused by the solid (Figure 4d), we would have $v_{s1} > v_{s3} > v_0$, a total stress of $\tau_1 > \tau_3 > \tau_0$, and a total viscosity of $\mu_1 > \mu_3 > \mu_0$, which corresponds to the experimental values in Figure 3g,h.

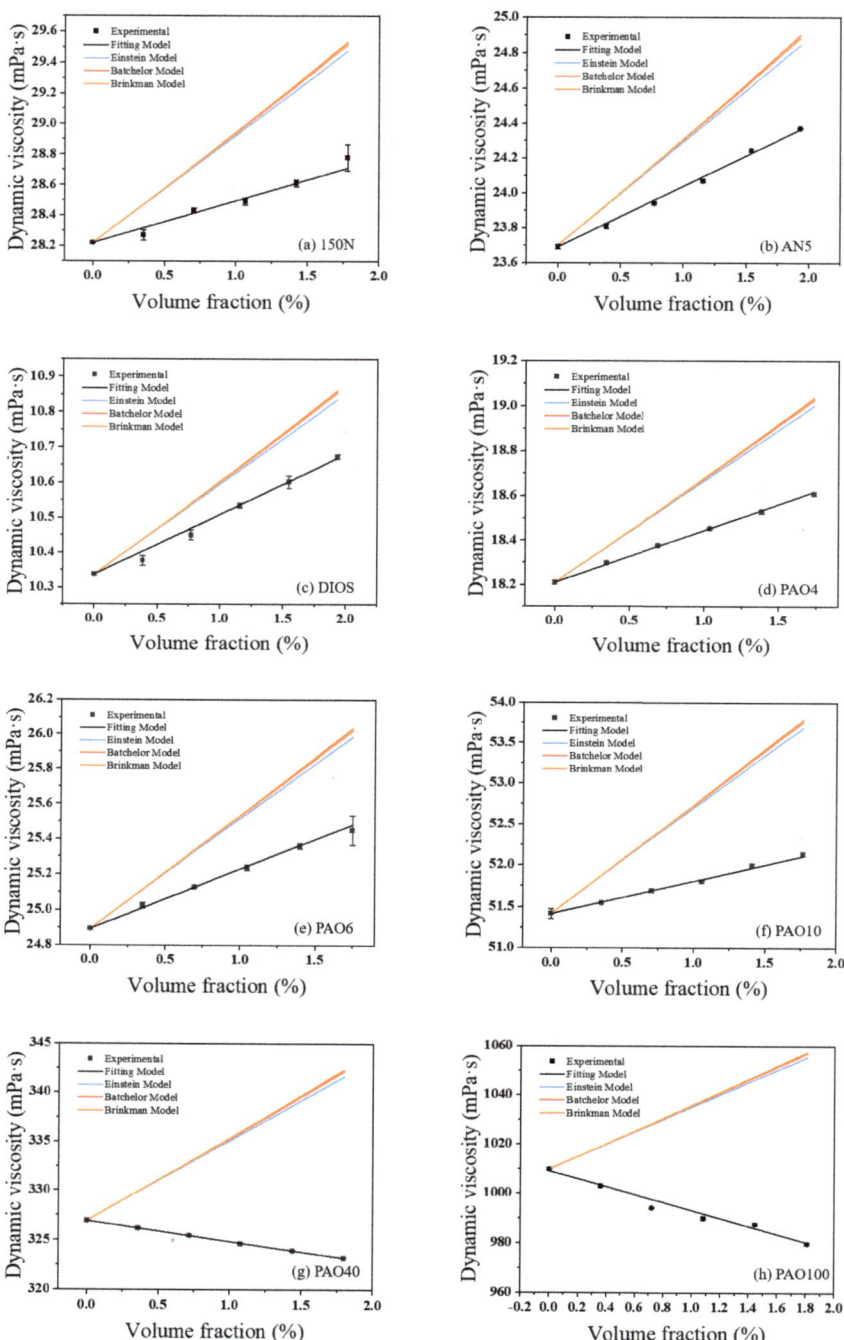

Figure 3. Comparison of experimental and predicted values of dynamic viscosity of various base oils with CuDDP.

Table 5. Values of coefficient α for different base oils.

Base Oils	Coefficient α
150N	0.98
AN5	1.47
DIOS	1.66
PAO4	1.28
PAO6	1.34
PAO10	0.76
PAO40	−0.64
PAO100	−1.64

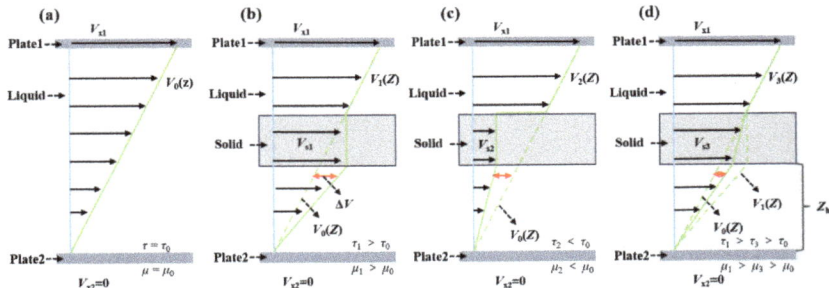

Figure 4. The influence of solids on fluid viscosity of (**a**) pure liquid, (**b**) liquid–solid blend without interface slip, (**c**) low-viscosity base oils containing CuDDP with a large interface slip, and (**d**) high-viscosity base oils containing CuDDP with a weak interface slip.

The aniline point is the lowest temperature required for the oil and an equal volume of aniline to dissolve into a single liquid phase with each other. It serves as a measure of the aromatic content in oil and can also reflect the polarity of lubricating oils [34]. Here, we use the aniline point to represent polarity and reflect the interfacial interactions. Table 6 displays the aniline point test outcomes for the analyzed base oils, and there are discernible correlations between the viscosity of polyalphaolefins and their aniline points. Through fitting the experimental data presented in Table 5 with the aniline point, the coefficient α can be articulated as follows:

$$\alpha = -0.0011 * e^{\frac{A}{21.46}} + 1.67 \tag{8}$$

where A is the aniline point of the base oil and e is the natural logarithmic base. The correlation coefficient is more than 0.98, which indicates that Equation (8) can be used to accurately express the relationship between the base oil aniline point and the coefficient α. When the aniline point is low (Figure 5), α basically remains unchanged with varying polarity of the high-polarity base oil, and in this case, the nanoadditive CuDDP would have a weak interfacial slip as well as an enhanced viscosity-increasing effect therein. When the base oil polarity decreases to a certain degree, α decreases dramatically therewith, and in this case, the interfacial slip is enhanced, while the viscosity-increasing effect of the nanoadditive would be negligible. When the base oil polarity decreases to a certain degree, α decreases dramatically therewith, and in this case, the interfacial slip is enhanced, while the viscosity-increasing effect of the nanoadditive would be negligible. When the base oil polarity continues to decrease, α will drop to 0 and even become negative. In this case, the viscosity-reduction effect caused by interfacial slip is greater than the viscosity-enhancement effect caused by the anti-shear effect of the nanoparticles, and the overall outcome would be a viscosity-reduction effect.

Table 6. Aniline point of various base oils.

Base Oils	Aniline Point (°C)
150N	126
DIOS	−29
AN5	29
PAO4	123
PAO6	137
PAO10	140
PAO40	164
PAO100	172

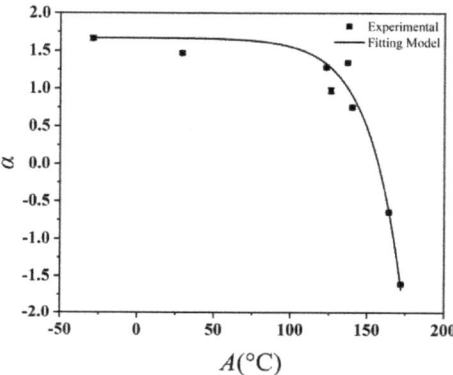

Figure 5. Variation in coefficient α of base oils with their aniline point.

By bringing Equation (8) into Equation (7), we have the final equation to describe the effect of CuDDP on the viscosity of different base oils:

$$\mu_{nf} = \mu_{bf}\left(1 + \left(-0.0011 * e^{\frac{A}{21.46}} + 1.67\right)\varphi\right) \quad (9)$$

Equation (9), which incorporates the viscosity and aniline point of the base oil, facilitates the computation of the viscosity for base oils infused with a specific concentration of CuDDP nanoadditive at 40 °C, yielding an error margin of less than 1.7%. This level of precision marks a notable enhancement over traditional nanofluid viscosity estimation models. It is important to note that viscosity is influenced by additional parameters, such as the surface polarity and geometric shape of the nanoparticles. Consequently, Equation (8) is tailored specifically for the viscosity determination of CuDDP nanolubricants. further investigative efforts are necessary to establish applicable predictive models.

4. Conclusions

Under the condition that the experimental temperature was 40 °C and the additive mass fraction ranged from 0.5% to 2.5%, we studied the effects of dialkyl dithiophosphate (HDDP) copper-modified (CuDDP) nanoparticles on the dynamic kinematic viscosity of mineral oils 150N, alkylated naphthalene (AN5), diisooctyl sebacate (DIOS), and polyalphaolefins (PAO4, PAO6, PAO10, PAO40, and PAO100). Based on classical formulae and experimental data, a novel equation was developed to quantify the interfacial interaction between the nanoparticles and base oil using aniline points, elucidating the impact of CuDDP nanoparticles on the viscosity of lubricating oil. The main findings are as follows.

CuDDP reduces the viscosity of higher-viscosity base oils, such as PAO40 and PAO100, and can increase the viscosity of lower-viscosity base oils, such as 150N, AN5, DIOS, PAO4, PAO6, and PAO10. The influence of CuDDP on the viscosity of base oils is governed by the interfacial slip effect and the nanoparticles' shear resistance. When the interfacial slip effect predominates, it can lead to a more significant decrease in viscosity compared to the viscosity-increasing anti-shear effect of the nanoparticles. Conversely, when shear resistance is more pronounced, an increase in viscosity occurs.

Experimental dynamic viscosity of eight base oils containing CuDDP was compared with values calculated using three classical nanofluid viscosity formulae. A significant deviation was observed between the predicted and experimental viscosity values, with the maximum deviation reaching 7.9%. This discrepancy indicates that traditional nanofluid viscosity equations cannot accurately characterize the effect of CuDDP on the viscosity of base oils.

A modified equation was developed by incorporating specific interfacial effects into Einstein's viscosity formula and quantifying these effects using aniline points. This new equation elucidates the relationship between the base oil's aniline points and the viscosity of CuDDP nanoparticles, yielding more accurate viscosity predictions for nanolubricants. The deviation in the predicted values from the experimental data is less than 1.7%, marking a significant improvement over traditional nanofluid viscosity models.

Author Contributions: X.W., conceptualization, methodology, and writing—original draft preparation; S.F. and N.S., investigation and data curation; L.Y., writing—review and editing; Y.Z. and S.Z., resources, writing—review and editing, project administration, and funding acquisition. All authors have read and agreed to the published version of the manuscript.

Funding: We acknowledge the financial support provided by the National Key Research Development Plan (grant 2023YFB3812104), National Natural Science Foundation of China (grants 52305189 and 52105180), and Henan Province Key Research and Development Project (grant 231111230600).

Data Availability Statement: The data that support the findings of this study are available from the corresponding author (S.Z.) upon reasonable request.

Conflicts of Interest: The authors declare no conflicts of interest.

Nomenclature

v_m	kinematic viscosity of oil blend (mm²/s)
v_1	kinematic viscosity of oil component 1 (mm²/s)
v_2	kinematic viscosity of oil component 2 (mm²/s)
ω_1	mass fraction of oil component 1
ω_2	mass fraction of oil component 2
φ	volume fraction
m_{np}	mass of nanoparticle (g)
m_{bf}	mass of base oil (g)
ρ_{np}	density of nanoparticle (g/m³)
ρ_{nf}	density of nanoparticle (g/m³)
μ_{nf}	nanofluid dynamic viscosity (mPa·s)
μ_{bf}	base fluid dynamic viscosity (mPa·s)
v_{x1}	plate 1 speed (m/s)
v_{x2}	plate 2 speed (m/s)
$V_0(z)$	flow velocity (m/s)
$V_1(z)$	flow velocity (no interface slip) (m/s)
$V_2(z)$	flow velocity (large interface slip) (m/s)
$V_3(z)$	flow velocity (weak interface slip) (m/s)
V_s	solid velocity (m/s)
τ_0–τ_3	shear stress (Pa)
μ_0–μ_3	nanofluid dynamic viscosity (Pa·s)
A	aniline point (°C)
α	coefficient

References

1. Du, F.; Li, C.; Li, D.; Sa, X.; Yu, Y.; Li, C.; Yang, Y.; Wang, J. Research Progress Regarding the Use of Metal and Metal Oxide Nanoparticles as Lubricant Additives. *Lubricants* **2022**, *10*, 196. [CrossRef]
2. Lu, Z.; Cao, Z.; Hu, E.; Hu, K.; Hu, X. Preparation and tribological properties of WS_2 and WS_2/TiO_2 nanoparticles. *Tribol. Int.* **2019**, *130*, 308–316. [CrossRef]
3. Padgurskas, J.; Rukuiza, R.; Prosyčevas, I.; Kreivaitis, R. Tribological properties of lubricant additives of Fe, Cu and Co nanoparticles. *Tribol. Int.* **2013**, *60*, 224–232. [CrossRef]
4. Duan, L.; Jia, D.; Zhan, S.; Zhang, W.; Yang, T.; Tu, J.; Liu, J.; Li, J.; Duan, H. Copper phosphate nanosheets as high-performance oil-based nanoadditives: Tribological properties and lubrication mechanism. *Tribol. Int.* **2023**, *179*, 108077. [CrossRef]
5. Guo, Z.; Zhang, Y.; Wang, J.; Gao, C.; Zhang, S.; Zhang, P.; Zhang, Z. Interactions of Cu nanoparticles with conventional lubricant additives on tribological performance and some physicochemical properties of an ester base oil. *Tribol. Int.* **2020**, *141*, 105941. [CrossRef]
6. Zhu, M.; Song, N.; Zhang, S.; Zhang, Y.; Yu, L.; Yang, G.; Zhang, P. Effect of micro nano-structured copper additives with different morphology on tribological properties and conductivity of lithium grease. *Tribol. Trans.* **2022**, *65*, 686–694. [CrossRef]
7. Zhang, Y.; Zhang, S.; Sun, D.; Yang, G.; Gao, C.; Zhou, C.; Zhang, C.; Zhang, P. Wide adaptability of Cu nano-additives to the hardness and composition of DLC coatings in DLC/PAO solid-liquid composite lubricating system. *Tribol. Int.* **2019**, *138*, 184–195. [CrossRef]
8. Shen, M.x.; Rong, K.j.; Li, C.h.; Xu, B.; Xiong, G.y.; Zhang, R.h. In situ Friction-Induced Copper Nanoparticles at the Sliding Interface Between Steel Tribo-Pairs and their Tribological Properties. *Tribol. Lett.* **2020**, *68*, 98. [CrossRef]
9. Martini, A.; Ramasamy, U.S.; Len, M. Review of Viscosity Modifier Lubricant Additives. *Tribol. Lett.* **2018**, *66*, 58. [CrossRef]
10. Froböse, E.; Murr, T.; Vdi. Viscosity reduction of driveline lubricants. In Proceedings of the VDI Congress on Drivetrain for Vehicles: Light—Compact—Efficient, Friedrichshafen, Germany, 19–20 June 2012; pp. 213–226.
11. Breki, A.; Nosonovsky, M. Einstein's Viscosity Equation for Nanolubricated Friction. *Langmuir* **2018**, *34*, 12968–12973. [CrossRef]
12. Dolatabadi, N.; Rahmani, R.; Rahnejat, H.; Garner, C.P.; Brunton, C. Performance of Poly Alpha Olefin Nanolubricant. *Lubricants* **2020**, *8*, 17. [CrossRef]
13. Esfe, M.H.; Motallebi, S.M.; Toghraie, D.; Hatami, H. Experimental study and viscosity modeling by adding oxide nanoparticles to oil to improve the performance. *Tribol. Int.* **2023**, *190*, 109031. [CrossRef]
14. Liu, X.; Xu, N.; Li, W.; Zhang, M.; Lou, W.; Wang, X. Viscosity modification of lubricating oil based on high-concentration silica nanoparticle colloidal system. *J. Dispers. Sci. Technol.* **2017**, *38*, 1360–1365. [CrossRef]
15. Wan, Q.; Jin, Y.; Sun, P.; Ding, Y. Rheological and tribological behaviour of lubricating oils containing platelet MoS_2 nanoparticles. *J. Nanoparticle Res.* **2014**, *16*, 2386. [CrossRef]
16. Mackay, M.E.; Dao, T.T.; Tuteja, A.; Ho, D.L.; van Horn, B.; Kim, H.C.; Hawker, C.J. Nanoscale effects leading to non-Einstein-like decrease in viscosity. *Nat. Mater.* **2003**, *2*, 762–766. [CrossRef]
17. Kotia, A.; Kumar, R.; Haldar, A.; Deval, P.; Ghosh, S.K. Characterization of Al_2O_3-SAE 15W40 engine oil nanolubricant and performance evaluation in 4-stroke diesel engine. *J. Braz. Soc. Mech. Sci. Eng.* **2018**, *40*, 38. [CrossRef]
18. Ma, J.; Shahsavar, A.; Al-Rashed, A.A.; Karimipour, A.; Yarmand, H.; Rostami, S. Viscosity, cloud point, freezing point and flash point of zinc oxide/SAE50 nanolubricant. *J. Mol. Liq.* **2020**, *298*, 112045. [CrossRef]
19. Hemmat Esfe, M.; Afrand, M.; Gharehkhani, S.; Rostamian, H.; Toghraie, D.; Dahari, M. An experimental study on viscosity of alumina-engine oil: Effects of temperature and nanoparticles concentration. *Int. Commun. Heat Mass Transf.* **2016**, *76*, 202–208. [CrossRef]
20. Hemmat Esfe, M.; Mosaferi, M. Effect of MgO nanoparticles suspension on rheological behavior and a new correlation. *J. Mol. Liq.* **2020**, *309*, 112632. [CrossRef]
21. Fahad, M.R.; Abdulmajeed, B.A. Experimental investigation of base oil properties containing modified TiO_2/CuO nanoparticles additives. *J. Phys. Conf. Ser.* **2021**, *1973*, 012089. [CrossRef]
22. Sui, T.; Ding, M.; Ji, C.; Yan, S.; Wei, J.; Wang, A.; Zhao, F.; Fei, J. Dispersibility and rheological behavior of functionalized silica nanoparticles as lubricant additives. *Ceram. Int.* **2018**, *44*, 18438–18443. [CrossRef]
23. Abdelhalim, M.A.; Mady, M.M.; Ghannam, M.M. Rheological and dielectric properties of different gold nanoparticle sizes. *Lipids Health Dis.* **2011**, *10*, 208. [CrossRef]
24. Dehghani, Y.; Abdollahi, A.; Karimipour, A. Experimental investigation toward obtaining a new correlation for viscosity of WO_3 and Al_2O_3 nanoparticles-loaded nanofluid within aqueous and non-aqueous basefluids. *J. Therm. Anal. Calorim.* **2018**, *135*, 713–728. [CrossRef]
25. Zhang, S.; Han, X. Effect of different surface modified nanoparticles on viscosity of nanofluids. *Adv. Mech. Eng.* **2018**, *10*, 1687814018762011. [CrossRef]
26. Zhou, J.; Yang, J.; Zhang, Z.; Liu, W.; Xue, Q. Study on the structure and tribological properties of surface-modified Cu nanoparticles. *Mater. Res. Bull.* **1999**, *34*, 1361–1367. [CrossRef]
27. Piwoński, I.; Kisielewska, A. Dialkyldithiophosphate Acids (HDDPs) as Effective Lubricants of Sol–Gel Titania Coatings in Technical Dry Friction Conditions. *Tribol. Lett.* **2011**, *45*, 237–249. [CrossRef]
28. Čoga, L.; Akbari, S.; Kovač, J.; Kalin, M. Differences in nano-topography and tribochemistry of ZDDP tribofilms from variations in contact configuration with steel and DLC surfaces. *Friction* **2021**, *10*, 296–315. [CrossRef]

29. Grunberg, L.; Nissan, A.H. Mixture law for viscosity. *Nature* **1949**, *164*, 799–800. [CrossRef]
30. Einstein, A. Eine neue Bestimmung der Moleküldimensionen. *Ann. Phys.* **1906**, *324*, 289–306. [CrossRef]
31. Batchelor, G.K. The effect of Brownian motion on the bulk stress in a suspension of spherical particles. *J. Fluid Mech.* **2006**, *83*, 97–117. [CrossRef]
32. Brinkman, H.C. The Viscosity of Concentrated Suspensions and Solutions. *J. Chem. Phys.* **1952**, *20*, 571. [CrossRef]
33. Yue, P.; Zhang, Y.; Zhang, S.; Jia, J.; Han, K.; Song, N. The key role of interfacial non-bonding interactions in regulating lubricant viscosity using nanoparticles. *Tribol. Int.* **2023**, *187*, 108716. [CrossRef]
34. Marie, H.; Rigol, S.; Deeg, H.P.; Philipp, H. Impact of Aniline Octane Booster on Lubricating Oil. In Proceedings of the SAE 2016 International Powertrains, Fuels & Lubricants Meeting, Baltimore, MD, USA, 24–26 October 2016; Volume 2273, p 8. [CrossRef]

Disclaimer/Publisher's Note: The statements, opinions and data contained in all publications are solely those of the individual author(s) and contributor(s) and not of MDPI and/or the editor(s). MDPI and/or the editor(s) disclaim responsibility for any injury to people or property resulting from any ideas, methods, instructions or products referred to in the content.

Article

Halloysite Reinforced Natural Esters for Energy Applications

Jose Jaime Taha-Tijerina [1,*], Karla Aviña [2], Victoria Padilla-Gainza [3] and Aditya Akundi [1]

1. Department of Informatics and Engineering Systems, The University of Texas Rio Grande Valley, Brownsville, TX 78520, USA
2. Engineering Department, Universidad de Monterrey, Av. Ignacio Morones Prieto 4500 Pte., San Pedro Garza García 66238, Mexico
3. Department of Mechanical Engineering, The University of Texas Rio Grande Valley, 1201 West University Drive, Edinburg, TX 78530, USA
* Correspondence: jose.taha@utrgv.edu

Abstract: Recently, environmentally friendly and sustainable materials are being developed, searching for biocompatible and efficient materials which could be incorporated into diverse industries and fields. Natural esters are investigated and have emerged as eco-friendly high-performance alternatives to mineral fluids. This research shows the evaluations on thermal transport and tribological properties of halloysite nanotubular structures (HNS) reinforcing natural ester lubricant at various filler fractions (0.01, 0.05, and 0.10 wt.%). Nanolubricant tribotestings were evaluated under two configurations, block-on-ring, and 4-balls, to obtain the coefficient of friction (COF) and wear scar diameter (WSD), respectively. Results indicated improvements, even at merely 0.01 wt.% HNS concentration, where COF and WSD were reduced by ~66% and 8%, respectively, when compared to pure natural ester. The maximum significant improvement was observed for the 0.05 wt.% concentration, which resulted in a reduction of 87% in COF and 37% in WSD. Thermal conductivity was analyzed under a temperature scan from room temperature up to 70 °C (343 K). Results indicate that thermal conductivity is improved as the HNS concentration and testing temperature are increased. Results revealed improvements for the nanolubricants in the range of 8–16% at 50 °C (323 K) and reached a maximum of 30% at 70 °C (343 K). Therefore, this research suggests that natural ester/HNS lubricants might be used in industrial applications as green lubricants.

Keywords: natural ester; halloysite; energy; wear; thermal conductivity

Citation: Taha-Tijerina, J.J.; Aviña, K.; Padilla-Gainza, V.; Akundi, A. Halloysite Reinforced Natural Esters for Energy Applications. *Lubricants* **2023**, *11*, 65. https://doi.org/10.3390/lubricants11020065

Received: 19 January 2023
Revised: 2 February 2023
Accepted: 3 February 2023
Published: 5 February 2023

Copyright: © 2023 by the authors. Licensee MDPI, Basel, Switzerland. This article is an open access article distributed under the terms and conditions of the Creative Commons Attribution (CC BY) license (https://creativecommons.org/licenses/by/4.0/).

1. Introduction

For more than a century, the primary source of energy has been fossil fuels. Petroleum-based fluids are used as lubricants and coolants in electronic or electrical equipment, machinery, transportation fields, and power transmission systems, among others [1–5]. Research and industrial focus are on critical challenges to reduce pollution and mitigate global warming and climate change, among others [6]. Due to their poor biodegradability, eco-toxicity, and extremely likely carcinogenic characteristics, these materials exhibit environmental concerns [7,8].

Among the diverse drawbacks to be resolved about these materials are how to properly dispose of them and how to hinder products from impacting negatively on both health and the environment [9–11]. Additionally, due to the forthcoming scarcity of petroleum reserves and the rise in lubricant disposal costs, among other factors, the importance of incorporating and applying renewable energies has grown significantly [12–14]. Consequently, interest in natural lubricants has increased recently.

Technology is moving forward very fast with high efficiency, miniaturization, and novel equipment requirements and developments. The increasing demand for higher-performance of conventional fluids and lubricants has been crucial for diverse industrial manufacturing processes. Energy applications such as power transmission systems, transportation vehicles, and battery subsystems require materials which would be lighter but with much higher mechanical properties, which require higher processing loadings, need a hefty lubrication regime due to higher wear and friction among base substrates and tooling.

Recently, several investigations have been performed by scientists and industry to find sustainable and eco-friendly alternatives [15–19]. The usage of these materials has been emerging and is more common in certain environmentally-sensitive fields [1–4,20–22]. Natural esters serve as an alternative solution against petroleum-based lubricants due to their non-toxic attributes and specific characteristics as renewable and eco-friendly. Furthermore, these fluids have been proven to diminish hydrocarbon and CO_2 emission levels [23], which is why their application in diverse systems and subsystems has recently increased in the industry [24]. Natural esters also possess excellent lubricity [25,26], high thermal conductivity [27,28], compatibility with additives, relatively low production costs [29], biodegradability, and high fire and flash points [30,31], comparable to mineral lubricants and fluids [26,32–34]. In general, these materials are defined as fatty acids, which contain an extended aliphatic chain in addition to the ester function, which determines the mechanical and chemical characteristics of the lubricants [35]. Unsaturated can display less viscous performance but are more susceptible to oxidation [35,36], leading to an increase in viscosity and degradation, therefore affecting the tribological, thermal, and other characteristics of the lubricant [37].

Regardless of all the excellent attributes of natural esters, the application of stand-alone material had hit their tribological and thermal transport limit. Advancements in science and nanotechnology provide the potential to enhance the performance of fluids and lubricants reinforced by nanostructures additives. Since the mid-1990s, the exploration of the incorporation of myriad solid ultrafine particles within conventional fluids and lubricants has been investigated [38]. These nanostructures possess superb characteristics compared to base fluids and lubricants, and when they are homogenously dispersed or formulated within conventional materials, there is a significant enhancement in their properties.

Lubricants play a paramount role in reducing the wear and friction of mechanical pairs or contacts as well as in internal components and mechanisms of machinery and devices. The most significant enhancements and advantages of applying nanoparticles as reinforcement are the reduction of coefficient of friction (COF) and wear. Therefore, reducing machining cutting forces and power consumption [39]. One more key feature is the improvement of surface roughness or finishing of processed components and products, having a significant impact on quality and secondary processing. An additional benefit for tools and components of machinery by applying nanoreinforced fluids and lubricants is that they also work as thermal dissipators of the heat generated by material interactions or mechanisms working in contact [40–43].

For instance, Rapoport et al. results demonstrated that WS_2 nanostructures reduced the COF of paraffin lubricant by 30% [44]. Kumar et al. [45] investigated natural oil blends (sunflower oil:rice bran oil) reinforced with CuO nanostructures as an additive, and it was observed that at 0.04 vol.% of nanoparticles resulted in the greatest improvements of 6% and 10% in WSD and COF, respectively when compared to conventional material. Omrani et al. [46] studied vegetable oil with the addition of nano-graphene by sliding contact, observing improvements in COF of ~84%. Karthikeyan et al. [47] performed pin-on-disk evaluations for olive- and castor-oil-based nanolubricants reinforced with 0.7 wt.% of molybdenum disulfide (MoS_2). A wear reduction of 21% and 37% for these vegetable nanolubricants, respectively. Potential uses for nanomaterials in the automotive industry with natural-based lubricants from bio-lubricants and nanostructures as additives were found by Yadav et al. [48]. Important wear scar and COF reductions in vegetable nanolubricants were also observed by Lim et al. [49], Gupta et al. [50], and Taha et al. [51]

The importance of thermal management as heat transport in industrial and energy segments is a key topic to work on with the aid of nanofluids. Since base fluids and lubricants have very limited thermal conductivity, researchers endeavor to strengthen this characteristic by incorporating diverse types, morphologies, and concentrations of nanostructures into conventional materials and have shown good enhancements. Nevertheless, a crucial aspect that must be addressed to preserve their thermophysical characteristics after production for an extended period is the stabilization of nanofluids and nanolubricants.

Jacob et al. [52] investigated the thermal conductivity of a natural ester (soybean lubricant) by adding Al_2O_3 nanostructures. They observed an enhancement of 6.3% and 10.3% at 0.04 wt.% and 0.1 wt% filler fraction, respectively. Karthikeyan et al. [47] studied MoS_2 nanostructure effects on vegetable lubricants (olive and castor oils). Both exhibited an improvement of 28% and 21%, respectively, at 0.7 wt% filler fraction. Farade et al. [53] developed cottonseed nanofluids with graphene oxide (GO) nanosheets at diverse concentrations varying from 0.01 wt.% to 0.05 wt.%. Thermal conductivity was improved at each level of filler fraction, and the value increased up to 36.4% at 0.05 wt.%. In another research on 2D nanostructures, Taha et al. [28] evaluated hexagonal boron nitride (h-BN), MoS_2, and their combination in a natural ester media. In their study, the improvement was increased with the nanostructure's filler fraction. The greatest improvement was at 323 K. It was shown that the incorporation of 2D nanomaterials enhanced the thermal transport performance in the 20% to 32% range. Similarly, Khan et al. [54] reported positive effects on the thermal transport behavior of TiO_2 and GO nanostructures within natural and synthetic ester lubricants.

Halloysite nanotubular structures (HNS) possess high modulus (140 GPa), allowing a high loading resistance under extreme pressure conditions [55]. These nanostructures are emerging as natural biocompatible minerals [56,57], readily available worldwide [58], with low-cost (about USD 3000 per ton) [59,60], corrosion protection [61,62], non-toxic and non-conductor resource as alternative "green" nanoreinforcement to improve tribological and thermal transport properties of conventional fluids and lubricants. They have been employed as mechanical reinforcing of polymer matrices and composites [63–65], improving tribological performance and other characteristics. Nevertheless, HNS has proven beneficial in characteristics of diverse systems and media as reinforcement; scarcity of research coupled with natural esters and lubricants has not been deeply explored.

For instance, for tribological studies, Ahmed et al. [66] studied the characteristics of castor oil by incorporating HNS. They were able to observe maximum reductions of 21.3% in wear scar diameter (WSD) and 28.2% in COF at 1 wt.% reinforcement. Qin et al. [67] investigated the effects of halloysite employing a ball-on-disk tribometer. Interesting results were shown where halloysite-containing lubricant exhibited good friction reduction of 34%, primarily attributed to a tribofilm formation of the nanostructures and also to the halloysite rolling role in the contact surface of the specimens [68]. In a similar work on pongamia lubricant, Suresha et al. [69] observed a wear and friction reduction of 10% and 14%, respectively, with the addition of 1.5 wt.% of HNS.

On the thermal aspect, Alberola et al. [70] evaluated the thermophysical behavior of halloysite nanofluids, showing a thermal conductivity increase of 8% for the 5 vol.% nanofluids at 80 °C. Similarly, Ba et al. [71] explored the effects on the thermal transport performance of halloysite nanofluids, where an increase of 18% for 1.5 vol.% was observed compared to the conventional fluid. In another investigation, Ba et al. [72] analyzed the pool boiling heat transport behavior of nano halloysite / deionized water (DI) fluids. Their findings revealed that increasing the reinforcement filler fraction leads to improved thermal performance. At 0.05 vol.% nanofluid, a greater improvement of ~6% at moderate heat flux (HF) was achieved, indicating that HNS is a promising material for applications that entail heat transfer performance.

In this work, the tribological behavior of HNS reinforcing natural ester lubricant at various filler fractions (by weight) is evaluated. Temperature-dependent measurements over ranges up to 70 °C (343 K) were also analyzed.

2. Materials and Methods

Natural ester—Envirotemp® FR3™ (Cargill Industrial Specialties—Minneapolis, MN, USA) (Table 1) served as base material to develop various nanolubricants reinforced with hollow rod-shaped HNS (Sigma-Aldrich Co., St. Louis, MO, USA. CAS #: 1332-58-7)

(Table 1) at various filler fractions: 0.01, 0.05, and 0.10 wt.%. Scanning electron microscopy (Carl Zeiss, Sigma VP, NY, USA) was employed to examine the size and morphology of HNS (Figure 1). From SEM images, it can be observed a tubular rod-type shape of the HNS. The measured lengths of these nanostructures have a maximum of 589 nm and 120 nm as a minimum. Average length was 302 nm with a standard deviation of 83 nm. The average measured diameter was 61 nm ± 18 nm, where the minimum and maximum diameters were 28 nm and 96 nm, respectively.

Table 1. Material characteristics.

Materials	Properties and Characteristics	
Conventional Lubricant	Density (20 °C)	Kinetic Viscosity (mm^2/s)
Natural ester	0.92 g/cm^3	190 @ 0 °C; 32–34 @ 40 °C; 7.7–8.3 @ 100 °C
Nanostructures	Properties	
Halloysite (HNS)	Chemical formula: $H_4Al_2O_9Si_2 \cdot 2H_2O$ Specific gravity: 2.57 g/cm^3 Molecular weight: 294.19 g/mol Size–Length: 302 nm ± 83 nm; Diameter: 61 nm ± 18 nm	

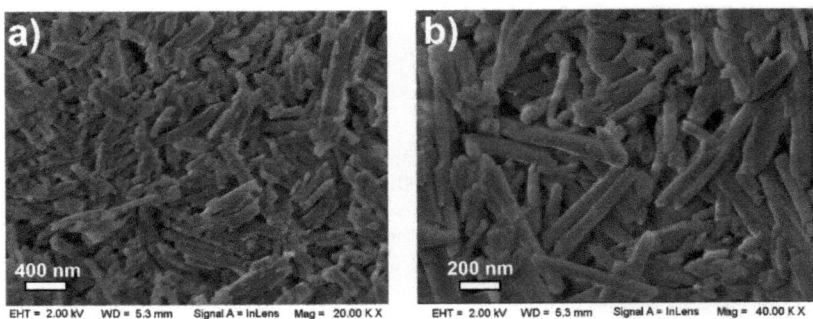

Figure 1. HNS morphology at (**a**) 20 KX, and (**b**) 40 KX magnifications.

Table 1 depicts the general properties and characteristics of the natural ester, nanostructures, and specimens for the tribological tests.

Nanolubricants Preparation

Homogeneous nanolubricants were prepared by a two-step methodology dispersing HNS within the natural ester lubricant. For each set of nanolubricants 42 mL glass containers were prepared at 0.01, 0.05, and 0.10 wt.%. Water bath ultrasonication (Branson ultrasonic homogenizer model 5510—Danbury, CT, USA, 40 kHz) was used for extensive time (8 h). According to the following methodology, a constant temperature (24 °C) was kept in the sonicator water bath to prevent the nanostructures from clumping together, causing fast sedimentation. All glass vials specimens were kept on a shelf for about 15 days without significant particle sedimentation (Supplementary Materials) Preliminary sonication of each sample for 15 min was performed before the experimental testing.

3. Experimental Details

3.1. Tribological Evaluations

For tribological measurements, COF and WSD were analyzed with a tribotester in a four-ball configuration based on the ASTM D 4172 methodology. This tribosystem employs three fixed steel balls on a cylindrical container (fluid cup) and one more on top of them (Figure 2), applying a load P = 40 kgf (392 N). A temperature of 75 °C was employed for the evaluated nanolubricants; the upper ball rotates at a specific rotational speed n = 1200 rpm during 1200 s. An AISI 52100 steel is used for the ball's material (60 HRC and 12.7 mm in diameter). This tribotester configuration is widely applied to determine the wear properties of lubricating fluids and oils in sliding metal-metal applications. The anti-wear performance of nanolubricants was tested with the block-on-ring configuration at room temperature (24 °C), 3000 N load, 200 rpms, and time of 1200 s. The evaluated nanolubricants were deposited in a fluid pool, allowing constant lubrication while the test ring was rotating. For these evaluations, the blocks used were AISI 1018 steel with 79 HRB hardness, and the rings were AISI D2 tool steel with 61 HRC hardness. To obtain statistically significant results, four replicas were measured for each set of nanolubricants. Worn scars of steel components were analyzed by a 3D surface profiler microscope measurement system (Keyence Corp Precision, Itasca, IL, USA) and a TescanVega 3SB Scanning Electron Microscope (SEM).

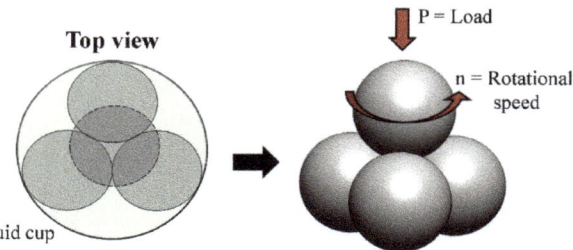

Figure 2. Schematic of tribotesting four-balls setup.

3.2. Thermal Conductivity Evaluations

Thermal conductivity evaluations of HNS nanolubricants were evaluated with a TEMPOS thermal analyzer device and KS-3 sensor probe (METER GROUP, Inc., Pullman, WA, USA) following the transient hot-wire (THW) methodology. A temperature-dependence scan up to 70 °C (343 K) was performed. Before thermal measurements, 20 min ultrasonication was applied to the samples. Above room temperature evaluations, before measurements, each set of specimens was maintained at least 12 min in thermal equilibrium. This procedure promotes the preparation of stable homogeneous nanolubricants, which are then evaluated. The obtained thermal conductivity values were compared with the conventional natural ester (k_0). The effective thermal conductivity of the nanolubricants is k_{eff}. For each set of specimens, at least six measurements were obtained, reporting the average values with error bars as standard deviation.

4. Results and Discussion

4.1. Tribological Performance

The tribological behavior of natural ester lubricant was analyzed with reinforcement of HNS. Figure 3 depicts COF curves recorded during block-on-ring evaluations being consistent with the testing conditions for all the samples. The tribotests were performed under lubrication with natural ester and HNS nanolubricants. It was observed that the addition of diverse filler fractions of HNS resulted in a profound effect of friction reduction during the tribological evaluations.

The initial measured COF for evaluated lubricants was in the vicinity of 0.085. After the first minutes of testing, the HNS nanolubricants showed a significant decrease in COF. Lower COF represents less energy loss caused by contact pairs friction. At merely 0.01 wt.% concentration, it was observed a COF value of 0.075, with maximum improvement for 0.05 wt.% of 0.029. As can be observed, after the complete run of experiments, the HNS nanolubricants showed superb behavior, reaching the greatest improvement of 87% at 0.05 wt.%, compared to pure lubricant.

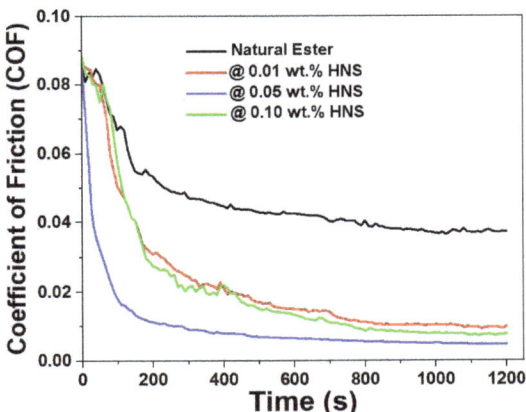

Figure 3. Block-on-ring evaluation for HNS nanolubricants showing COF versus time.

Figure 4 shows the average COF values for nanolubricants. COF was reduced from 0.042 for the natural ester to 0.015 with 0.01 wt.% HNS. The best improvement was observed at 0.05 wt.% concentration, where COF is 0.006. It is also shown an increase in these properties at higher filler fractions because, at higher concentrations and increasing temperature, nanostructures tend to agglomerate, decreasing the lubrication performance [73,74]. The lower improvement was shown at 0.10 wt.% filler fraction; this can be attributed to a higher concentration of nanostructure agglomeration. Nanostructures deposit in the surface depressions; hence, as smaller the size of the nanostructures, the more likely to creep into those gaps, minimizing contact and friction is mend to reduce the roughness and also to withstand higher loads [39]. Previous research of sunflower and soybean nanolubricants with SiO_2 displayed a reduction in the COF of 10% and 26% in comparison. [34].

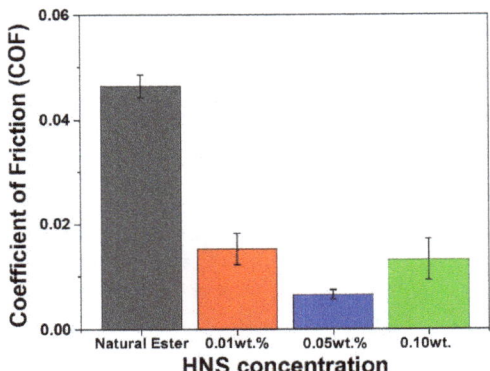

Figure 4. Average COF results for natural ester reinforced with HNS nanostructures.

Figure 5 depicts SEM of the scars of worn steel balls for conventional natural ester and HNS nanolubricants at various filler fractions. The average diameters of the three bottom steel balls were used to calculate the WSD, as shown in Table 2. When HNS were incorporated into natural ester, 188 μm scar diameter was measured, showing a decrease of 8.29%, with the highest reduction of 37.56% (128 μm) at 0.05 wt.%. The highest concentration, 0.10 wt.% of HNS, showed a reduction of 28.29%. This minor decrease in the lubrication properties for higher nanostructure filler fraction could be attributed to the tendency of agglomeration of nanostructures, as was also observed in the COF evaluations.

In accordance with Wu et al. [75], the experimental temperature has a direct impact on the lubricating mechanisms; higher evaluation temperatures result in a decrease in viscosity, which could affect the formation of the lubricant tribofilm between contact surfaces.

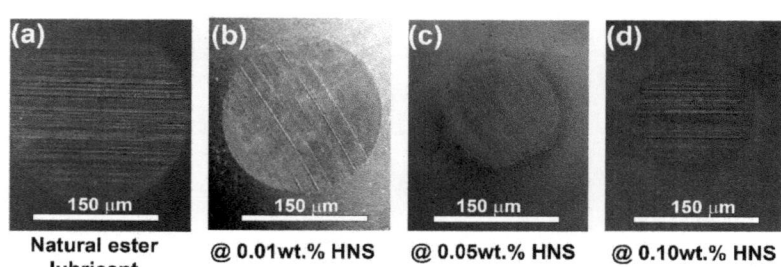

Figure 5. SEM of steel ball wear for: (**a**) natural ester, and (**b**–**d**) HNS nanolubricants.

Table 2. WSD of HNS nanolubricants by ASTM D5183.

Lubricant	WSD (μm) (Reduction %)
Natural ester	205
@ 0.01 wt.% HNS	188 (8.29%)
@ 0.05 wt.% HNS	128 (37.56%)
@ 0.10 wt.% HNS	147 (28.29%)

For anti-wear conditions, tribological mechanisms for HNS play a paramount role in the lubricant characteristics and properties. It must be mentioned that the characteristics and performance of the lubricant are significantly influenced by these nanostructures. The size and morphology of HNS have considerable effects on the properties and characteristics of conventional lubricants, such as have been observed with other rod-type structures such as single/multi-wall carbon nanotubes (CNTs). These peculiar structures act by rolling and sliding under transversal sliding forces action, playing a bearing-like function among contact components and pairs in friction, changing the sliding friction into rolling friction and enhancing the anti-wear characteristics of the material. According to Zhang et al. [76], friction reduction is not promoted by the action of the nanorods themselves working as molecular bearings. Instead, a vortex structure is developed during the friction process, causing the rolling friction behavior. In this case, HNS has a load-bearing mechanism that promotes the reduction of wear and lowers COF [77,78]. Additionally, the smoothing and mending effects of nanostructures after being deposited on contact areas (COF reduction [39]) and the rolling-sliding effect by rod-type nanostructures (decreasing WSD [79,80]) could affect the tribological performance.

4.2. Thermal Performance

The Brownian motion of nanostructures within a fluid or lubricant is mostly governed by the nanofluid thermal transport properties (thermal conductivity). Shafi et al. [81] described how the Brownian motion of nanostructures enhances the thermal transport behavior of a fluid or lubricant; first, nanostructures collide and create a solid- -solid conduction mode of heat transfer (percolation channel formation). Then, thermal conductivity is enhanced by a convective heat transfer mode.

Figure 6 depicts the temperature-scanning evaluations for the thermal conductivity of HNS nanolubricants at various concentrations. The natural ester did not exhibit significant affectation in thermal conductivity (only a 2.4% increase) as the temperature was raised up to 70 °C (343 K), compared to room temperature. Moreover, the nanolubricant's thermal performance was observed to be gradually enhanced with the HNS concentration increase.

Furthermore, as the filler fraction of nanofluids is increased and the testing temperature is also elevated, the thermal conductivity improves, indicating the contribution of the thermal transport characteristics. For instance, it can be observed that at room temperature, a slight increase in thermal conductivity is shown for nanolubricants, achieving a maximum of 4% at 0.10 wt.%. Elevating the test temperature to 50 °C (323 K) showed enhancements of 8, 11, and 16% for 0.01, 0.05, and 0.10 wt.%, respectively, compared to conventional natural ester. The maximum evaluated temperature of 70 °C (343 K) showed a maximum enhancement of 30% at 0.10 wt.%. It is important to mention that as the testing temperature is elevated, there is more deviation from the data obtained. This is mainly due to the properties of the lubricant and its interaction with the nanostructures.

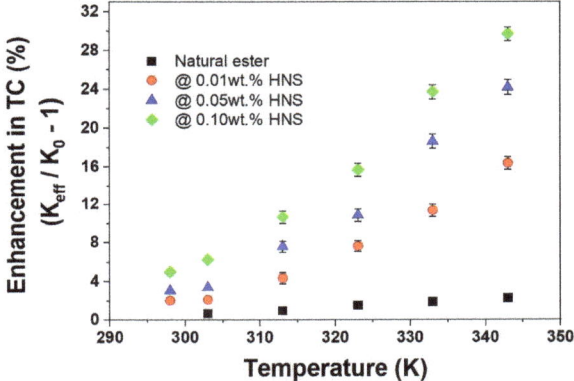

Figure 6. Temperature-dependence evaluation for thermal conductivity.

According to Guo et al. [82], it is suggested that due to the small quantity of concentration of nanostructures, the improvement in thermal conductivity is triggered by molecular interactions between the natural ester and the HNS. Additionally, the percolation mechanism, as well as the liquid layering interface at the nanostructures—lubricant also promotes this enhancement [51,83,84]. The lubricant molecules interacting with the nanotubular particles are prone to form a systematic layered structure around the nanostructures, which is associated with improved lubricant thermal transport [81].

5. Conclusions

The lubrication and thermal transport performance of natural ester lubricant reinforced with HNS were broadly addressed in this research. In general, HNS nanolubricants resulted in a significant decrease in COF and WSD. Even at merely 0.01 wt.% concentration of HNS, results showed a ~66% and 8% reduction, respectively, when compared to natural ester. The maximum enhancement was shown for the 0.05 wt.% filler fraction, resulting in a reduction of 87% in COF and 37% in WSD. Results indicate that at higher HNS concentration (0.10 wt.%), a minor impact was observed, which could be attributed to the tendency of agglomeration of nanostructures. The morphology and size of HNS promote the antiwear behavior of the nanolubricants, as load bearing and rolling-sliding mechanism of the nanostructures, as well as the contribution from the smoothing and mending effects in the contact areas, reduce WSD and COF.

Nanolubricants have a good effect on thermal conductivity as well. A temperature scanning evaluation was performed. At room temperature, reinforcing HNS has a slight impact on thermal conductivity, but as the temperature is elevated, the effect is more significant. For instance, at 50 °C (323 K), thermal conductivity measurements resulted in enhancements of 8%, 11%, and 16% for 0.01, 0.05, and 0.10 wt.%, respectively, compared to conventional natural ester. Achieving the greatest improvement of 30% with 0.01 wt.%

HNS concentration at 70 °C (343 K). This behavior, additionally from the Brownian motion effect on the natural ester, is attributed to the percolation mechanism as well as the liquid layering interface contributing to the enhancement in thermal transport characteristics.

A key driver is increasing environmental awareness of novel scientific and technological developments. HNS has been demonstrated to be a good eco-friendly alternative to petroleum-based fluids, mainly in sensitive industrial fields where these materials have the potential to succeed.

Supplementary Materials: The following supporting information can be downloaded at: https://www.mdpi.com/article/10.3390/lubricants11020065/s1. Figure S1. Observation of nanostructure's sedimentation after 15 days on the shelf.

Author Contributions: J.J.T.-T. contributed to the conceptualization, project administration, methodology, literature research, measuring campaign, data analysis, data interpretation, validation, formal analysis, resources, investigation, figures, study design, supervision, and writing—original. K.A. contributed to the measuring campaign methodology, resources, validation, formal analysis, investigation and writing—original. V.P.-G. contributed to the measuring campaign, data interpretation and investigation. A.A. contributed to the investigation, literature research and formal analysis. All authors have read and agreed to the published version of the manuscript.

Funding: This research received no external funding.

Data Availability Statement: Not applicable.

Acknowledgments: Authors acknowledge Cargill Industrial Specialties, The University of Texas Rio Grande Valley and Karen Lozano for their support given in this research.

Conflicts of Interest: The authors declare no conflict of interest.

References

1. Wong, K.V.; De Leon, O. Applications of Nanofluids: Current and Future. *Adv. Mech. Eng.* **2010**, *2*, 519659. [CrossRef]
2. Salehirad, M.; Nikje, M.M.A. Properties of Modified Hexagonal Boron Nitride as Stable Nanofluids for Thermal Management Applications. *Russ. J. Appl. Chem.* **2019**, *92*, 78–86. [CrossRef]
3. Fofana, I. 50 years in the development of insulating liquids. *IEEE Electr. Insul. Mag.* **2013**, *29*, 13–25. [CrossRef]
4. Taha-Tijerina, J.; Narayanan, T.N.; Gao, G.; Rohde, M.; Tsentalovich, D.A.; Pasquali, M.; Ajayan, P.M. Electrically insulating thermal nano-oils using 2D fillers. *ACS Nano* **2012**, *6*, 1214–1220. [CrossRef]
5. Contreras, J.E.; Rodriguez, E.A.; Taha-Tijerina, J. Chapter 39: Recent Trends of Nanomaterials for High-Voltage Applications. In *Handbook of Nanomaterials for Industrial Applications*; Elsevier: Amsterdam, The Netherlands, 2018; pp. 724–738.
6. Abas, N.; Kalair, A.; Khan, N. Review of fossil fuels and future energy technologies. *Futures* **2015**, *69*, 31–49. [CrossRef]
7. Madanhire, I.; Mbohwa, C.; Madanhire, I.; Mbohwa, C. Development of Biodegradable Lubricants. In *Mitigating Environmental Impact of Petroleum Lubricants*; Springer: Berlin/Heidelberg, Germany, 2016; pp. 85–101.
8. Nowak, P.; Kucharska, K.; Kamiński, M. Ecological and health effects of lubricant oils emitted into the environment. *Int. J. Environ. Res. Public Health* **2019**, *16*, 3002. [CrossRef]
9. Krolczyk, G.M.; Maruda, R.W.; Krolczyk, J.B.; Wojciechowski, S.; Mia, M.; Nieslony, P. Ecological trends in machining as a key factor in sustainable production—A review. *J. Clean. Prod.* **2019**, *218*, 601–615. [CrossRef]
10. Benedicto, E.; Carou, D.; Rubio, E.M. Technical, Economic and Environmental Review of the Lubrication/Cooling Systems used in Machining Processes. *Procedia Eng.* **2017**, *184*, 99–116. [CrossRef]
11. Taha-Tijerina, J.; Narayanan, T.N.; Avali, S.; Ajayan, P.M. 2D Structures-based Energy Management Nanofluids. In Proceedings of the ASME 2012 International Mechanical Engineering Congress & Exposition (IMECE 2012), Houston, TX, USA, 9–15 November 2012; p. 87890.
12. Liu, R. The challenges and opportunities of nanofluids. In Proceedings of the 2019 2nd International Conference on Electrical Materials and Power Equipment (ICEMPE), Guangzhou, China, 7–10 April 2019; pp. 110–114.
13. Sharma, P.; Said, Z.; Kumar, A.; Nizetic, S.; Pandey, A.; Hoang, A.T.; Huang, Z.; Afzal, A.; Li, C.; Le, A.T.; et al. Recent Advances in Machine Learning Research for Nanofluid-Based Heat Transfer in Renewable Energy System. *Energy Fuels* **2022**, *36*, 6626–6658. [CrossRef]
14. Rafiq, M.; Lv, Y.Z.; Zhou, Y.; Ma, K.B.; Wang, W.; Li, C.R. Use of vegetable oils as transformer oils-A review. *Renew. Sustain. Energy Rev.* **2015**, *52*, 308–324. [CrossRef]
15. Kumar, S.; Vo, D.-V.N.; Selvaraju, N.; Ramalingam, S.; Sikdar, S.; Rahman, H. Synergistic Study of Solid Lubricant Nano-Additives Incorporated in canola oil for Enhancing Energy Efficiency and Sustainability. *Sustainability* **2022**, *14*, 290.

16. Fernández-Silva, S.D.; García-Morales, M.; Ruffel, C.; Delgado, M.A. Influence of the Nanoclay Concentration and Oil Viscosity on the Rheological and Tribological Properties of Nanoclay-Based Ecolubricants. *Lubricants* **2021**, *9*, 8. [CrossRef]
17. Narayanan, G.A.A.; Babu, R.; Vasanthakumari, R. Studies on Halloysite Nanotubes (HNT) Natural Rubber Nanocomposites for Mechanical Thermal and Wear Properties. *Int. J. Eng. Res. Technol.* **2016**, *5*, 152–156.
18. Ahmed Abdalglil Mustafa, W.; Dassenoy, F.; Sarno, M.; Senatore, A. A review on potentials and challenges of nanolubricants as promising lubricants for electric vehicles. *Lubr. Sci.* **2022**, *34*, 1–29. [CrossRef]
19. Kadirgama, G.; Kamarulzaman, M.K.; Ramasamy, D.; Kadirgama, K.; Hisham, S. Classification of Lubricants Base Oils for Nanolubricants Applications—A Review. In *ICMER 2021: Technological Advancement in Mechanical and Automotive Engineering*; Ismail, M.Y., Mohd Sani, M.S., Kumarasamy, S., Hamidi, M.A., Shaari, M.S., Eds.; Lecture Notes in Mechanical Engineering; Springer: Singapore, 2023; pp. 205–213.
20. Huang, Z.; Li, J.; Yao, W.; Wang, F.; Wan, F.; Tan, Y. Electrical and thermal properties of insulating oil-based nanofluids: A comprehensive overview. *IET Nanodielectr.* **2019**, *2*, 27–40. [CrossRef]
21. Taha-Tijerina, J.J. Thermal transport and Challenges on Nanofluids Performance. In *Microfluidics and Nanofluidics*, 1st ed.; Kandelousi, M.S., Ed.; IntechOpen: Rijeka, Croatia, 2018; pp. 215–256.
22. Nagendramma, P.; Kaul, S. Development of ecofriendly/biodegradable lubricants: An overview. *Renew. Sustain. Energy Rev.* **2012**, *16*, 764–774. [CrossRef]
23. Pawar, R.V.; Hulwan, D.B.; Mandale, M.B. Recent advancements in synthesis, rheological characterization, and tribological performance of vegetable oil-based lubricants enhanced with nanoparticles for sustainable lubrication. *J. Clean. Prod.* **2022**, *378*, 134454. [CrossRef]
24. Ab Ghani, S.; Muhamad, N.A.; Noorden, Z.A.; Zainuddin, H.; Abu Bakar, N.; Talib, M.A. Methods for improving the workability of natural ester insulating oils in power transformer applications: A review. *Electr. Power Syst. Res.* **2018**, *163*, 655–667. [CrossRef]
25. Thampi, A.D.; Prasanth, M.A.; Anandu, A.P.; Sneha, E.; Sasidharan, B.; Rani, S. The effect of nanoparticle additives on the tribological properties of various lubricating oils—Review. *Mater. Today Proc.* **2021**, *47*, 4919–4924. [CrossRef]
26. Kazeem, R.A.; Fadare, D.A.; Ikumapayi, O.M.; Adediran, A.A.; Aliyu, S.J.; Akinlabi, S.A. Advances in the Application of Vegetable-Oil-Based Cutting Fluids to Sustainable Machining Operation: A Review. *Lubricants* **2022**, *10*, 69. [CrossRef]
27. Dombek, G.; Nadolny, Z.; Marcinkowska, A. Thermal properties of natural ester and low viscosity natural ester in the aspect of the reliable operation of the transformer cooling system. *Eksploat. Niezawodn. Maint. Reliab.* **2019**, *21*, 384–391. [CrossRef]
28. Taha-Tijerina, J.; Ribeiro, H.; Aviña, K.; Martínez, J.M.; Godoy, A.P.; Cremonezzi, J.M. Thermal Conductivity Performance of 2D h-BN/MoS2/-Hybrid Nanostructures Used on Natural and Synthetic Esters. *Nanomaterials* **2020**, *10*, 1160. [CrossRef] [PubMed]
29. Zainal, N.A.; Zulkifli, N.W.M.; Gulzar, M.; Masjuki, H.H. A review on the chemistry, production, and technological potential of bio-based lubricants. *Renew. Sustain. Energy Rev.* **2018**, *82*, 80–102. [CrossRef]
30. Zhang, C.; Yang, Z.; Lu, Z.; Wang, X.; Jia, L.; Wang, J.; Gao, Q.; Li, L.; Zhou, C.; Chen, G.; et al. Synthesis and tribological properties of bio-inspired green dopamine oil soluble additive. *Tribol. Int.* **2022**, *174*, 107697. [CrossRef]
31. Durango-Giraldo, G.; Zapata-Hernandez, C.; Santa, J.F.; Buitrago-Sierra, R. Palm oil as a biolubricant: Literature review of processing parameters and tribological performance. *J. Ind. Eng. Chem.* **2022**, *107*, 31–44. [CrossRef]
32. Abdalla, H.S.; Patel, S. The performance and oxidation stability of sustainable metalworking fluid derived from vegetable extracts. *Proc. Inst. Mech. Eng. Part B J. Eng. Manuf.* **2006**, *220*, 2027–2040. [CrossRef]
33. Petlyuk, A.M.; Adams, R.J. Oxidation stability and tribological behavior of vegetable oil hydraulic fluids. *Tribol. Trans.* **2004**, *47*, 182–187. [CrossRef]
34. Taha-Tijerina, J.; Aviña, K.; Diabb, J.M. Tribological and Thermal Transport Performance of SiO_2-Based Natural Lubricants. *Lubricants* **2019**, *7*, 71. [CrossRef]
35. Eberhardt, R.; Muhr, H.M.; Lick, W.; Baumann, F.; Pukel, G. Comparison of alternative insulating fluids. In Proceedings of the Annual Report Conference on Electrical Insulation and Dielectric Phenomena, Quebec, QC, Canada, 26–29 October 2008; IEEE: Piscataway, NJ, USA, 2008; pp. 591–593.
36. Fox, N.J.; Stachowiak, G.W. Vegetable oil-based lubricants—A review of oxidation. *Tribol. Int.* **2007**, *40*, 1035–1046. [CrossRef]
37. IEEE Power Engineering Society. *C57.147-2018—IEEE Guide for Acceptance and Maintenance of Natural Ester Insulating Liquid in Transformers*; IEEE: Piscataway, NJ, USA, 2018.
38. Choi, S.U.S.; Eastman, J.A. Enhancing thermal conductivity of fluids with nanoparticles. In *Developments and Applications of Non-Newtonian Flows*; ASME: New York, NY, USA, 1995; Volume 231, pp. 99–105.
39. Li, H.; Zhang, Y.; Li, C.; Zhou, Z.; Nie, X.; Chen, Y.; Cao, H.; Liu, B.; Zhang, N.; Said, Z.; et al. Extreme pressure and antiwear additives for lubricant: Academic insights and perspectives. *Int. J. Adv. Manuf. Technol.* **2022**, *120*, 1–27. [CrossRef]
40. Sharma, A.K.; Tiwari, A.K.; Dixit, A.R. Progress of Nanofluid Application in Machining: A Review. *Mater. Manuf. Process.* **2015**, *30*, 813–828. [CrossRef]
41. Ben Said, L.; Kolsi, L.; Ghachem, K.; Almeshaal, M.; Maatki, C. Application of nanofluids as cutting fluids in machining operations: A brief review. *Appl. Nanosci.* **2021**, 1–32. [CrossRef]
42. Kishawy, H.A.; Hegab, H.; Deiab, I.; Eltaggaz, A. Sustainability assessment during machining Ti-6Al-4V with nano-additives-based minimum quantity lubrication. *J. Manuf. Mater. Process.* **2019**, *3*, 61. [CrossRef]
43. Kadirgama, K. A comprehensive review on the application of nanofluids in the machining process. *Int. J. Adv. Manuf. Technol.* **2021**, *115*, 2669–2681. [CrossRef]

44. Rapoport, L.; Leshchinsky, V.; Lapsker, I.; Volovik, Y.; Nepomnyashchy, O.; Lvovsky, M. tribological properties of WS$_2$ nanoparticles under mixed lubrication. *Wear* **2003**, *255*, 785–793. [CrossRef]
45. Sunil Kumar, D.; Garg, H.C.; Kumar, G. Tribological analysis of blended vegetable oils containing CuO nanoparticles as an additive. *Mater. Today Proc.* **2022**, *51*, 1259–1265. [CrossRef]
46. Omrani, E.; Siddaiah, A.; Moghadam, A.D.; Garg, U.; Rohatgi, P.; Menezes, P.L. Ball Milled Graphene Nano Additives for Enhancing Sliding Contact in Vegetable Oil. *Nanomaterials* **2021**, *11*, 610. [CrossRef]
47. Karthikeyan, K.M.B.; Vijayanand, J.; Arun, K.; Rao, V.S. Thermophysical and wear properties of eco-friendly nano lubricants. *Mater. Today Proc.* **2021**, *39*, 285–291. [CrossRef]
48. Yadav, A.; Singh, Y.; Negi, P. A review on the characterization of bio based lubricants from vegetable oils and role of nanoparticles as additives. *Mater. Today Proc.* **2021**, *46*, 10513–10517. [CrossRef]
49. Lim, S.K.; Azmi, W.H.; Jamaludin, A.S.; Yusoff, A.R.; Lim, S.K.; Azmi, W.H. Characteristics of Hybrid Nanolubricants for MQL Cooling Lubrication Machining Application. *Lubricants* **2022**, *10*, 350. [CrossRef]
50. Gupta, H.S.; Sehgal, R.; Wani, M.F. Tribological characterization of eco-friendly bio-based mahua and flaxseed oil through nanoparticles. *Biomass Convers. Biorefin.* **2022**, 1–13. [CrossRef]
51. Taha-Tijerina, J.; Shaji, S.; Sharma Kanakkillam, S.; Mendivil Palma, M.I.; Aviña, K. Tribological and Thermal Transport of Ag-Vegetable Nanofluids Prepared by Laser Ablation. *Appl. Sci.* **2020**, *10*, 1779. [CrossRef]
52. Jacob, J.; Preetha, P.; Sindhu, T.K. Stability analysis and characterization of natural ester nanofluids for transformers. *IEEE Trans. Dielectr. Electr. Insul.* **2020**, *27*, 1715–1723. [CrossRef]
53. Farade, R.A.; Wahab, N.I.A.; Mansour, D.E.A.; Azis, N.B.; Jasni, J.B.; Soudagar, M.E.M. Development of Graphene Oxide-Based Nonedible Cottonseed Nanofluids for Power Transformers. *Materials* **2020**, *13*, 2569. [CrossRef] [PubMed]
54. Khan, S.A.; Tariq, M.; Khan, A.A.; Alamri, B.; Mihet-Popa, L. Assessment of Thermophysical Performance of Ester-Based Nanofluids for Enhanced Insulation Cooling in Transformers. *Electronics* **2022**, *11*, 376. [CrossRef]
55. Lecouvet, B.; Horion, J.; D'Haese, C.; Bailly, C.; Nysten, B. Elastic modulus of halloysite nanotubes. *Nanotechnology* **2013**, *24*, 105704. [CrossRef] [PubMed]
56. Yuan, P.; Tan, D.; Annabi-Bergaya, F. Properties and applications of halloysite nanotubes: Recent research advances and future prospects. *Appl. Clay Sci.* **2015**, *112–113*, 75–93. [CrossRef]
57. Yang, S.; Li, S.; Yin, X.; Wang, L.; Chen, D.; Zhou, Y.; Wang, H. Preparation and characterization of non-solvent halloysite nanotubes nanofluids. *Appl. Clay Sci.* **2016**, *126*, 215–222. [CrossRef]
58. Joussein, E.; Petit, S.; Churchman, J.; Theng, B.; Righi, D.; Delvaux, B. Halloysite clay minerals—A review. *Clay Miner.* **2005**, *40*, 383–426. [CrossRef]
59. Lv, Y.; Sun, X.; Yan, S.; Xiong, S.; Wang, L.; Wang, H. Solvent-free halloysite nanotubes nanofluids based polyacrylonitrile fibrous membranes for protective and breathable textiles. *Compos. Commun.* **2022**, *33*, 101211. [CrossRef]
60. Massaro, M.; Noto, R.; Riela, S. Past, Present and Future Perspectives on Halloysite Clay Minerals. *Molecules* **2020**, *25*, 4863. [CrossRef]
61. Udoh, I.I.; Shi, H.; Daniel, E.F.; Li, J.; Gu, S.; Liu, F. Active anticorrosion and self-healing coatings: A review with focus on multi-action smart coating strategies. *J. Mater. Sci. Technol.* **2022**, *116*, 224–237. [CrossRef]
62. Massaro, M.; Noto, R.; Riela, S. Halloysite Nanotubes: Smart Nanomaterials in Catalysis. *J. Catal.* **2022**, *12*, 149. [CrossRef]
63. Paul, A.; Augustine, R.; Hasan, A.; Zahid, A.A.; Thomas, S.; Agatemor, C. Halloysite nanotube and chitosan polymer composites: Physicochemical and drug delivery properties. *J. Drug Deliv. Sci. Technol.* **2022**, *72*, 103380. [CrossRef]
64. Same, S.; Nakhjavani, S.A.; Samee, G.; Navidi, G.; Jahanbani, Y.; Davaran, S. Halloysite clay nanotube in regenerative medicine for tissue and wound healing. *Ceram. Int.* **2022**, *48*, 31065–31079. [CrossRef]
65. Liao, J.; Wang, H.; Liu, N.; Yang, H. Functionally modified halloysite nanotubes for personalized bioapplications. *Adv. Colloid Interface Sci.* **2023**, *311*, 102812. [CrossRef] [PubMed]
66. Ahmed, M.S.; Nair, K.P.; Tirth, V.; Elkhaleefa, A.; Rehan, M. Tribological evaluation of date seed oil and castor oil blends with halloysite nanotube additives as environment friendly bio-lubricants. *Biomass Convers. Biorefin.* **2021**, 1–10. [CrossRef]
67. Qin, Y.; Wu, M.; Yang, Y.; Yang, Y.; Yang, G. Enhanced ability of halloysite nanotubes to form multilayer nanocrystalline tribofilms by thermal activation. *Tribol. Int.* **2022**, *174*, 107718. [CrossRef]
68. Yuan, Q.; Yang, Y.; Yang, Y.; Wu, M.; Yang, G. Formation mechanism of wear-resistant composite film by Span 80-decorated halloysite nanotubes. *Ceram. Int.* **2022**, *48*, 23897–23907. [CrossRef]
69. Suresha, B.; Hemanth, G.; Rakesh, A.; Adarsh, K.M. Tribological behaviour of pongamia oil as lubricant with and without halloysite nanotubes using four-ball tester. *AIP Conf. Proc.* **2019**, *2128*, 030011.
70. Alberola, J.A.; Mondragón, R.; Juliá, J.E.; Hernández, L.; Cabedo, L. Characterization of halloysite-water nanofluid for heat transfer applications. *Appl. Clay Sci.* **2014**, *99*, 54–61. [CrossRef]
71. Ba, T.; Alkurdi, A.Q.; Lukács, I.E.; Molnár, J.; Wongwises, S.; Gróf, G. A Novel Experimental Study on the Rheological Properties and Thermal Conductivity of Halloysite Nanofluids. *Nanomaterials* **2020**, *10*, 1834.
72. Ba, T.; Baqer, A.; Saad Kamel, M.; Gróf, G.; Odhiambo, V.O.; Wongwises, S. Experimental Study of Halloysite Nanofluids in Pool Boiling Heat Transfer. *Molecules* **2022**, *27*, 729.
73. Guimarey, M.J.G.; Liñeira del Río, J.M.; Fernández, J. Improvement of the lubrication performance of an ester base oil with coated ferrite nanoadditives for different material pairs. *J. Mol. Liq.* **2022**, *350*, 118550. [CrossRef]

74. Chen, Y.; Renner, P.; Liang, H. Dispersion of Nanoparticles in Lubricating Oil: A Critical Review. *Lubricants* **2019**, *7*, 7. [CrossRef]
75. Wu, P.R.; Liu, Z.; Cheng, Z.L. Growth of MoS$_2$ Nanotubes Templated by Halloysite Nanotubes for the Reduction of Friction in Oil. *ACS Omega* **2018**, *3*, 15002–15008. [CrossRef]
76. Zhang, Z.; Yu, G.; Geng, Z.; Tian, P.; Ren, K.; Wu, W.; Gong, Z. Ultra low friction of conductive carbon nanotube films and their structural evolution during sliding. *Diam. Relat. Mater.* **2021**, *120*, 108617. [CrossRef]
77. Wang, B.; Qiu, F.; Barber, G.C.; Zou, Q.; Wang, J.; Guo, S. Role of nano-sized materials as lubricant additives in friction and wear reduction: A review. *Wear* **2022**, *490–491*, 204206. [CrossRef]
78. Gara, L.; Zou, Q. Friction and Wear Characteristics of Water-Based ZnO and Al$_2$O$_3$ Nanofluids. *Tribol. Trans.* **2012**, *55*, 345–350. [CrossRef]
79. Wu, Y.Y.; Tsui, W.C.; Liu, T.C. Experimental analysis of tribological properties of lubricating oils with nanoparticle additives. *Wear* **2007**, *262*, 819–825. [CrossRef]
80. Azman, N.F.; Samion, S. Dispersion Stability and Lubrication Mechanism of Nanolubricants: A Review. *Int. J. Precis. Eng. Manuf. Technol.* **2019**, *6*, 393–414. [CrossRef]
81. Shafi, W.K.; Charoo, M.S. An overall review on the tribological, thermal and rheological properties of nanolubricants. *Tribol. Mater. Surf. Interfaces* **2020**, *15*, 20–54. [CrossRef]
82. Guo, W.; Li, G.; Zheng, Y.; Dong, C. Measurement of the thermal conductivity of SiO$_2$ nanofluids with an optimized transient hot wire method. *Thermochim. Acta* **2018**, *661*, 84–97. [CrossRef]
83. Taha-Tijerina, J.; Aviña, K.; Martínez, J.M.; Arquieta-Guillén, P.Y.; González-Escobedo, M. Carbon Nanotori Structures for Thermal Transport Applications on Lubricants. *Nanomaterials* **2021**, *11*, 1158. [CrossRef] [PubMed]
84. Taha-Tijerina, J.; Cadena-de la Peña, N.; Cue-Sampedro, R.; Rivera-Solorio, C. Thermo-physical evaluation of dielectric mineral oil-based nitride and oxide nanofluids for thermal transport applications. *J. Therm. Sci. Technol.* **2019**, *14*, JTST0007. [CrossRef]

Disclaimer/Publisher's Note: The statements, opinions and data contained in all publications are solely those of the individual author(s) and contributor(s) and not of MDPI and/or the editor(s). MDPI and/or the editor(s) disclaim responsibility for any injury to people or property resulting from any ideas, methods, instructions or products referred to in the content.

Article

Preparation and Tribological Behavior of Nitrogen-Doped Willow Catkins/MoS₂ Nanocomposites as Lubricant Additives in Liquid Paraffin

Yaping Xing [1], Ebo Liu [1], Bailin Ren [1], Lisha Liu [1], Zhiguo Liu [1], Bocheng Zhu [1], Xiaotian Wang [1], Zhengfeng Jia [1,*], Weifang Han [1,*] and Yungang Bai [2]

[1] School of Materials Science and Engineering, Liaocheng University, Liaocheng 252059, China
[2] Shandong Qichanxintu Composite Material Co., Ltd., Liaocheng 252800, China; bai9521@hotmail.com
* Correspondence: jiazhfeng@lcu.edu.cn (Z.J.); hanweifang@lcu.edu.cn (W.H.)

Abstract: In this study, willow catkins/MoS₂ nanoparticles (denoted as WCMSs) have been prepared using a hydrothermal method. The WCMSs were modified with oleic acid (OA) to improve dispersion in base oil. The friction and wear properties of WCMSs in liquid paraffin (LP) for steel balls were investigated using a four-ball wear tester. The results have shown that at a high reaction temperature, willow catkins (being used as a template) and urea (being used as a nitrogen resource) can effectively decrease the wear scar diameters (WSDs) and coefficients of friction (COFs). At a concentration of 0.5 wt.%, the WSD and COF of steel balls, when lubricated using LP containing modified WCMS with urea, decreased from 0.65 mm and 0.175 of pure LP to 0.46 mm and 0.09, respectively. The addition of urea and hydroxylated catkins can generate a significant number of loose nano-sheets and even graphene-like sheets. The weak van der Waals forces, decreasing the shear forces that the steel balls must overcome, provide effective lubrication during rotation. On the other hand, the tribo-films containing MoS₂, FeS, azide, metal oxides and other compounds play important roles in reducing friction and facilitating anti-wear properties.

Keywords: willow catkins; MoS₂; hydrothermal; urea; wear

Citation: Xing, Y.; Liu, E.; Ren, B.; Liu, L.; Liu, Z.; Zhu, B.; Wang, X.; Jia, Z.; Han, W.; Bai, Y. Preparation and Tribological Behavior of Nitrogen-Doped Willow Catkins/MoS₂ Nanocomposites as Lubricant Additives in Liquid Paraffin. *Lubricants* **2023**, *11*, 524. https://doi.org/10.3390/lubricants11120524

Received: 2 November 2023
Revised: 6 December 2023
Accepted: 7 December 2023
Published: 10 December 2023

Copyright: © 2023 by the authors. Licensee MDPI, Basel, Switzerland. This article is an open access article distributed under the terms and conditions of the Creative Commons Attribution (CC BY) license (https://creativecommons.org/licenses/by/4.0/).

1. Introduction

Nanomaterials are widely studied because of their excellent properties with respect to tribology, notably in reducing friction, enabling wear resistance, with significant implications for environmental protection [1–3]. Two-dimensional (2D) nanomaterials, including BN, graphene, and MoS₂ exhibit superior tribological properties because of their distinct layer structures and self-lubricating ability [3,4]. Weak van der Waals forces between the adjacent lamellae decrease the shear stress during sliding [2]. Liquid exfoliation and hydrothermal reaction are two standard procedures in the fabrication of MoS₂ nanosheets [5,6]. Ion intercalation and exchange by sonication in solution are often used in exfoliation methods [5]. In the case of the hydrothermal route, the nature of the solution, temperature, and hybrid elements can be adjusted to improve material properties [2,7,8]. With the objective of improving the tribological properties of MoS₂, composites including MoS₂/carbon [7,9], metal oxide/MoS₂ [10], and boron nitride/MoS₂ [1] have been investigated to tune possible synergistic effects. Wu et al. [7] have synthesized a series of MoS₂@carbon nanocomposites using a hydrothermal method, and noted a synergism between the MoS₂ and carbon nanocomposites that resulted in improved dispersibility and tribological properties relative to the individual components. Furthermore, MoS₂ nanoparticles grown on carbon nanotubes, graphene, and fullerene C60 were executed using a simple solvothermal method. On the other hand, surface functionalization and elemental doping of nanoparticles can further improve the dispersion stability and chemical activity in base oil (mineral oil, such as engine oil and/or liquid paraffin), and their tribological behaviors in lubricants can be enhanced due to the tribochemical reactions between these functional groups/doped atoms and the friction surface [1,9]. Various dispersants and surface-active chemicals, including alkylamines, alkanethiols, organic acid, and amine were used for the functionalization of nanostructured and doped elements, respectively [9].

Willow catkins (denoted as WCs), the willow flowers in spring, are widely dispersed by wind and cause damage to the environment due to their high specific surface area and facile spontaneous combustion [11,12]. Direct combustion of catkins results in environmental pollution with the release of CO_2 and NO_x [11]. Many researchers have addressed the practical use of renewable biomass materials, including catkins, celtuce leaves, and seed shells [11,13,14]. The pyrolysis of these biomass materials generates carbon materials and the possible fabrication of composites represents a reuse of catkins [15–17]. Zhang et al. [18] generated N/S co-doped carbon micro-tubes by the pyrolysis of catkins in an inert atmosphere. The resultant hybrids exhibited superior electrocatalytic properties when compared with commercial products. While this work offers a means of reusing catkins in treating environmental pollution and water contamination, the associated energy consumption represents a challenge in terms of sustainability. So far, authors did not find any research on the composites of willow catkins and MoS_2 being used as lubricant additives.

In this study, willow catkins and urea have been used as matrix and nitrogen resource, respectively, to prepare nitrogen-doped willow catkins/MoS_2 nanocomposites using hydrothermal methods. The composites have been modified with OA to enhance the dispersibility in LP to evaluate tribological properties using a four-ball wear tester [19]. The wear mechanism of the composites has also been investigated.

2. Materials and Methods

2.1. Materials and Preparation

Sodium molybdate dihydrate, thioacetamide, oleic acid and NaOH were purchased from the Aladdin Biochemical Technology Company (Shanghai, China) and used as supplied. The LP (viscosity: 258.56 mm^2/s (40 °C), flash point: 279 °C, pour point: −15 °C) was obtained from Liaocheng Manxiandi lubricating oil company (Liaocheng, China) and used without treatment. AISI-52100 steel balls (Ø12.7 mm) were employed in the four-ball wear tester. The willow catkins were collected from the campus of Liaocheng University.

The willow catkins were washed sequentially with tap water and deionized water, transferred into a NaOH solution (2.0 wt.%) with vigorous stirring for 24 h, then centrifuged and washed to achieve well-dispersed hydroxylated willow catkins solutions. Sodium molybdate dihydrate (1 g) and thioacetamide (1 g) were added to the hydroxylated willow catkins solution (30 mL) with stirring, then transferred into the hydrothermal reactor. The solution was heated above 100 °C for 24 h to obtain the composites ($WC_cM_1S_1$). For the purposes of comparison, particles different percentages of molybdate dihydrate and thioacetamide, with/without willow catkins and urea were also synthesized (see Table S1).

The modification and wear experiments using WCMSs have been outlined in previous papers [19]. The LP containing OA-modified WCMSs (code as OAWCMSs and with concentrations of 0.2, 0.5, 1.0, 1.5, and 2.0 wt%, respectively) were stable within 7 days (see Figure S1). Wear experiments were conducted using a MRS-10A four-ball wear tester of Jinan Shunmao Experimental Technology Company in China (392 N load, 1450 rpm, and 30 min).

2.2. Characterization

The microstructure of the composites was assessed using a JEM-2100 high-resolution transmission electron microscope (HRTEM) and a SIGMA500/VP field-emission scanning electron microscope (FE-SEM) equipped with energy-dispersive X-ray analysis (EDXA, Kevex Sigma, USA). The X-ray diffraction (XRD) patterns were collected on a Bruker D8 Advanced X-ray Diffractometer using Cu Kα radiation in the 2θ range between 10° and 80° with a scanning rate of 8° min^{-1}. The ESCA LAB Xi+ X-ray photoelectron spectrometer

(XPS) and Bruck IFs66v spectrometer were used to collect X-ray photoelectron spectra and Fourier transform infrared (FT-IR) spectra of the composites and/or worn surfaces, respectively. Al-Ka radiation was used as the excitation source to determine the binding energies of the target elements at a pass energy of 29.4 eV and a resolution of ±0.2 eV. The binding energy of C1s (284.5 eV) was used as the internal reference for XPS analysis. Thermogravimetric analysis (TGA) was conducted using a NETZSCHSTA499 simultaneous thermal analyzer, operating from room temperature to 800 °C at a heating rate of 10 °C/min in N_2.

3. Results and Discussion

3.1. Results

Representative FE-SEM and TEM images of WCMSs produced at different hydrothermal temperatures are presented in Figure 1. It can be seen that the MoS_2 nano-sheets increase in size and become closely clustered on raising the hydrothermal temperature from 120 °C to 200 °C, where the thickness of the nano-sheets was less than 100 nm (Figure 1a,b). The mapping images show that MoS_2 sheets are strongly attached on the surfaces of willow catkins (Figure S2). The elliptic holes exhibit a major axis of 5 μm and both sides of the entocoele are strongly attached with MoS_2 nano-sheets (Figure 1c). The thickness of these nano-sheets is still less than 100 nm (Figure 1c,d). The TEM analysis has established that the willow catkins were covered with MoS_2 and the interlayer spacing between the nano-sheets was ca. 0.65 nm (Figure 1e,f).

Figure 1. Cont.

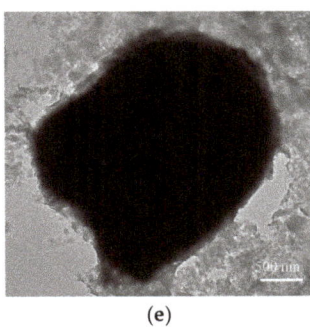
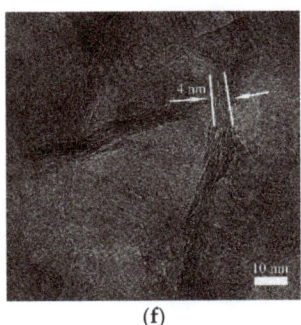

(e)   (f)

Figure 1. FE-SEM images of $WC_0M_2S_1$ produced at a hydrothermal temperature of 120 °C (**a**) and 200 °C (**b**). FE-SEM (**c**,**d**) and TEM (**e**,**f**) images of $WC_2M_2S_1$ at a hydrothermal temperature of 200 °C.

The XRD patterns and Raman spectra of $WC_2M_2S_1$ generated at a hydrothermal temperature of 120 °C and 200 °C, respectively, are presented in Figure 2. The diffraction peaks at 13.89°, 33.42°, 38.87°, 48.32°, and 58.91° can be attributed to (002), (100), (103), (105), and (008) crystal planes, respectively, establishing a hexagonal MoS_2 structure (card No. 37-1492) [20]. An increase in hydrothermal temperature is accompanied by sharper diffraction peaks with increased intensity, indicating improved the growth of MoS_2 crystals along (002) plane and (100) plane, respectively [21]. The Raman spectra presented in Figure 2b–d were obtained using a laser irradiation wavelength of 532 nm. The peaks at 376.48 cm^{-1} and 402.26 cm^{-1} are attributed to the $E^1{}_{2g}$ and $A^1{}_g$ modes of MoS_2, suggesting in-layer vibration of S and Mo atoms and outer layer vibration of S atoms along the c axis, respectively [21,22]. Furthermore, the distance value of the $E^1{}_{2g}$ and $A^1{}_g$ peaks is 25.65 cm^{-1}, which is smaller than the value (26.3 cm^{-1}) for bulk MoS_2. The lower $E^1{}_{2g}$ and $A^1{}_g$ peak distances indicate a decrease in the number of layers [21], which results from the use of willow catkins. The Raman peak at 819.88 cm^{-1} is assigned to the vibration of MoO_3, possibly suggesting a transfer of MoO_3 into MoS_2 as the hydrothermal temperature was increased from 120 °C to 200 °C [20]. The D and G peaks of carbon (at 1380 cm^{-1} and 1600 cm^{-1}) were not observed after hydrothermal treatment [21].

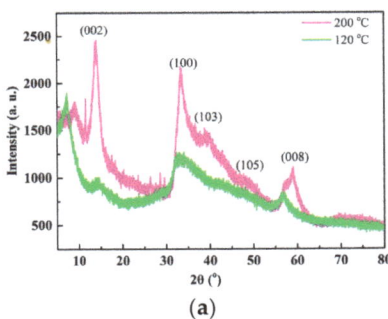
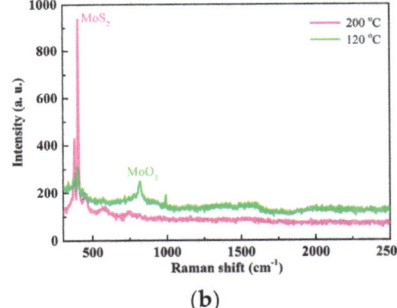

(a)   (b)

Figure 2. *Cont.*

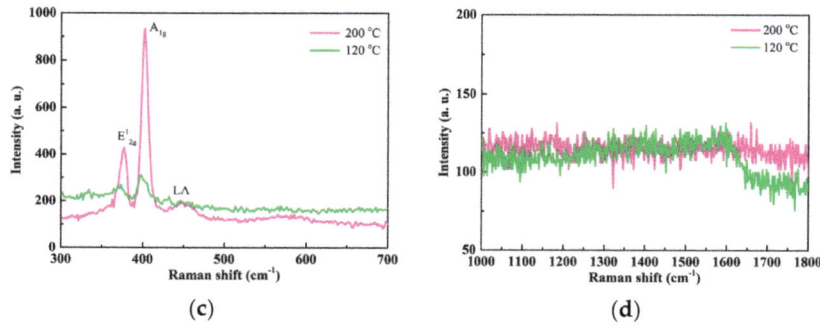

Figure 2. XRD patterns (**a**) and Raman spectra (**b–d**) for WC$_2$M$_2$S$_1$ produced at a hydrothermal temperature of 120 °C and 200 °C, respectively.

The FT-IR spectrum of WC$_2$M$_2$S$_1$ (Figure 3a) shows a broad peak at ca. 3200 cm^{-1}, attributed to -OH and/-NH$_2$ on the surface of willow catkins [23]. Peaks at ca. 1139 cm^{-1} and 888 cm^{-1} are assigned to Mo-O and S-S bonds, respectively [8]. The peaks at ca. 1017 cm^{-1} and 1418 cm^{-1} are due to C-O-C and -COO- groups, respectively [23]. A weak peak at ca. 695 cm^{-1} is assigned to the Mo-S bond [24]. The peak at ca. 1611 cm^{-1}, attributed to C=N and/or C=C groups, suggests the presence of aromatic rings associated with willow catkins [11]. The TGA curves (Figure 3b) have revealed weight losses for WC$_2$M$_2$S$_1$ before and after modification of 8.5% and 25.1%, respectively [19].

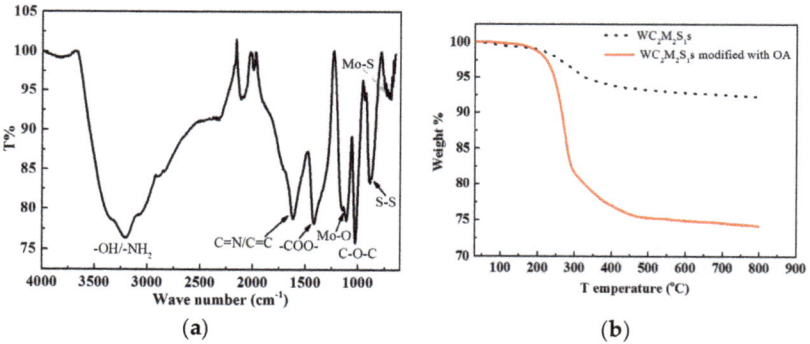

Figure 3. (**a**) FT-IR spectrum of WC$_2$M$_2$S$_1$, and (**b**) TGA curves of WC$_2$M$_2$S$_1$ before and after modification.

The WSD-concentration and COF-concentration curves for LP containing OA modified MSs (denoted as OAMSs) are presented in Figure 4 for different reaction temperatures and percentages of sodium molybdate dihydrate and thioacetamide. Figure 4a,b show the WSD-concentration and COF-concentration curves for steel balls with lubrication by LP-OAMSs synthesized at different temperatures. In the case of the three lubricants, lower WSDs and COFs were recorded relative to pure LP. Increasing the synthesis temperature to 200 °C resulted in a WSD decrease from 0.65 mm for pure LP to 0.48 mm, and a decrease in COF from 0.178 to 0.121. The steel balls using lubricants containing additives synthesized at 200 °C exhibit better anti-wear and friction reducing ability than observed at other temperatures. Consequently, the synthesis temperature was fixed at 200 °C.

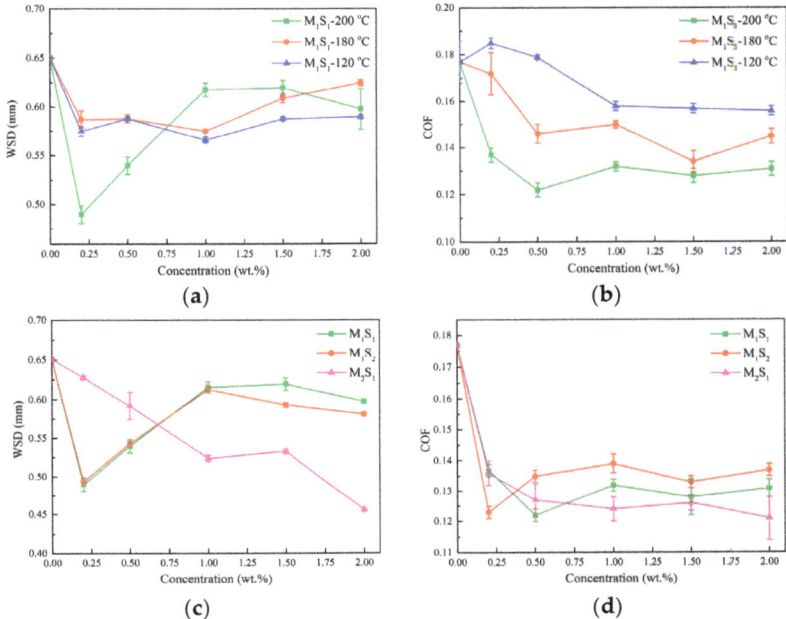

Figure 4. (a) WSD concentration curves and (b) COF concentration curves of LP-OA-modified M_1S_1 at different reaction temperatures. (c) WSD concentration curves and (d) COF concentration curves of LP-OA-modified MS with different percentages at 200 °C.

The tribological behavior of MSs with different percentages of sodium molybdate dihydrate and thioacetamide was investigated; the results are presented in Figure 4c,d. As the weights of sodium molybdate dihydrate and thioacetamide were 2 g and 1 g, respectively, the samples are denoted by M_2S_1. The steel balls lubricated with LP-OA-modified M_2S_1 exhibit less friction reduction and anti-wear capability than achieved with the other two lubricants. At a concentration of 2.0 wt.%, a WSD of 0.46 mm was recorded for LP-OAM_2S_1, which is lower than that for LP-M_1S_1 (0.60 mm) and LP-M_1S_2 (0.59 mm), respectively. The weights of sodium molybdate dihydrate and thioacetamide were fixed at 2 g and 1 g, respectively. The non-linearity of COFs with concentrations might be attributed to the changes of viscosity index, adsorption–desorption ability and/or the spread ability of additives in base oil, respectively [25–27].

The WSD concentration curves and COF concentration curves presented in Figure 5a and 5b, respectively, show the results generated for LP-OAWCM_2S_1 using different volumes of willow catkins solution. It can be seen that the steel balls exhibit smaller WSDs for lubrication by LP-OAWC$_3M_2S_1$ with concentrations less than 1.5 wt.% relative to samples without willow catkins. The WSD of steel balls lubricated with LP-OAWC$_3M_2S_1$ at a concentration of 0.2 wt.% is 0.50 mm, which is much smaller than that with LP (0.65 mm) and LP-OAWC$_0M_2S_1$ (0.60 mm). In the case of the COFs of steel balls with LP containing the three additives, there is no marked dependence where use of LP containing the three additives at a concentration of 1.0 wt.% resulted in a COF of 0.121, much lower than that obtained with pure LP (0.178).

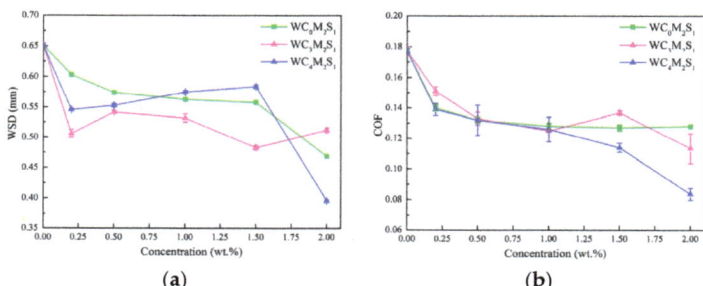

Figure 5. (a) WSD concentration curves and (b) COF concentration curves of LP-OA-modified WCM$_2$S$_1$ with different volumes of Willow catkins solutions.

In a further investigation of the tribological properties of WC$_3$M$_2$S$_1$, urea was added to the solution during hydrothermal reaction (see Figure 6). When 8 g urea was included in the hydrothermal solution, both the WSDs and COFs for the steel balls lubricated with LP-OA modified WC$_3$M$_2$S$_1$ exhibited a marked drop. At a concentration of 0.5 wt.%, a WSD of 0.46 mm was calculated for the balls lubricated with LP containing WC$_3$M$_2$S$_1$ and urea (8 g), which is ca. 70% and 87% of that achieved with pure LP and LP containing WC$_3$M$_2$S$_1$ without urea, respectively. In the case of samples with other amounts of urea (12 g and 4 g), the measured WSDs showed no obvious differences relative to the addition of 8 g urea. Furthermore, the COFs for the balls lubricated with LP containing WC$_3$M$_2$S$_1$ with urea were much lower than lubrication without urea. For example, at a concentration of 0.5 wt.%, a COF of 0.09 mm was recorded for the balls lubricated with LP containing WC$_3$M$_2$S$_1$ with urea (8 g), which is ca. 50% and 69% of that recorded for lubrication with pure LP and LP containing WC$_3$M$_2$S$_1$ without urea, respectively. From an overview of the results, WC$_3$M$_2$S$_1$ with 8 g urea exhibited the lowest coefficients of friction.

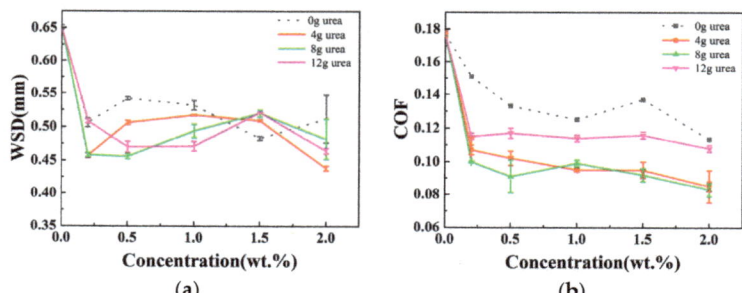

Figure 6. (a) WSD concentration and (b) COF concentration curves of LP-OA-modified WC$_3$M$_2$S$_1$ with different volumes of urea.

3.2. Discussion

The above experiments serve to demonstrate that the combination of willow catkins and urea can improve the tribological properties of MoS$_2$. Generally speaking, the low friction coefficient of MoS$_2$ was related to its easy sliding between layers under the shearing force of weak van der Waals force between molecular layers. The possible lubrication mechanisms were generally suggested as rolling, deformation, exfoliation-transferring and easier interlayer sliding, and the related factors about improving lubrication performance are also explored. Many reports also suggest that both weak Van der Waals forces between the MoS$_2$ layers and the protective tribo-films containing Mo, S, Fe, C, O and other elements can decrease friction forces and wear rate [7,9,28–31]. For example, Sun et al. synthesized N-doped carbon quantum dots as lubricant additive to investigate the tribological behavior of

MoS$_2$ nanofluid [9]. The tribo-films containing amorphous substances, ultrafine crystalline nanoparticles, and self-lubricating FeSO$_4$/Fe$_2$(SO$_4$)$_3$ were formed during rotation. Non-equilibrium molecular dynamics simulation results indicated the interaction between S atoms in MoS$_2$ as well as these O- and N-containing functional groups in nitrogen-doped carbon quantum dots with steel surfaces enhanced the stability and strength of tribo-films. Liu studied the tribological properties of coral-like MoS$_2$ as additives of liquid paraffin using four-ball testing machine [30]. The results shows that MoS$_2$ sheets stack randomly under weak Van der Waals force and large layer space can weaken the attractive force, and the nanostructure will be easily distorted, exfoliated and even oxidized during friction. On the other hand, the surface of the steel ball is formed by the boundary lubrication and protective film formed by newly generated Fe-based compounds, molybdenum oxide and other products. So the friction reduction and anti-wear mechanism were focused on the tribo-films and/or the Van der Waals forces.

The morphologies and corresponding elemental analysis of worn surfaces are presented in Figure 7 for lubrication using LP and LP-OA-modified WC$_3$M$_2$S$_1$ with 8 g urea. The WSD of the worn surface using LP-OA modified WC$_3$M$_2$S$_1$ with 8 g urea is 0.45 mm, which is much smaller than the WSD (0.65 mm) using pure LP. The SEM and associated EDXA have revealed that the worn surface of the steel balls lubricated with LP-OA modified WC$_3$M$_2$S$_1$ with urea is smooth with evidence of a Fe, C, O, Mo, S and N content. It should be noted that the percentage of nitrogen on the worn surface increased from 0.06 at.% to 0.92 at.% when 8 g urea was included in the treatment.

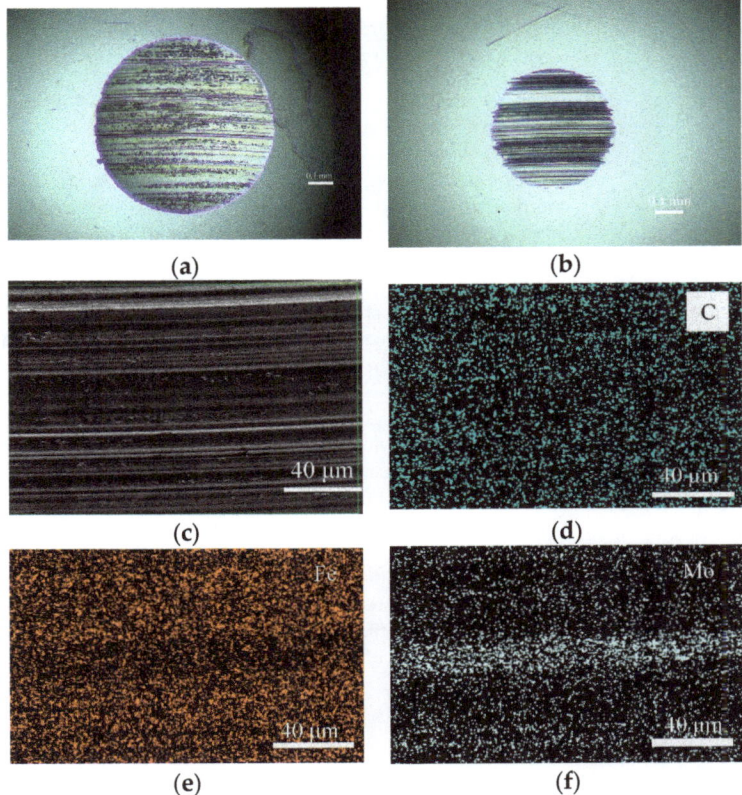

Figure 7. *Cont.*

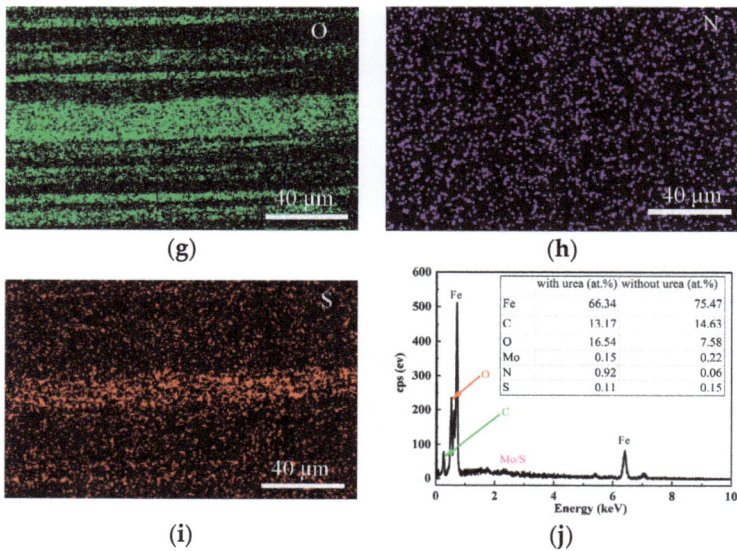

Figure 7. (**a**) Optical image of the steel ball wear scar following lubrication with LP. (**b**) Optical image, (**c**) SEM image, (**d–i**) EDS element mapping, and (**j**) EDS spectrum of the steel ball wear scar following lubrication with LP containing WC$_3$M$_2$S$_1$ with 8 g urea.

The XPS analysis of the worn surfaces following lubrication using LP-WC$_3$M$_2$S$_1$, with and without urea, is presented in Figure 8. Peaks due to C1s at 283.1 eV, 284.7 eV, 286.0 eV, and 288.1 eV can be attributed to carbide, carbon, carbon with N/S, and O-C=O, respectively (see Figure 8a and 8g) [19,32]. A comparison of Figure 8g with Figure 8a reveals a greater component of carbide and carbon with N/S on the worn surfaces, which must result from the use of urea during composite treatment. The O1s peaks following lubrication with LP-WC$_3$M$_2$S$_1$ are assigned to metal oxides, carbonates, and sulfates [9,19,29]. The signals generated for the worn sample with lubrication using WC$_3$M$_2$S$_1$ with urea may be attributed to metal oxides, carbonates and nitrates (see Figure 8b,h) [9,19]. The occurrence of N1s peaks has revealed the formation of nitrides, organic matrix, and azide on the worn surface [9,32]. A consideration of Figure 8c,i suggests the presence of azide on the tribo-film for lubrication with LP containing WC$_3$M$_2$S$_1$ with urea. A Mo3d peaks following lubrication with LP-WC$_3$M$_2$S$_1$ are attributed to the MoO$_3$ or FeMoO$_4$ (232.2–235.4 eV), MoS$_2$ (229.1–232.3 eV), and oxysulfide compound of type MoOxSy (230.7–233.8 eV), respectively (shown as: Figure 8d and 8j) [33]. Furthermore, the weak S2p signal at 161.70 eV and 169.15 eV are due to sulfide and sulfate, respectively, suggesting the presence of FeS, MoS$_2$ and FeSO$_4$ on the worn surfaces [7,34]. The Fe2p signal was deconvoluted into two peaks at 710.44 eV and 713.04 eV, which may be ascribed to FeS and Fe$_2$O$_3$ on the worn surfaces, respectively [7,34]. The results indicate that a tribo-film containing carbon, C-O/N/S, nitride, and sulfide was formed on the worn surfaces during rotation. In the case of the samples lubricated with LP-WC$_3$M$_2$S$_1$ and urea, the formation of surface azide, nitride, and FeS may help improve the friction reducing and anti-wear capability of the steel balls.

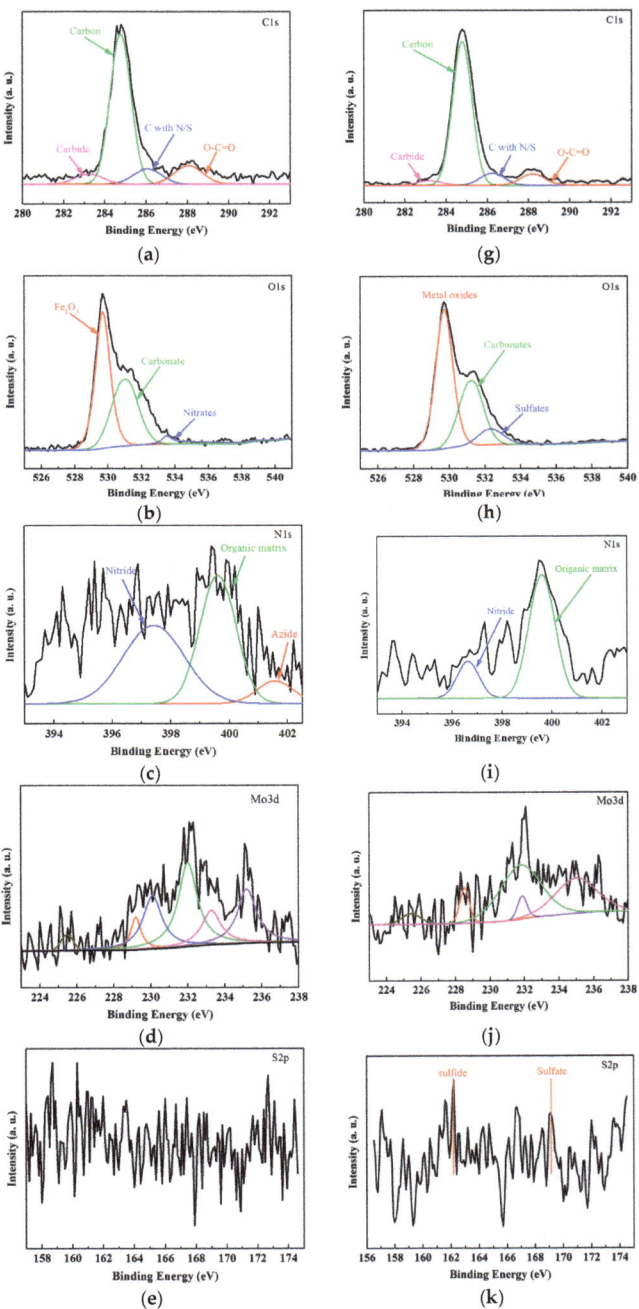

Figure 8. *Cont.*

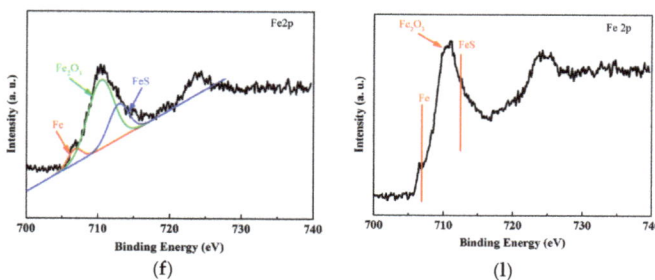

Figure 8. XPS spectra of the worn surfaces lubricated using LP-WC$_3$M$_2$S$_1$ with (**a–f**) and without (**g–l**) urea.

In order to arrive at a relationship between the wear mechanism and structure of the composites, XRD and the EDXA measurements were conducted for WC$_3$M$_2$S$_1$ with and without urea, and the results are shown in Figure 9. Taking the WC$_3$M$_2$S$_1$ sample, the XRD peaks at 14.10°, 33.32°, 59.02° are attributed to (002), (100), and (008) crystal planes, respectively. In the case of the samples with urea, the 2θ values associated with the (100) and (008) planes were shifted to 32.38° and 57.26°, respectively. The peak due to (002) was split into two peaks at 18.40° and 9.27°, which indicates that graphene-like MoS$_2$ formed due to the inclusion of urea in the solution [35]. The FE-SEM analysis has revealed the presence of loose MoS$_2$ nano-sheets on the samples of WC$_3$M$_2$S$_1$ with urea (Figure 9a). Possibly, the large spaces of these sheets decreased the shearing force of weak van der Waals force between molecular layers and friction coefficient of MoS$_2$ composites. The urea in the solution may decrease the interaction between the solution and MoS$_2$ and increase the inter-layer spacing of the MoS$_2$ nano-sheets [36]. The mapping analysis (shown in Figure S3) and the EDXA spectrum (Figure 9b) have established that the composites contain Mo, S, C, O, and N, where the ratio of Mo to S is 1:2, confirming that MoS$_2$ nano-sheets covered the surfaces of willow catkins. The images and mapping suggest that the MoS$_2$ filled the willow catkins cavity. The nitrogen content increased from 3.84 at.% to 5.5 at.% due to the addition of urea, which suggests that the hybrid of nitrogen with WCMS was formed.

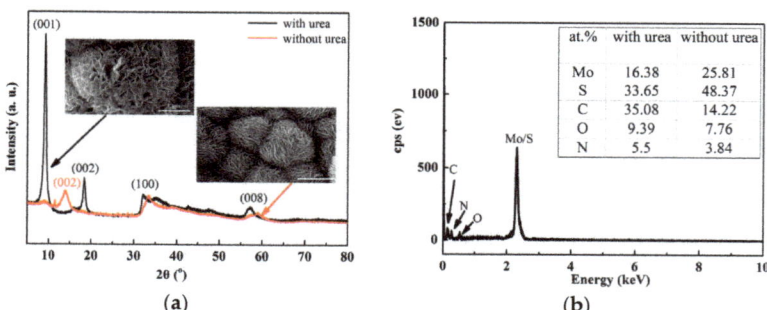

Figure 9. (**a**) XRD pattern and (**b**) EDXA spectra of the WC$_3$M$_2$S$_1$ with and without urea.

The results of the XPS analysis of WC$_3$M$_2$S$_1$ with and without urea are presented in Figure S4. The C1s peaks at ca. 285 eV and 286.1 eV are attributed to free carbon and carbon bonded to nitrogen/oxygen/sulfur, respectively [19,24]. The weak N1s peaks at 407.1 eV, 404.8 eV, 403.7 eV, 400.9 eV, and 401.2 eV are assigned to nitrates, nitrites, the organic matrix, C-NH$_2$, and ammonium salt, respectively [24,31]. Comparing the nitrogen signal for samples with and without urea, the ammonium salt was replaced by organic compounds when urea was added to the solution. The O1s peaks at 533.7 eV, 532.1 eV, and 530.4 eV are attributed to N-O, S-O, and Mo-O, respectively [24,31]. It is possible that Mo and S on the surface of MoS_2 nano-sheets bonded with oxygen on the surface of hydroxylated willow catkins [34]. Furthermore, the Mo3d and S2p curves have revealed that MoS_2 and MoO_3 were formed, confirming the presence of Mo-O on the nanosheet surface [24,31]. It can be deduced that the MoS_2 sheets were anchored to the hydroxide of the hydroxylated catkins in alkaline solution with the formation of loose MoS_2 [37]. The N_2 adsorption/desorption isotherms (at 77 K) of M_2S_1 and $WC_3M_2S_1$ (presented in Figure S5) have established that the BET surface areas increased from 8.6236 m^2/g to 15.3994 m^2/g as willow catkins were transferred into the composites, which is consistent with the formation of loose composites. Many researchers suggested that ultrathin MoS_2 could significantly improve the lubrication property of oils, because they could be penetrated into the contact area easily [29,30]. Notably, MoS_2 plays an important role in friction reduction and abrasion resistance. Nano-scale MoS_2 possesses high specific surface area, which is more advantageous to be adsorbed on the contact surface. In friction, the shedded MoS_2 flakes can be replenished and updated immediately by other adsorbed MoS_2 film, which plays an important role in lubrication function [30]. Research has shown that the addition of urea to the precursor solution can influence nucleation and growth of crystalline MoS_2, and generate a greater number of loose nano-sheets [38]. These possess weak van der Waals forces of the loose sheets, decreasing the shear forces that the steel balls must overcome, providing effective lubrication during rotation [19].

The above analysis might give some clues of mechanism. First, the lager catkins surface and the significant hydroxide can anchor the MoS_2 sheets with chemical bonds and hinder aggregation of MoS_2 sheets. Second, the addition of urea can generate a significant number of loose nano-sheets and even graphene-like sheets. The weak van der Waals forces between the loose sheets decrease the shear forces of steel balls during rotation, providing effective lubrication. Third, the tribo-film containing MoS_2, FeS, azide, metal oxides and other compounds plays an important role in reducing friction and the anti-wear properties of the composites (see Figure 10).

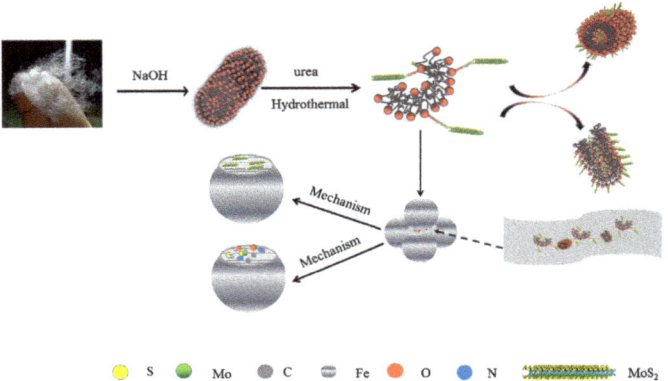

Figure 10. The growth and wear mechanism of WCMS with urea.

4. Conclusions

In this study, willow catkins/MoS_2 nanoparticles have been prepared using a hydrothermal method. By controlling the reaction temperature, the combination of willow catkins and urea can effectively improve tribological properties. The WSD of steel balls lubricated with LP-OAW$C_3M_2S_1$ at a concentration of 0.2 wt.% is 0.50 mm, which is much smaller than that with LP (0.65 mm) and LP-OAW$C_0M_2S_1$ (0.60 mm), respectively. At a concentration of 0.5 wt.%, the WSD and COF of steel balls when lubricated using LP containing modified $WC_3M_2S_1$ with urea (8g) were 0.46 mm and 0.09, representing ca. 70% and 50% of that observed for lubrication with pure LP, respectively. The tribo-film contained MoS$_2$, FeS, azide, metal oxides and other compounds that play an important role in reducing friction and the anti-wear properties of the composites. The catkins surface contains a significant hydroxide component that serves to anchor the MoS$_2$ sheets and hinder aggregation of MoS$_2$. The addition of urea to the precursor solution can alter the nucleation and growth of crystalline MoS$_2$, generating a significant number of loose nano-sheets and even graphene-like sheets. The weak van der Waals forces between the loose sheets decrease the shear forces of steel balls during rotation, providing effective lubrication.

Supplementary Materials: The following supporting information can be downloaded at https://www.mdpi.com/article/10.3390/lubricants11120524/s1. Figure S1. The stability of lubricants with additives. Figure S2. The mapping analysis of the WC$_2$M$_2$S$_1$s with the hydrothermal temperature of 200 °C. Figure S3. The mapping analysis of the WC$_2$M$_2$S$_1$s with urea at a hydrothermal temperature of 200 °C. Figure S4. The XPS spectra of the WC$_3$M$_2$S$_1$ with (a–e) and without (f–j) urea. Figure S5. N$_2$ adsorption/desorption isotherms (77K) of M$_2$S$_1$s (A) and WC$_3$M$_2$S$_1$s (B), respectively. Table S1. The catalogue and component content of the composites.

Author Contributions: Y.X.: data curation, formal analysis, investigation, methodology, writing—original draft. E.L. and X.W.: data curation, investigation, methodology. B.R.: visualization, investigation. L.L.: investigation. Z.L.: supervision. B.Z.: software and validation. Z.J. and Y.B.: conceptualization, investigation, funding acquisition, project administration, supervision, writing—review and editing. W.H.: conceptualization, investigation, funding acquisition, writing—reviewing and editing. All authors have read and agreed to the published version of the manuscript.

Funding: This research was funded by the Project of Natural Science Foundation of China (Grant No. 51905247; 52105190), Shan Dong Province Science and technology smes innovation ability improvement project (Grant No. 2022TSGC2564) and Shan Dong Province Nature Science Foundation (Grant ZR2020ME133 and ZR2020QE044).

Data Availability Statement: Data are contained within the article.

Conflicts of Interest: Author Yungang Bai was employed by the company Shandong Qichanxintu Composite Material Co., Ltd., China. The remaining authors declare that the research was conducted in the absence of any commercial or financial relationships that could be construed as a potential conflict of interest.

References

1. Kumari, S.; Chouhan, A.; Konathala, L.S.K.; Sharma, O.P.; Ray, S.S.; Ray, A.; Khatri, O.P. Chemically functionalized 2D/2D hexagonal boron nitride/molybdenum disulfide heterostructure for enhancement of lubrication properties. *Appl. Surf. Sci.* **2022**, *579*, 152157. [CrossRef]
2. Zhang, W.; Demydov, D.; Jahan, M.P.; Mistry, K.; Erdemir, A.; Malshe, A.P. Fundamental understanding of the tribological and thermal behavior of Ag-MoS$_2$ nanoparticle-based multi-component lubricating system. *Wear* **2012**, *288*, 9–16. [CrossRef]
3. Su, Y.; Zhang, Y.; Song, J.; Hu, L. Novel approach to the fabrication of an alumina-MoS$_2$ self-lubricating composite via the in situ synthesis of nanosized MoS$_2$. *ACS Appl. Mater Interfaces* **2017**, *9*, 30263–30266. [CrossRef] [PubMed]
4. Pan, S.; Jin, K.; Wang, T.; Zhang, Z.; Zheng, L.; Umehara, N. Metal matrix nanocomposites in tribology: Manufacturing, performance, and mechanisms. *Friction* **2022**, *10*, 1596–1634. [CrossRef]
5. Coleman, J.N.; Lotya, M.; O'Neill, A.; Bergin, S.D.; King, P.J.; Khan, U.; Young, K.; Gaucher, A.; De, S.; Smith, R.J.; et al. Two-dimensional nanosheets produced by liquid exfoliation of layered materials. *Science* **2011**, *331*, 568–571. [CrossRef]
6. Nicolosi, V.; Chhowalla, M.; Kanatzidis, M.G.; Strano, M.S.; Coleman, J.N. Liquid exfoliation of layered materials. *Science* **2013**, *340*, 1226419. [CrossRef]
7. Gong, K.; Lou, W.; Zhao, G.; Wu, X.; Wang, X. MoS$_2$ nanoparticles grown on carbon nanomaterials for lubricating oil additives. *Friction* **2021**, *9*, 747–757. [CrossRef]
8. Qiu, S.; Hu, Y.; Shi, Y.; Hou, Y.; Kan, Y.; Chu, F.; Sheng, H.; Yuen, R.K.; Xing, W. In situ growth of polyphosphazene particles on molybdenum disulfide nanosheets for flame retardant and friction application. *Compos. Part A Appl. Sci. Manuf.* **2018**, *114*, 407–417. [CrossRef]
9. He, J.; Sun, J.; Choi, J.; Wang, C.; Su, D. Synthesis of N-doped carbon quantum dots as lubricant additive to enhance the tribological behavior of MoS$_2$ nanofluid. *Friction* **2023**, *11*, 441–459. [CrossRef]

10. Zheng, X.; Xu, Y.; Geng, J.; Peng, Y.; Olson, D.; Hu, X. Tribological behavior of Fe_3O_4/MoS_2 nanocomposites additives in aqueous and oil phase media. *Tribol. Int.* **2016**, *102*, 79–87. [CrossRef]
11. Wang, K.; Zhao, N.; Lei, S.; Yan, R.; Tian, X.; Wang, J.; Song, Y.; Xu, D.; Guo, Q.; Liu, L. Promising biomass-based activated carbons derived from willow catkins for high performance supercapacitors. *Electrochim. Acta* **2015**, *166*, 1–11. [CrossRef]
12. Zhang, S.; Zang, L.; Dou, T.; Zou, J.; Zhang, Y.; Sun, L. Willow catkins-derived porous carbon membrane with hydrophilic property for efficient solar steam generation. *ACS Omega* **2020**, *5*, 2878–2885. [CrossRef] [PubMed]
13. Wang, M.; Zhou, J.; Wu, S.; Wang, H.; Yang, W. Green synthesis of capacitive carbon derived from platanus catkins with high energy density. *J. Mater. Sci. Mater. Electron.* **2019**, *30*, 4184–4195. [CrossRef]
14. Yan, Y.; Fan, C.; Yang, Y.; Xie, Y.; Cao, Y.; Lin, J.; Zou, Y.; You, C.; Xu, Y.; Yang, R. Effects of structural feature of biomass raw materials on carbon products as matrix in cathode of Li-S battery and its electrochemical performance. *Ionics* **2020**, *26*, 6035–6047. [CrossRef]
15. Tong, M.; Cao, B.; Li, Y.; Chen, L.; Fu, Y. Biomass carbon combined antimony sulfide with various contents as anodes with improved cycle stability in the sodium ion batteries. *J. Alloys Compd.* **2023**, *936*, 168270. [CrossRef]
16. Gouda, M.S.; Shehab, M.; Helmy, S.; Soliman, M.; Salama, R.S. Nickel and cobalt oxides supported on activated carbon derived from willow catkin for efficient supercapacitor electrode. *J Energy Storage* **2023**, *61*, 106806. [CrossRef]
17. Zhao, D.; Wang, L.; Qiu, M.; Zhang, N. Amorphous Se restrained by biomass-derived defective carbon for stable Na-Se batteries. *ACS Appl. Energy Mater.* **2021**, *4*, 7219–7225. [CrossRef]
18. Song, L.; Chang, J.; Ma, Y.; Jiang, W.; Xu, Y.; Liang, C.; Chen, Z.; Zhang, Y. Biomass-derived nitrogen and sulfur co-doped carbon microtubes for the oxygen reduction reaction. *Mater. Chem. Front.* **2020**, *4*, 3251–3257. [CrossRef]
19. Zang, C.; Yang, M.; Liu, E.; Qian, Q.; Zhao, J.; Zhen, J.; Zhang, R.; Jia, Z.; Han, W. Synthesis, characterization and tribological behaviors of hexagonal boron nitride/copper nanocomposites as lubricant additives. *Tribol. Int.* **2022**, *165*, 107312. [CrossRef]
20. Li, W.; Yan, Z.; Shen, D.; Zhang, Z.; Yang, R. Microstructures and tribological properties of MoS_2 overlayers on MAO Al alloy. *Tribol. Int.* **2023**, *1181*, 108348. [CrossRef]
21. Sarwar, S.; Karamat, S.; Bhatti, A.S.; Aydinol, M.K.; Oral, A.; Hassan, M.U. Synthesis of graphene-MoS_2 composite based anode from oxides and their electrochemical behavior. *Chem. Phys. Lett.* **2021**, *781*, 138969. [CrossRef]
22. Shen, P.; Yang, X.; Du, M.; Zhang, H. Temperature and laser-power dependent Raman spectra of MoS_2/RGO hybrid and few-layered MoS_2. *Phys. B Condens. Matter* **2021**, *604*, 412693. [CrossRef]
23. Xuan, D.; Zhou, Y.; Nie, W.; Chen, P. Sodium alginate-assisted exfoliation of MoS_2 and its reinforcement in polymer nanocomposites. *Carbohydr. Polym.* **2017**, *155*, 40–48. [CrossRef] [PubMed]
24. Mall, V.K.; Ojha, R.P.; Tiwari, P.; Prakash, R. Immunosuppressive drug sensor based on MoS_2-polycarboxyindole modified electrodes. *Results Chem.* **2022**, *4*, 100345. [CrossRef]
25. Hirayama, S.; Kurokawa, T.; Gong, J.P. Non-linear rheological study of hydrogel sliding friction in water and concentrated hyaluronan solution. *Tribol. Int.* **2020**, *147*, 106270. [CrossRef]
26. Sharma, A.K.; Katiyar, J.K.; Bhaumik, S.; Roy, S. Influence of alumina/MWCNT hybrid nanoparticle additives on tribological properties of lubricants in turning operations. *Friction* **2019**, *7*, 153–168. [CrossRef]
27. Mousavi, S.B.; Heris, S.Z.; Estellé, P. experimental comparison between ZnO and MoS_2 nanoparticles as additives or performance of diesel oil-based nano lubricant. *Sci. Rep.* **2020**, *10*, 5813. [CrossRef]
28. Guimarey, M.J.; Viesca, J.L.; Abdelkader, A.M.; Thomas, B.; Battez, A.H.; Hadfield, M. Electrochemically exfoliated graphene and molybdenum disulfide nanoplatelets as lubricant additives. *J. Mol. Liq.* **2021**, *342*, 116959. [CrossRef]
29. Chen, J.; Xu, Z.; Hu, Y.; Yi, M. PEG-assisted solvothermal synthesis of MoS_2 nanosheets with enhanced tribological property. *Lubr. Sci.* **2020**, *32*, 273–282. [CrossRef]
30. Liu, L.; Huang, Z.; Huang, P. Fabrication of coral-like MoS_2 and its application in improving the tribological performance of liquid paraffin. *Tribol. Int.* **2016**, *104*, 303–308. [CrossRef]
31. Wu, X.; Gong, K.; Zhao, G.; Lou, W.; Wang, X.; Liu, W. Surface modification of MoS_2 nanosheets as effective lubricant additives for reducing friction and wear in poly-α-olefin. *Ind. Eng. Chem. Res.* **2018**, *57*, 8105–8114. [CrossRef]
32. Kathiravan, D.; Huang, B.R.; Saravanan, A.; Prasannan, A.; Hong, P.D. Highly enhanced hydrogen sensing properties of sericin-induced exfoliated MoS_2 nanosheets at room temperature. *Sens. Actuators B Chem.* **2019**, *279*, 138–147. [CrossRef]
33. Garcia, I.; Galipaud, J.; Kosta, I.; Grande, H.; Garcia-Lecina, E.; Dassenoy, F. Influence of the organic moiety on the tribological properties of MoS_2: Glycol hybrid nanoparticles-based dispersions. *Tribol. Int.* **2020**, *68*, 104. [CrossRef]
34. Xu, W.; Fu, C.; Hu, Y.; Chen, J.; Yang, Y.; Yi, M. Synthesis of hollow core-shell MoS_2 nanoparticles with enhanced lubrication performance as oil additives. *Bull. Mater. Sci.* **2021**, *44*, 88. [CrossRef]
35. Chen, T.; Xia, Y.; Jia, Z.; Liu, Z.; Zhang, H. Synthesis, characterization, and tribological behavior of oleic acid capped graphene oxide. *J. Nanomater.* **2014**, *2014*, 654145. [CrossRef]
36. Pal, K.; Majumder, T.P.; Schirhagl, R.; Ghosh, S.; Roy, S.K.; Dabrowski, R. Efficient one-step novel synthesis of ZnO nanospikes to nanoflakes doped OAFLCs (W-182) host: Optical and dielectric response. *Appl. Surf. Sci.* **2013**, *280*, 405–417. [CrossRef]

37. Liu, J.; Zhao, S.; Wang, C.; Ma, Y.; He, L.; Liu, B.; Zhang, Z. Catkin-derived mesoporous carbon-supported molybdenum disulfide and nickel hydroxyl oxide hybrid as a bifunctional electrocatalyst for driving overall water splitting. *J. Colloid Interface Sci.* **2022**, *608*, 1627–1637. [CrossRef]
38. Liu, Y.; Li, X.; Liu, T.; Zheng, Z.; Liu, Q.; Wang, Y.; Qin, Z.; Guo, H.; Liang, Y. Alcohols assisted in-situ growth of MoS_2 membrane on tubular ceramic substrate for nanofiltration. *J. Membr. Sci.* **2022**, *659*, 120777. [CrossRef]

Disclaimer/Publisher's Note: The statements, opinions and data contained in all publications are solely those of the individual author(s) and contributor(s) and not of MDPI and/or the editor(s). MDPI and/or the editor(s) disclaim responsibility for any injury to people or property resulting from any ideas, methods, instructions or products referred to in the content.

Article

Study on the Influence of the MoS$_2$ Addition Method on the Tribological and Corrosion Properties of Greases

Can Zhu [1], Zhongyi He [1,*], Liping Xiong [1], Jiusheng Li [2], Yinglei Wu [3,*] and Lili Li [1]

[1] School of Materials Science and Engineering, East China Jiaotong University, Nanchang 330013, China; zczy990117@163.com (C.Z.); helijia666@163.com (L.X.); lilshd18@gmail.com (L.L.)
[2] Laboratory of Advanced Lubricating Materials, Shanghai Advanced Research Institute, Chinese Academy of Sciences, Shanghai 201210, China; lijs@sari.ac.cn
[3] School of Chemistry and Chemical Engineering, Shanghai University of Engineering Science, Shanghai 201210, China
* Correspondence: hzy220567@163.com (Z.H.); wuyl@sues.edu.cn (Y.W.)

Abstract: MoS$_2$ lithium-based grease is suitable for lubrication protection between bearings at high temperatures and loads due to its excellent tribological properties. However, there is little research on the influence of different addition methods of MoS$_2$ additive on its tribology and corrosion properties. In this work, eco-friendly vegetable oil was selected as the base oil, with MoS$_2$ powder as the additive to synthesize lithium-based grease. The effects of different adding modes of MoS$_2$ on the tribology and corrosion properties of the grease were studied. The experimental results showed that adding 0.01 wt% MoS$_2$ before thickening (Method D) was more conducive to improving the tribological properties of lithium grease. The average friction coefficient was reduced by 26.1%, and the average wear scar diameter was reduced by 0.16 mm. After grinding and adding (Method B) 0.01 wt% MoS$_2$, the corrosion inhibition efficiency of the steel sheet was as high as 96.97%. The reason was that the tribochemical reaction of MoS$_2$ evenly distributed throughout the grease during friction, forming a thin friction film, reducing friction and wear. The protective film formed by MoS$_2$ and GCr15-bearing steel improved the corrosion inhibition performance of the grease.

Keywords: MoS$_2$; lithium-base grease; addition method; friction; corrosion

---

Citation: Zhu, C.; He, Z.; Xiong, L.; Li, J.; Wu, Y.; Li, L. Study on the Influence of the MoS$_2$ Addition Method on the Tribological and Corrosion Properties of Greases. *Lubricants* **2023**, *11*, 517. https://doi.org/10.3390/lubricants11120517

Received: 13 November 2023
Revised: 1 December 2023
Accepted: 5 December 2023
Published: 8 December 2023

Copyright: © 2023 by the authors. Licensee MDPI, Basel, Switzerland. This article is an open access article distributed under the terms and conditions of the Creative Commons Attribution (CC BY) license (https://creativecommons.org/licenses/by/4.0/).

1. Introduction

Grease is a common lubricant composed of base oil and thickener and is widely used in machinery lubrication due to its unique properties. It is often the lubricant of choice for rolling bearings, plain bearings, slider bearings, gears, pivots, couplings, guides, pin bushings, and sliding contacts [1]. Greases have a good sealing ability, are leak-resistant, have corrosion resistance, and require little maintenance [2]. Grease occupies an important position in the global economy and plays an important role in maintaining the normal operation of various mechanical equipment, reducing friction and wear during operation, and extending the service life of mechanical equipment [3].

At present, there are five main types of base oils for the production of biodegradable fats and oils, highly unsaturated vegetable oil [4], low-viscosity polyalphaolefin, polyethylene glycol, dibasic acid ester, and polyol ester. The main advantages of vegetable oil are low toxicity, high biodegradation rate, low cost, and renewable [5].

In general, various types of greases can be used for bearing lubrication [6]. The low-performance calcium and sodium grease is cheap [7], but the lubrication effect is not as good as the high-performance grease, and the grease change cycle is short. Ball-bearing grease can be used for equipment with general speed and heavy working load. Its mechanical stability and colloidal stability are better than calcium and sodium base greases [8]. Compared with other soap-based greases, lithium-based greases can still exert excellent lubrication performance under extremely harsh operating conditions. Lithium-based greases improve the reliability and durability of mechanical components by improving lubrication performance, reducing friction, reducing wear on mating surfaces, and preventing sintering [9–11].

MoS_2 is a two-dimensional layered structure formed by an S-Mo-S atomic covalent bond; the layers are connected by weak van der Waals forces; the molecular layers are easy to slide so that they show excellent frictional properties [12] and have been widely used in the field of solid lubrication [13]. In addition, compared with other lubricants, a series of lubricants containing MoS_2 displays tribological advantages. Research has found that MoS_2 is prone to react with non-ferrous metals such as iron substrates, generating friction films containing iron sulfide and its oxides, metal oxides, and disulfides [14], thus exhibiting excellent low friction and wear resistance. Therefore, MoS_2 is commonly used as an additive and is widely used in lubricating oils or greases to reduce friction and wear [15,16].

In the field of lubrication engineering, the tribological behavior of grease is largely determined by the performance of additives. The main function of additives in grease is to improve and enhance the performance [17], and the addition of different additives will lead to differences in the performance of the same grease. Additives are generally added during the preparation of the grease [18] or are directly mixed with the additive. However, the effect of additive adding method has not been reported. In this paper, the effects of trace amounts of MoS_2 addition methods on the tribology and corrosion properties of lithium grease were studied by us.

2. Experimental Part

2.1. Experimental Materials and Characterization

Thickening agent (lithium dodecyl stearate, white powder) purchased from Xinwang Chemical Co., Ltd. (Huzhou, China). Longevity Flower corn oil (edible grade) purchased from Samsung Corn Industry Technology Co., Ltd. (Binzhou, China) was employed as a base oil. Petroleum Ether (60–90) and ethyl alcohol (AR) were purchased from Xilong Science Co., Ltd. (Shantou, China).

MoS_2 powder (AR), produced by Tianjin Zhiyuan Chemical Reagent Co., Ltd. (Tianjin, China). The morphology of the samples was observed by Hitachi SU8010 scanning electron microscope (SEM, HitachiHigh-Technologies Corporation, Tokyo, Japan). The powder samples were ultrasonically dispersed on the silicon wafer with anhydrous ethanol, and then gold was sprayed after drying. Figure 1 shows the SEM image of MoS_2 powder. It can be seen from SEM that MoS_2 is a lamellar structure. Because the layered structure of MoS_2 is only affected by weak van der Waals forces between layers, it is prone to sliding. This results in lubricants containing MoS_2 typically exhibiting good friction reduction and wear resistance [19].

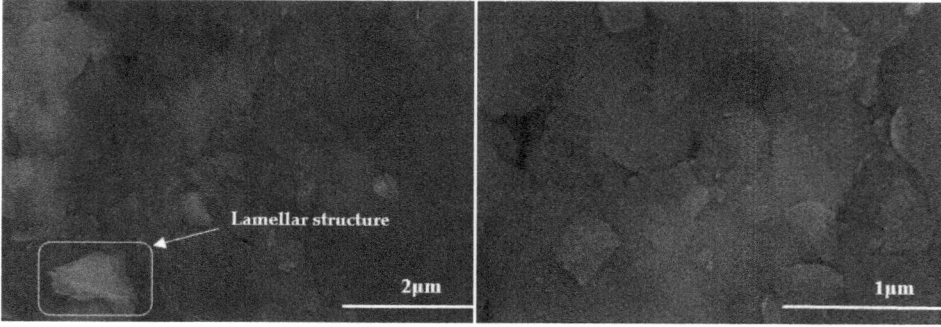

Figure 1. SEM image of MoS_2 powder.

2.2. Preparation of Grease

Based on the preparation method of He et al. [20], the specific preparation method of grease is as follows: 150 g of corn oil was heated in an oil bath to 110 °C. In the atmosphere of nitrogen, 25 g of thickening agent (12-hydroxystearate lithium) was gradually added to corn oil for thickening treatment. After stirring the reactant for 1 h, the temperature was raised to 180 °C for 1 h, and finally, the reaction temperature was raised to 210 °C until the magnetic stirrer could not stir the grease, then mechanical stirring was used for 1 h. After that, the reaction was cooled naturally in the oil bath. The cooled grease was then ground 5–10 times by a three-roll grinder (S65, China Changzhou Caobao Machinery Co., Ltd., Changzhou, China) to obtain the samples of the greases used in the following experiments. The additive concentration of MoS_2 were 0.01 wt%, 0.03 wt%, 0.05 wt% and 0.07 wt%.

In this study, MoS_2 additives with different mass fractions (ω) were added to the prepared grease samples in different ways. The addition methods were divided into the following four types: (1) adding additives directly during the thickening process; the method of adding additives was called "Method A". (2) Additives were added during three-roll grinding; this method was referred to as "Method B". (3) After adding additives during the thickening process, the sample was ground with three-rollers for 5–10 times; the method of adding additives was labeled as "Method C". (4) The MoS_2 additive was added 20 min before thickening, and the resulting grease was ground 5–10 times with a three-roll grinder; this method was recorded as "Method D".

2.3. Performance Test

2.3.1. Physical and Chemical Properties of Grease

Following the test method specified in the GB/T4929 standard [21], a drop point tester (Shanghai Fine Analysis Instrument Manufacturing Co., Ltd., SYD-4929, Shanghai, China) was used to determine the drop point of the prepared lithium greases. According to the national standard GB/T269 [22], a cone penetration tester (Beijing Luchen Weiye Instrument Equipment Co., Ltd., ZRD-3, Beijing, China) was used to measure the cone penetration of the prepared lithium greases.

2.3.2. Tribological Properties

The tribological properties of MoS_2 in grease were investigated by a vertical universal friction and wear testing machine (MMW-1B of Jinan Shunmao Instrument Co., Ltd., Jinan, China). According to the national standard GB-T3142-1982 [23], the high-load and high-speed test conditions of 392 N, 1450 rpm, and 60 min were adopted, and the friction coefficient was measured and recorded by the mechanical sensor in the friction and wear testing machine in real-time. The GCr15-bearing steel ball has a diameter of 10 mm and a hardness of 62–65 HRC. Before the test, the steel ball and the oil box were ultrasonically cleaned with petroleum ether for 30 min, and the test room temperature was 25 ± 5 °C. The experiment was repeated three times under the same conditions to ensure the accuracy of the test results. After the friction experiment, the wear scar diameter was measured by the optical microscope (M203, Aosiwei Optical Instrument Co., Ltd., Shenzhen, China).

2.4. Analysis of Wore Surfaces

The surface morphology of the steel ball was analyzed by scanning electron microscopy (SEM) of Hitachi SU8010. Before the SEM examination, the test balls were ultrasonically cleaned twice with petroleum ether for about 15 min each time. X-ray photoelectron spectroscopy (XPS, Thermo Fisher Scientific K-Alpha, Waltham, MA, USA) is a highly sensitive ultramicroscopic surface analysis technology; it can be used to achieve a qualitative and semi-quantitative analysis of the composition elements, chemical states, chemical bonds, and charge distribution on the surface of samples according to the peak shape, position, and intensity of different characteristic peaks in XPS spectra. Therefore, XPS was also used for the analysis of worn surfaces in this study. The XPS excitation source used was Al Kα rays (hv = 1486.68 eV), with a working voltage of 15 kV and a filament current of 10 mA.

The signal was accumulated for 5–10 cycles. The test passing-energy was 50 eV, with a step size of 0.05 eV, and the energy of C1s = 284.80 eV was used as the reference standard.

2.5. Corrosion Test

This study investigated the corrosion resistance of lubricating greases prepared by adding different amounts of MoS_2 additives in different ways on iron sheets. The performance of corrosion resistance was quantified by the weight loss of the steel plate.

Before the test, the 20 × 15 × 3 mm GCr15-bearing steel plate was polished with 320#, 800#, 1000#, 1500# and 2000# sandpaper in turn. Then ultrasonic cleaning was carried out with deionized water and anhydrous alcohol, respectively, to remove the iron filings on the surface of the steel plate. After the steel plates were naturally air-dried, they were buried in different greases. After 15 days, the steel plates were removed and cleaned with petroleum ether and anhydrous ethanol and then dried in air. The front and back mass of the steel plate was recorded, and the corrosion resistance of different greases can be analyzed according to the quality difference of each steel plate. The corrosion surface of GCr15 bearing steel sheet was evaluated via scanning electron microscopy (SEM) using a Hitachi SU8010, and three-dimensional (3D) morphology was performed using equipment from Zygo Corporation (ZGP OPTICAL PROFILER).

3. Results and Discussion

3.1. Physical and Chemical Properties of Grease

Table 1 shows the drop point and cone penetration of lithium grease after adding MoS_2 additive with different mass fractions in different ways. It can be seen from the data in Table 1 that after adding MoS_2 additives with different mass fractions (ω) in different ways, the drop point of the grease was improved. The coning degree of the prepared lithium grease meets the coning degree requirements of No.4 greases (175~205). Table 1 shows that when MoS_2 was added, the drop point and cone penetration of the grease changed, indicating that the addition of MoS_2 may have a certain effect on the colloidal structure of lithium grease [24].

Table 1. Dropping point and cone penetration of greases produced by different additional methods of MoS_2 with different mass fractions (ω).

	Addition Method	\multicolumn{5}{c}{ω (wt%)}			
		0	0.01	0.03	0.05	0.07
Dropping point (°C)	Method A	196.5	197	197	202	202
	Method B		202	202	200	199
	Method C		197	197.5	198	198
	Method D		203	199	200	201
Cone penetration (0.1 mm)	Method A	180.6	186.1	202.8	198.1	181.6
	Method B		202.4	201.2	180	192.2
	Method C		175.7	202.3	186.9	180.4
	Method D		184.6	187.5	198.2	180.5

3.2. Tribological Properties

3.2.1. Friction Reduction Performance

Under the experimental conditions of 392 N, 1450 rpm, room temperature, and 60 min, the friction reduction and anti-wear properties of MoS_2 greases prepared in different ways were estimated. Figure 2 shows the friction coefficient relationship curve and average friction coefficient of MoS_2 greases with different mass fractions prepared by different addition methods. As can be seen from the friction coefficient curve of Figure 2a, when the content of MoS_2 was 0.01% and 0.03%, the friction coefficient relationship curve changes like a peak after a stable period, which may be because the additive content was small, and cannot play the role of lubrication for a long time during the friction process. The friction coefficient curve corresponding to 0.05 wt% MoS$_2$ content begins to stabilize after about 500 s. And the average friction coefficient (Figure 3a) was the smallest, with an average friction coefficient of 0.037, which decreases by 19.6% compared with the blank group. The results showed that 0.05 wt% was the best content when MoS$_2$ was added by "Method A".

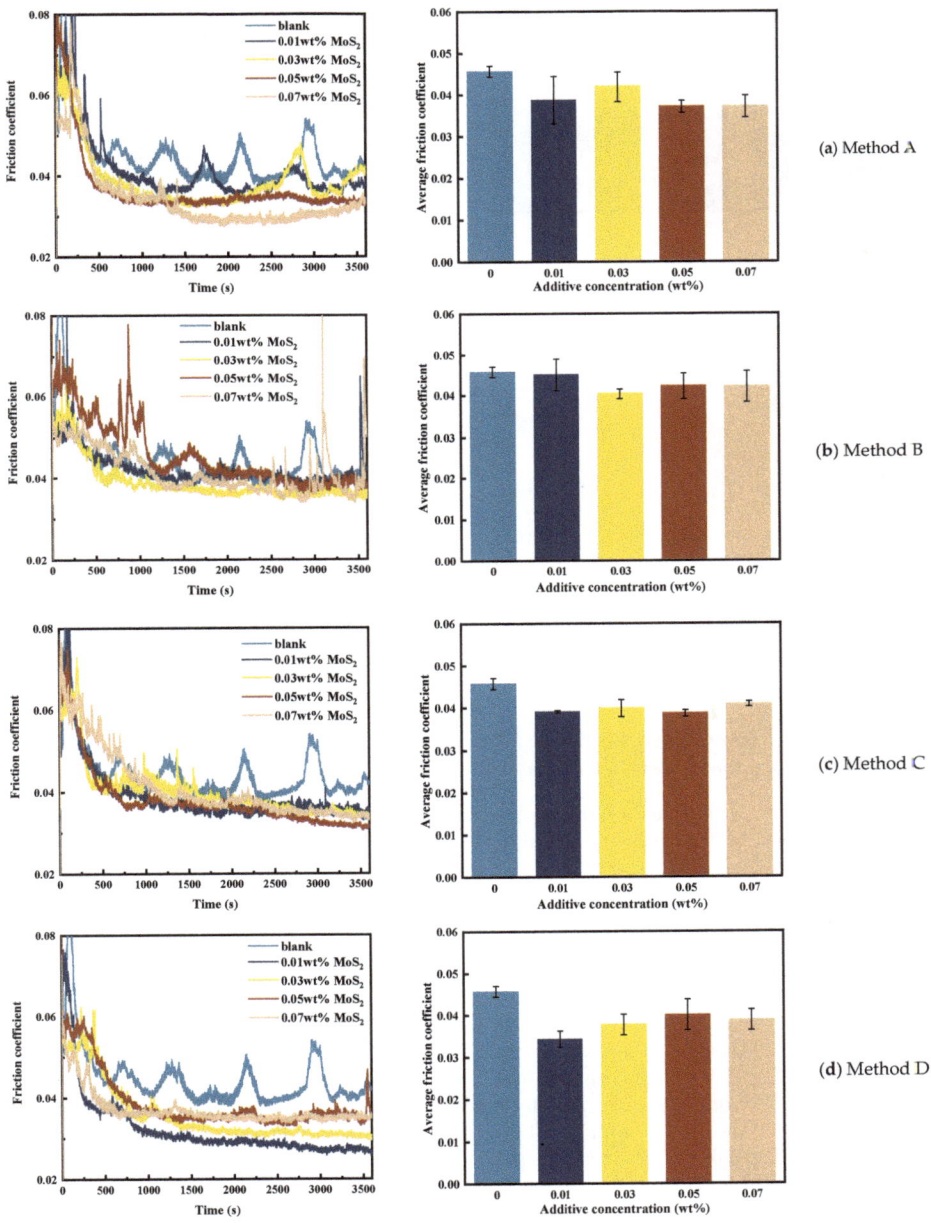

Figure 2. Friction coefficient curves and average friction coefficient of MoS$_2$ greases prepared by different addition methods with variable amounts of MoS. (392 N, 1450 rpm, 60 min).

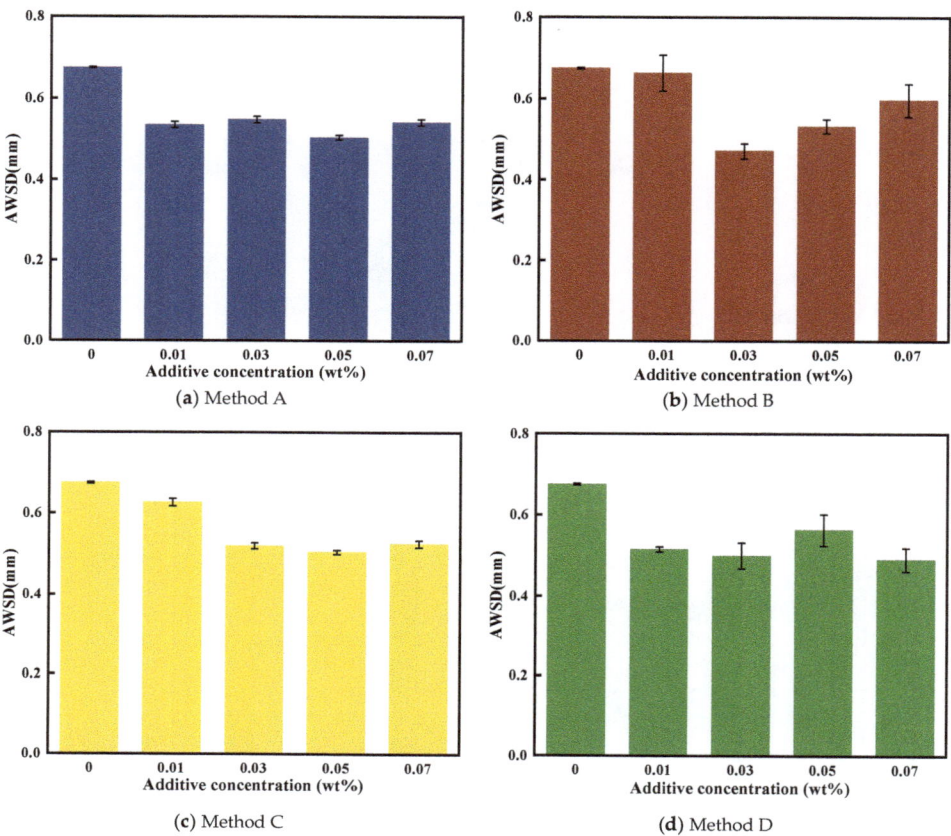

Figure 3. Wear scar diameter of MoS$_2$ greases prepared by different addition methods with variable amounts of MoS$_2$. (392 N, 1450 rpm, 60 min).

Figure 2b shows that when MoS$_2$ was added via "Method B", the average friction coefficient of the prepared grease under experimental conditions first decreased and then increased. By comparing the friction coefficient curves, it can be seen that the optimal addition amount of MoS$_2$ additive in the grease prepared using this method was 0.03 wt%. The average friction coefficient of MoS$_2$ grease at the optimal concentration is 0.040, which decreases by 13.0%.

The above two methods can improve the anti-friction performance of grease, but "Method B" requires less optimal additive content. The average friction coefficient corresponding to the optimal content of "Method A" is smaller than that of "Method B". Therefore, "Method C" combines the above two methods.

It can be seen in Figure 2c that the friction coefficient of MoS$_2$ grease prepared via "Method C" was generally stable under experimental conditions, and gradually decreased over time. The three-roll grinding process in "Method C" may help the MoS$_2$ to be more evenly distributed in the grease, thereby reducing friction for longer. By comparing, it can be found that the optimal MoS$_2$ additive content of MoS$_2$ grease prepared via "Method C" was 0.05 wt%, and the corresponding average friction coefficient was 0.039. When compared with the average friction coefficient of 0.046 in the blank group, it can be found that the average friction coefficient was reduced by 15.2%.

"Method C" and "Method D" are different in the addition time of additives. It can be seen in Figure 2d that when MoS_2 was added via "Method D", the friction coefficient curves of MoS_2 grease with different mass fractions were stable, and the optimal content of MoS_2 corresponding to this method was 0.01 wt%, and its average friction coefficient was 0.034. This was a 26.1% reduction compared to the blank group.

Compared with four different MoS_2 adding methods, it can be found that the friction coefficient of grease obtained via "Method D" was the most stable, the average friction coefficient was reduced the most, and the required optimal MoS_2 content was lowest. This may be due to the longest reaction time between the additive and the base oil, which was more conducive to the uniform mixing of the additive in the grease so that the friction reduction effect can be effectively achieved for a longer time. The lubricating grease prepared by this method can achieve a good lubrication effect based on a very low additive content of 0.01 wt%.

3.2.2. Anti-Wear Performance

Figure 3 shows the average wear scar diameter of MoS_2 greases with different mass fractions prepared by different addition methods under the test conditions of 392 N and 1450 rpm. In Figure 3a, it can be seen that the average wear scar diameter corresponding to the lubricating grease prepared via "Method A" shows that the optimal MoS_2 content was 0.05 wt%, at which point the minimum average wear scar diameter was 0.50 mm. In Figure 3b, it can be seen that the wear scar diameter caused by MoS_2 lubricating grease prepared via "Method B" shows a trend of first decreasing and then increasing with the increase of MoS_2 content, and the optimal content of additives required for the lubricating grease prepared via this method was 0.03 wt%. From the average wear scar diameter in Figure 3c, it can be observed that the optimal amount of MoS_2 added to the lubricating grease prepared via "Method C" was also 0.05 wt%. When the MoS_2 content was 0.01 wt%, a small amount of MoS_2 could not effectively resist wear during the friction process. When the amount of MoS_2 exceeded 0.03 wt%, due to the low surface tension and hydrophilic properties of the MoS_2 sliding layer [25], MoS_2 particles may be unevenly dispersed in the lubricating grease, resulting in a gradual increase in wear scar diameter. Figure 3d shows the corresponding wear scar diameter of the lubricating grease prepared via "Method D". From Figure 3d, it can be seen that the lubricating grease configured with 0.01 wt% MoS_2 exhibits the smallest average wear scar diameter, which was reduced by 0.16 mm compared to the blank group without MoS_2.

3.3. SEM Analysis of Worn Surfaces

Figure 4 shows the SEM images of worn surfaces corresponding to blank lubricating grease and lubricating grease with the optimal mass fraction of MoS_2 added in different ways. In Figure 4a, it can be seen that the worn surface corresponding to the blank grease exhibits obvious wear scar tracks, with wide and large furrows on the wear scar tracks. The wear scar tracks on the steel ball surface (Figure 4b–e) corresponding to the lubricating grease containing the MoS_2 additive become relatively less obvious, and the surface becomes smoother. This may be because MoS_2 has a layered structure that is easy to slide, so it can form a stable film on the friction surface under friction conditions, thereby achieving an anti-wear effect.

By comparing Figure 4b–e, it can be found that the wear scar tracks (300 μm) of "Method A" are relatively uniform. The wear scar tracks (300 μm) of "Method B" are concentrated. "Method C" and "Method D" have a similar range of wear scar tracks. The wear scar tracks of steel balls become shallower in turn, and the MoS_2 grease prepared via "Method D" shows the shallowest wear scar tracks and the smoothest grinding surface, indicating that the grease prepared by this addition method has the best anti-wear effect. This may be because the addition of MoS_2 before thickening allows MoS_2 and the base oil to be in contact and mixed for a longer period, and may also have a longer chemical reaction. Moreover, this method adopted a three-roller grinder for grinding, which allows MoS$_2$ to be more evenly distributed in the lubricating grease.

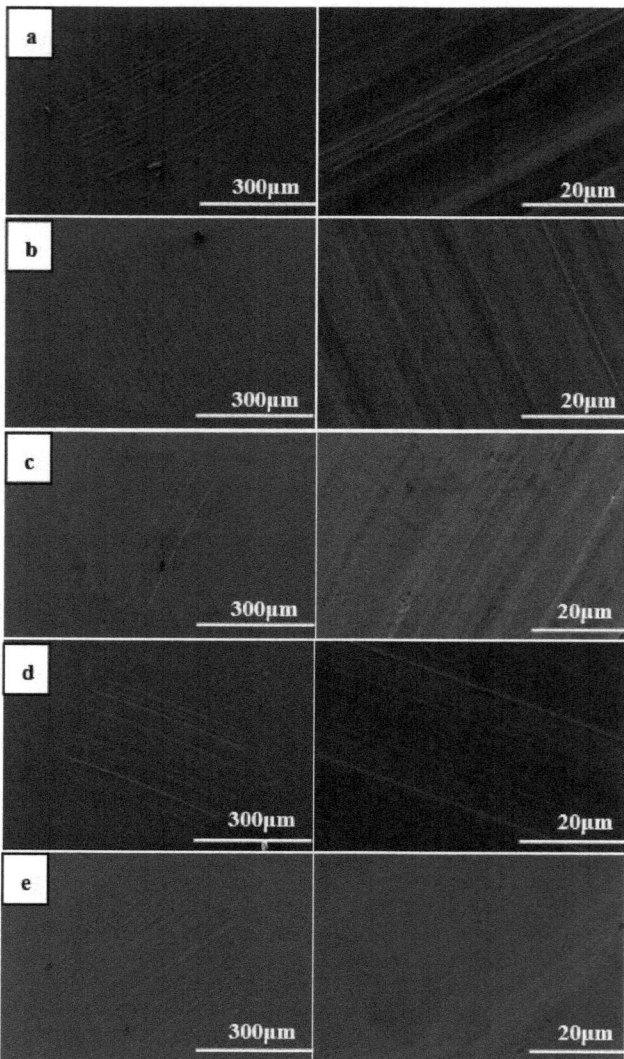

Figure 4. SEM images of worn surfaces: (**a**) blank group, (**b**) Method A with 0.05 wt% MoS$_2$, (**c**) Method B with 0.03 wt% MoS$_2$, (**d**) Method C with 0.05 wt% MoS$_2$, (**e**) Method D with 0.01 wt% MoS$_2$. (392 N, 1450 rpm, 60 min).

3.4. Soap Fiber Structure of Lithium Lipids

Figure 5 shows the SEM images of the soap fiber structure of blank grease and lithium-based lubricating grease prepared by adding MoS$_2$ through "Method D" with the best tribological performance under the friction test conditions of 392 N, 1450 r/min, and 60 min before and after friction. By comparison, it can be found that the network structure of soap fiber of blank lithium-based grease was sparse before and after the friction. The more compact the soap fiber structure, the higher the drop point of the grease, the better the colloidal stability [26]. The soap base structure of "Method D" had small pores and strong oil absorption capacity, so it can act on the friction pair for a long time and play a role in reducing friction and anti-wear. The soap fiber structure of the blank grease and the grease of "Method D" did not change between pre-friction and post-friction, respectively. This shows that the performance of lithium-base grease is excellent, and the addition of MoS_2 has no obvious effect on the structure of soap fiber.

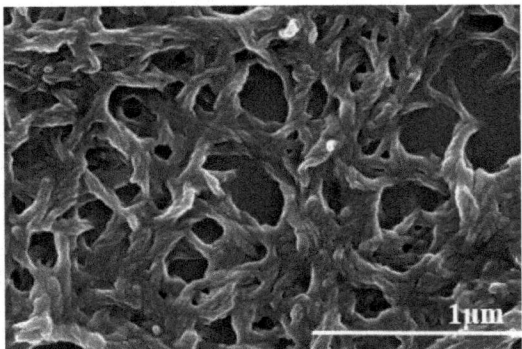
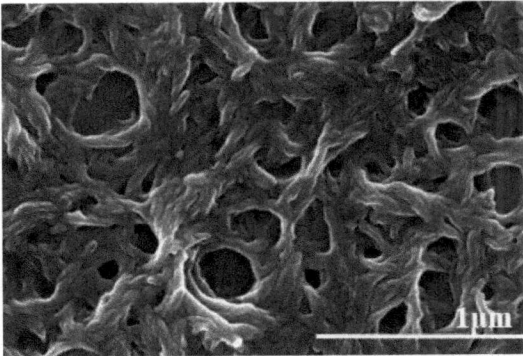

Pre-friction   Post-friction

(**a**) Blank grease

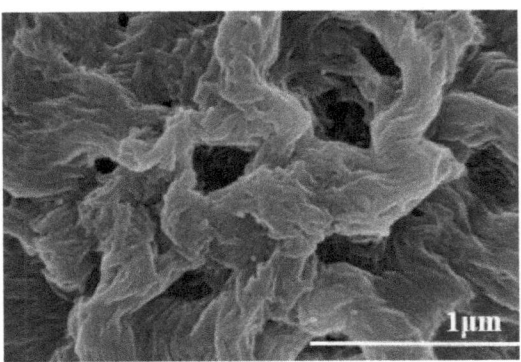
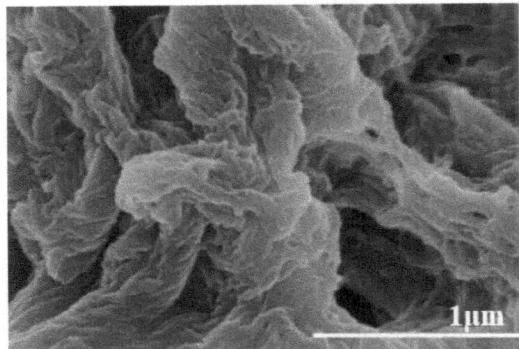

Pre-friction   Post-friction

(**b**) 0.01 wt% MoS2 by "Method D"

Figure 5. SEM image of soap fiber structure of blank grease and lithium grease with 0.01 wt% MoS_2 via "Method D" before and after friction. (392 N, 1450 rpm, 60 min).

3.5. XPS Analysis of Worn Surfaces

In order to reveal the lubrication mechanism of MoS_2 as an additive in lithium grease, the chemical state of elements on the surface of the wear scar was analyzed by XPS. Figure 6 shows the XPS spectra of C1s, O1s, Fe2p, S2p, and Mo3d on the worn surface of steel balls caused by lithium-based grease with 0.01% MoS_2.

The peaks with binding energies of 284.8, 284.98 and 287.18 eV in the C1s spectra (Figure 6b) correspond to C-C, C-O-C [27] and C=O [28] bonds, respectively. For the O1s (Figure 6c) spectrogram, two photoelectron contribution are found at 529.37 eV and 531.81 eV, attributing to metal oxides (Fe_2O_3) and C-O bonds [29]. The binding energies of Fe2p (Figure 6d) at 706.30 and 711.76 eV peaks correspond to FeS$_2$ and Fe$_2$O$_3$, respectively [30]. The 709.64 and 723.19 eV correspond to Fe 2p$_{3/2}$ and Fe 2p$_{1/2}$ of FeO. As for the S2p-XPS spectra (Figure 6e), two peaks belonging to MoS$_2$ [31]/FeS$_2$ and metal sulfate are found at 161.50 eV and 168.60 eV. The peaks at 232.68 and 235.83 eV in the Mo3d-XPS spectra (Figure 6f) correspond to MoO$_2$ and MoO$_3$, respectively [32]. The appearance of MoO$_2$ and MoO$_3$ indicates that the MoS$_2$ additive has been oxidized during the friction process; that is, MoS$_2$ may have partially caused tribochemical reactions during the friction process.

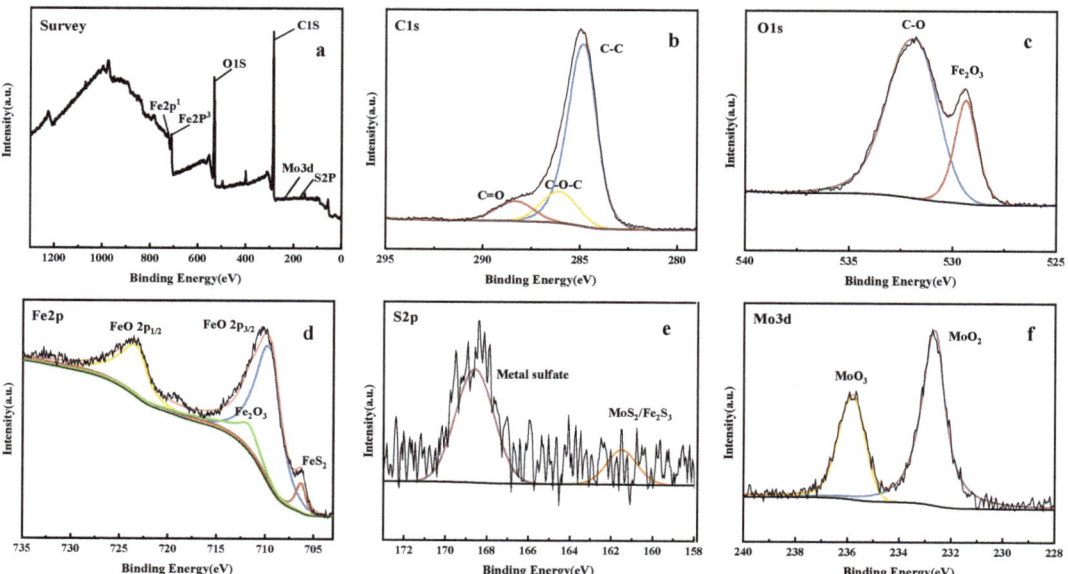

Figure 6. XPS spectra of the worn surface caused by 0.01 wt% MoS$_2$ lithium grease prepared via "Method D", (**a**) XPS survey spectrum, (**b**) C 1s, (**c**) O 1s, (**d**) Fe 2p, (**e**) S 2p, (**f**) Mo 3d.

3.6. Results of Corrosion Test

To explore the corrosion performance of MoS$_2$-lithium grease on metal under natural conditions, the corrosion test was conducted with the GCr15−bearing steel sheet. Table 2 shows the quality changes of the steel sheets before and after 15 days.

Table 2. Mass change (Δm, g) of GCr15-bearing steel plate after 15 days of exposure to MoS$_2$ lubricating grease.

Addition Method	Mass Change (Δm, g)				
	ω (wt%)				
	0	0.01	0.03	0.05	0.07
Method A	0.0198	0.0034	0.0109	0.0059	0.0040
Method B	0.0198	0.0006	0.0083	0.0034	0.0070
Method C	0.0198	0.0089	0.0032	0.0089	0.0017
Method D	0.0198	0.0161	0.0158	0.0167	0.0177

The corrosion rate of MoS$_2$ grease on bearing steel can be calculated according to the following formula [33]:

$$CR = \frac{\Delta m}{At} \quad (1)$$

$$\eta_{wL} = \frac{CR_0 - CR_1}{CR_0} \times 100\% \qquad (2)$$

where Δm represents weight loss, A is the surface area of the bearing steel, t is the time of the steel plate buried in the grease, CR represents the corrosion rate, CR_0 and CR_1 represent the corrosion rate with and without additives, respectively.

Figure 7 shows the corrosion rate of GCr15−bearing steel sheets buried in MoS_2 lubricating grease with different mass fractions prepared by different addition methods for 15 days. Compared with the blank group without MoS_2, it can be found that the addition of MoS_2 additive reduces the corrosion rate of bearing steel in the natural environment, indicating that the addition of MoS_2 effectively slows down the corrosion process of bearing steel.

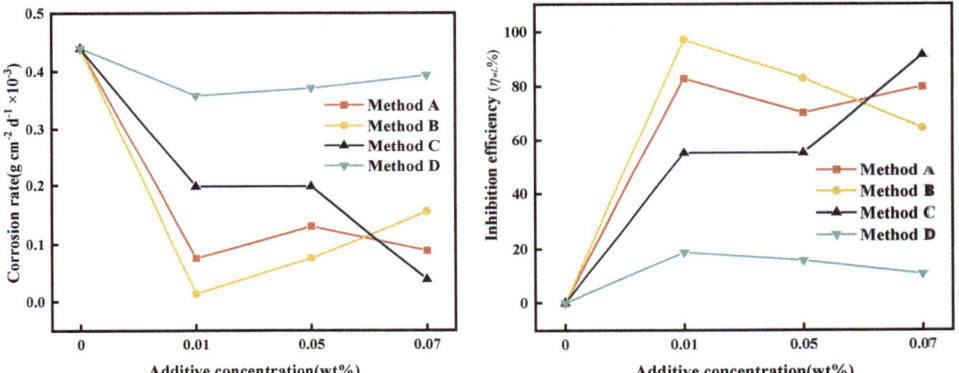

Figure 7. Corrosion rate and corrosion inhibition efficiency of GCr15−bearing steel plate buried in MoS_2 greases with different mass fractions prepared by varying addition methods.

However, as the addition of additives changes, the content of MoS_2 with the best corrosion inhibition efficiency also changes. The optimal corrosion inhibition efficiency of MoS_2 via "Method A" was found at 0.01 wt% MoS_2 content. When MoS_2 was added via "Method B", the optimal corrosion inhibition efficiency occurred when the MoS_2 content was 0.01 wt%. When MoS_2 was added via "Method C", the optimal additive content for corrosion inhibition efficiency was 0.07 wt%. When "Method D" was adopted to add MoS_2, 0.03 wt% MoS_2 exhibited the best corrosion inhibition efficiency. This change may be due to the varying degree of uniform distribution of MoS_2 in the lubricating grease when added in different ways, resulting in different corrosion results. The data in Table 2 and Figure 7 both show that when "Method B" was used to add 0.01 wt% MoS_2, the configured grease had the lowest corrosion rate on the bearing steel; that is, the corresponding corrosion inhibition efficiency was the highest, reaching 96.97%.

The SEM image of the GCr15−bearing steel sheet embedded in 0.01% MoS_2 lubricating grease for 15 days is shown in Figure 8. Figure 8a shows that no obvious surface coverings or corrosion pits were found on the surface of the original bearing steel sheet. However, the surface of the bearing steel corresponding to the blank grease without MoS_2 showed obvious corrosion, with an unsmooth surface and severe corrosion pits (Figure 8b). Figure 8c is the SEM image of the surface of the steel sheet caused by 0.01 wt% MoS_2 grease prepared via "Method A". It can be seen that corrosion pits appear in a small part of the surface of the steel sheet under such circumstances. When 0.01 wt% MoS_2 was added via "Method B", the resulting grease causes the slightest corrosion on the surface of the steel sheet (Figure 8d). When 0.01 wt% MoS_2 grease was prepared via "Method C", the corrosion test results in uneven corrosion on the surface of the steel sheet, as shown in Figure 8e. As can be seen from Figure 8f, the grease prepared by adding MoS$_2$ additive via the process of "Method D" will cause local corrosion on the surface of the steel sheet. From Figure 8c–f, it can be found that MoS$_2$ grease has a certain corrosion effect on the steel surface, which may be because the S element in MoS$_2$ can combine with the Fe element in the bearing steel to produce iron sulfide and other products [14], so MoS$_2$ has a certain corrosion effect, and the surface of the steel sheet was therefore pitted. However, by comparing Figures 8b and 8c–f, it can be found that after the addition of MoS$_2$ additive, the corrosion phenomenon on the surface of the bearing steel was slowed down in different forms, indicating that MoS$_2$ has a certain corrosion inhibition effect.

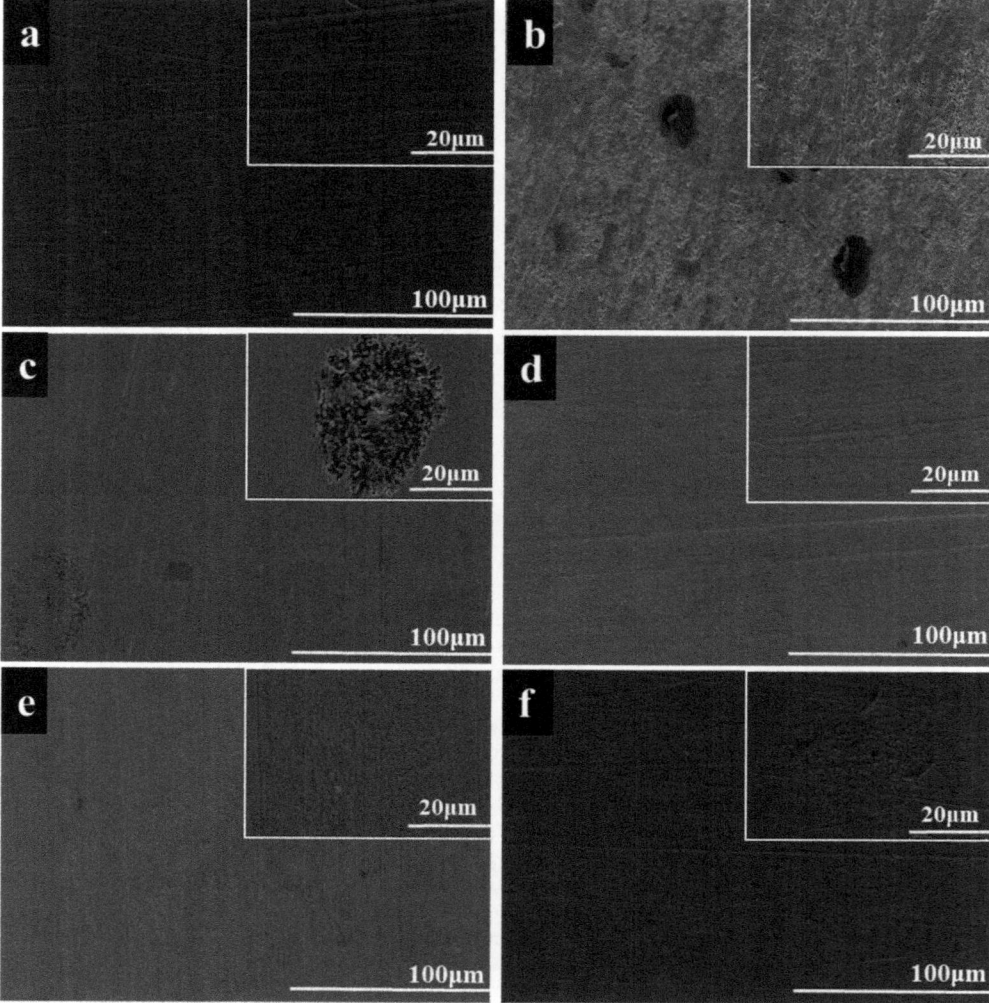

Figure 8. SEM image of GCr15 steel sheet embedded with 0.01 wt% MoS$_2$ lubricating grease prepared via different addition methods after 15 days: (**a**) original steel surface, (**b**) blank grease, (**c**) Method A, (**d**) Method B, (**e**) Method C, (**f**) Method D.

To better demonstrate the morphology and roughness of GCr15−bearing steel sheets after corrosion experiments, three-dimensional (3D) morphology studies were also conducted on the corroded steel sheets. Figure 9 shows the 3D morphology of the steel sheet after being embedded in 0.01% MoS_2 grease for 15 days.

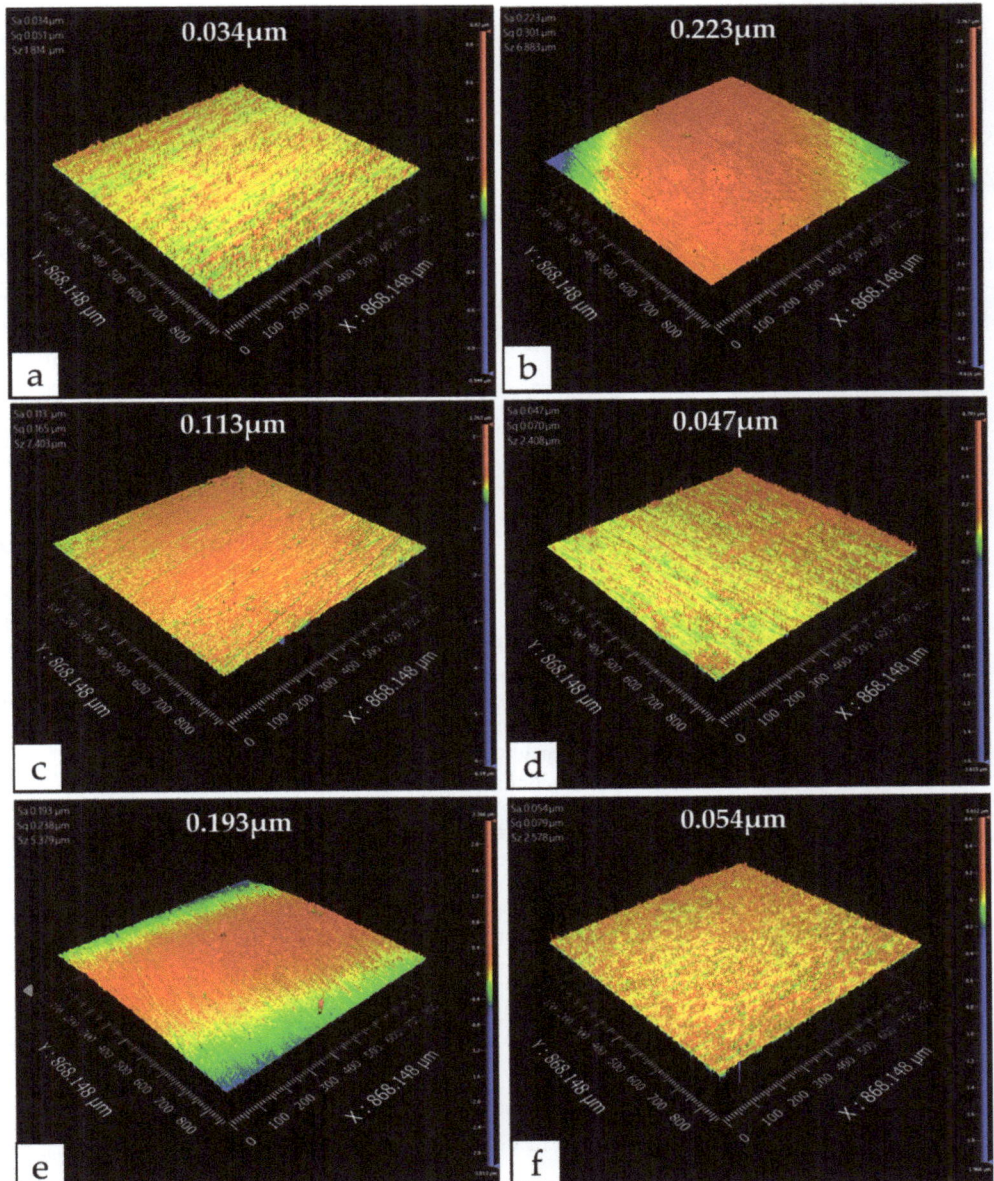

Figure 9. 3D morphology of GCr15−bearing steel sheet embedded in 0.01 wt% MoS_2 lubricating grease prepared via different methods for 15 days: (**a**) original steel sheet, (**b**) blank lubricating grease without MoS_2, (**c**) Method A, (**d**) Method B, (**e**) Method C, (**f**) Method D.

In Figure 9a, it can be observed that the surface roughness of the original bearing steel was 0.034 μm. After 15 days of embedding with blank lubricating grease without MoS_2, the surface roughness of the bearing steel was the highest, reaching 0.223 μm (Figure 9b). Figure 9c illustrates that the surface roughness of the corroded steel plate corresponding to the lubricating grease prepared via "Method A" was 0.113 μm. Figure 9d shows that the roughness of the corroded steel surface corresponding to the lubricating grease prepared via "Method B" was 0.047 μm. Figure 9e,f represent the 3D images of the corroded steel surface corresponding to the lubricating grease prepared via "Method C" and "Method D", respectively, with a surface roughness of 0.193 μm and 0.054 μm.

Comparing Figures 9b and 9c–f, it can be found that regardless of the method of adding 0.01 wt% MoS_2, the formulated lubricating grease will reduce the surface roughness of bearing steel after corrosion. The adoption of the "Method B" resulted in the maximum reduction of surface roughness (Figure 9d) by 0.176 μm. The process of "Method C" (Figure 9e) having the least reduction in surface roughness resulted in a reduction of only 0.03 μm. The 3D morphology in Figure 9 is in good agreement with the SEM results in Figure 8. This further indicates that adding MoS_2 additive can slow down the corrosion rate of bearing steel and can protect the corrosion of the device for a long time under natural conditions.

4. Conclusions

The effects of MoS_2 addition methods on the tribological and corrosion properties of MoS_2-containing lithium grease were investigated in this study. Based on the above analysis, the following conclusions can be drawn:

1. When MoS_2 was used as an additive in lithium-based greases, it exhibited ideal friction reduction and anti-wear effects and corrosion inhibition performance;
2. The content of additive MoS_2 was 0.01%, the friction coefficient of "Method D" was the most stable, the average friction coefficient was 0.034, and the average wear scar diameter was reduced by 0.16 mm. "Method B" had the highest corrosion inhibition efficiency (96.97%);
3. The soap base porosity of "Method D" is small, and the ability to wrap the base oil and MoS_2 is strong, so that it can play the role of anti-friction and anti-wear for a long time. In addition, MoS_2 is oxidized to MoO_2 and MoO_3 during the friction process. The tribochemical reaction occurred between MoS_2 and the rubbing pair, forming a thin friction film, which reduces friction and wear;
4. MoS_2 lithium grease has good anti-corrosion properties, which maybe "Method B" can evenly distribute MoS_2 on the surface of the grease, so that the protective film formed by MoS_2 and GCr15−bearing steel can effectively slow down the corrosion effect of lithium-based grease on steel.

Author Contributions: Writing—original draft, C.Z.; Writing—review & editing, C.Z., Z.H. and Y.W.; Investigation, C.Z.; Funding acquisition, Z.H. and L.X.; Resources, Z.H. and L.X.; Project administration, Z.H.; Supervision, J.L.; Formal analysis, L.L. All authors have read and agreed to the published version of the manuscript.

Funding: Financial support provided by the National Natural Science Foundation of China (51965020; 52201160); Jiangxi Natural Science Foundation of China (20232ACB204001; 20232BAB204008; 20232BAB212007; 20212BBE53041, 20224BAB204048, 20224ACB204014).

Data Availability Statement: Data are contained within the article.

Conflicts of Interest: The authors declare no conflict of interest.

References

1. Meijer, R.J.; Lugt, P.M. The Grease Worker and Its Applicability to Study Mechanical Aging of Lubricating Greases for Rolling Bearings. *Tribol. Trans.* **2021**, *65*, 32–45. [CrossRef]
2. Gurt, A.; Khonsari, M. The Use of Entropy in Modeling the Mechanical Degradation of Grease. *Lubricants* **2019**, *7*, 82. [CrossRef]

3. Wu, C.; Hong, Y.; Ni, J.; Teal, P.D.; Yao, L.; Li, X. Investigation of mixed hBN/Al$_2$O$_3$ nanoparticles as additives on grease performance in rolling bearing under limited lubricant supply. *Colloids Surf. A Physicochem. Eng. Asp.* **2023**, *659*, 130811. [CrossRef]
4. Uppar, R.; Dinesha, P.; Kumar, S. A critical review on vegetable oil-based bio-lubricants: Preparation, characterization, and challenges. *Environ. Dev. Sustain.* **2022**, *25*, 9011–9046. [CrossRef]
5. Yadav, A.; Singh, Y.; Negi, P. A review on the characterization of bio based lubricants from vegetable oils and role of nanoparticles as additives. *Mater. Today Proc.* **2021**, *46*, 10513–10517. [CrossRef]
6. Lugt, P.M. Modern advancements in lubricating grease technology. *Tribol. Int.* **2016**, *97*, 467–477. [CrossRef]
7. Wu, L.; Yang, B.; Zhao, F.; Zhang, Z. Tribological Properties of Complex Calcium Sulfonate Grease with Ultrafine SiO$_2$/MoS$_2$ Powders. *J. Nano Res.* **2021**, *66*, 35–44. [CrossRef]
8. Chen, X.; Chen, G. Analysis of lithium molybdenum disulfide base grease and its application in roll bearing. *Sci. Technol. Inf.* **2013**, *21*, 119–121. (In Chinese)
9. Wang, Y.; Wu, B. Friction Characteristics and Mechanisms of Two Lithium Greases in Elastohydrodynamic Lubrication. *J. Fail. Anal. Prev.* **2020**, *20*, 1266–1273. [CrossRef]
10. Kanazawa, Y.; Sayles, R.S.; Kadiric, A. Film formation and friction in grease lubricated rolling-sliding non-conformal contacts. *Tribol. Int.* **2017**, *109*, 505–518. [CrossRef]
11. De Laurentis, N.; Cann, P.; Lugt, P.M.; Kadiric, A. The Influence of Base Oil Properties on the Friction Behaviour of Lithium Greases in Rolling/Sliding Concentrated Contacts. *Tribol. Lett.* **2017**, *65*, 128. [CrossRef]
12. Rapoport, L.; Bilik, Y.; Feldman, Y.; Homyonfer, M.; Cohen, S.R.; Tenne, R. Hollow nanoparticles of WS$_2$ as potential solid-state lubricants. *Nature* **1997**, *387*, 791–793. [CrossRef]
13. Li, J.; Wang, Y.; Liu, D.; He, K.; Lu, L. Research progress on the properties and application of molybdenum disulfide. *Powder Metall. Technol.* **2021**, *39*, 471–478. (In Chinese) [CrossRef]
14. Niste, V.B.; Ratoi, M. Tungsten dichalcogenide lubricant nanoadditives for demanding applications. *Mater. Today Commun.* **2016**, *8*, 1–11. [CrossRef]
15. Wu, Z. Synthesis of Molybdenum (Tungsten) Disulfide Nanostructures and Their Properties. Ph.D. Thesis, Central South University, Changsha, China, 2012.
16. Wen, P.; Lei, Y. Preparation and Tribological Properties of Covalent Organic Framework Nanomaterials. *Tribology* **2022**, *42*, 123–130. (In Chinese) [CrossRef]
17. Cheng, S.; Guo, X.; Jiang, M.; He, Y. Research progress of lithium complex grease. *Contemp. Chem. Ind.* **2018**, *47*, 152–158. (In Chinese) [CrossRef]
18. Xie, L.; Lu, Z.; Kong, X.; Chen, L.; Luo, H. Study on the effect of ZnS nano additive on the lubrication performance of complex calcium-based grease and its mechanism. *J. Baoji Univ. Arts Sci. (Nat. Sci.)* **2023**, *43*, 41–47. (In Chinese) [CrossRef]
19. Cheng, Y. Preparation and Tribological Properties of Space Lubricating Grease Containing MoS$_2$ Nanoparticles. Master's Thesis, Hefei University of Technology, Hefei, China, 2012.
20. He, Z.; Zhu, X.; Tang, J.; Qiu, W.; Rao, S. Tribological Properties of Nano Lanthanum Oxide as Additives in Lithium Base Greases with Rapeseed Oil. *Chin. Rare Earths* **2017**, *38*, 23–29. (In Chinese) [CrossRef]
21. GB-T4929-1985; Lubricating Grease-Determination of Dropping Point. Standards Press of China: Beijing, China, 1985.
22. GB-T269-1991; Lubricating Grease and Petrolatum-Determination of Cone Penetration. Standards Press of China: Beijing, China, 1992.
23. GB/T 3142-1982; Lubricants-Determination of Load-Carrying Capacity (Four Balls Method). Standards Press of China: Beijing, China, 1990.
24. Wang, W.; Tian, S.; Sun, H. Influence of base oil and additives on microstructure of grease. *Pet. Prod. Appl. Res.* **2015**, *33*, 26–34. (In Chinese) [CrossRef]
25. Yuan, S. Study on the surface modification and friction properties of molybdenum disulfide. *Mod. Salt Chem. Ind.* **2019**, *46*, 50–54. (In Chinese) [CrossRef]
26. Zhang, E.; Li, W.; Zhao, G.; Wang, Z.; Wang, X. A Study on Microstructure, Friction and Rheology of Four Lithium Greases Formulated with Four Different Base Oils. *Tribol. Lett.* **2021**, *69*, 98. [CrossRef]
27. Kosynkin, D.V.; Higginbotham, A.L.; Sinitskii, A.; Lomeda, J.R.; Dimiev, A.; Price, B.K.; Tour, J.M. Longitudinal unzipping of carbon nanotubes to form graphene nanoribbons. *Nature* **2009**, *458*, 872–876. [CrossRef] [PubMed]
28. Liang, Y.; He, X.; Chen, L.; Zhang, Y. Preparation and characterization of TiO$_2$–Graphene@Fe$_3$O$_4$ magnetic composite and its application in the removal of trace amounts of microcystin-LR. *RSC Adv.* **2014**, *4*, 56883–56891. [CrossRef]
29. Horikawa, T.; Sakao, N.; Sekida, T.; Hayashi, J.I.; Do, D.D.; Katoh, M. Preparation of nitrogen-doped porous carbon by ammonia gas treatment and the effects of N-doping on water adsorption. *Carbon* **2012**, *50*, 1833–1842. [CrossRef]
30. Xiang, S.; Long, X.; Zhang, Q.; Ma, P.; Yang, X.; Xu, H.; Lu, P.; Su, P.; Yang, W.; He, Y. Enhancing Lubrication Performance of Calcium Sulfonate Complex Grease Dispersed with Two-Dimensional MoS$_2$ Nanosheets. *Lubricants* **2023**, *11*, 336. [CrossRef]
31. Fan, X.; Li, W.; Li, H.; Zhu, M.; Xia, Y.; Wang, J. Probing the effect of thickener on tribological properties of lubricating greases. *Tribol. Int.* **2018**, *118*, 128–139. [CrossRef]

32. Jiang, H.; Hou, X.; Ma, Y.; Su, D.; Qian, Y.; Ahmed Ali, M.K.; Dearn, K.D.J.W. The tribological performance evaluation of steel-steel contact surface lubricated by polyalphaolefins containing surfactant-modified hybrid MoS_2/h-BN nano-additives. *Wear* **2022**, *504–505*, 204426. [CrossRef]
33. Zhu, Y.; Sun, Q.; Wang, Y.; Tang, J.; Wang, Y.; Wang, H. Molecular dynamic simulation and experimental investigation on the synergistic mechanism and synergistic effect of oleic acid imidazoline and l-cysteine corrosion inhibitors. *Corros. Sci.* **2021**, *185*, 109414. [CrossRef]

Disclaimer/Publisher's Note: The statements, opinions and data contained in all publications are solely those of the individual author(s) and contributor(s) and not of MDPI and/or the editor(s). MDPI and/or the editor(s) disclaim responsibility for any injury to people or property resulting from any ideas, methods, instructions or products referred to in the content.

Article

Preparation and Tribological Behavior of Nitrogen-Doped Carbon Nanotube/Ag Nanocomposites as Lubricant Additives

Shaokun Jia [1,2], Jiahuan Zhao [1], Guangzhen Hao [1], Jifeng Feng [1], Chuanbo Zhang [1], Zhihui Wang [1], Zhengfeng Jia [1,*] and Yungang Bai [3]

[1] School of Materials Science and Engineering, Liaocheng University, Liaocheng 252059, China; h15864371243@163.com (G.H.)
[2] School of Materials Science and Engineering, Wuhan University of Technology, Wuhan 430070, China
[3] Shandong Qichanxintu Composite Material Co., Ltd., Liaocheng 252059, China
* Correspondence: jiazhfeng@lcu.edu.cn

Abstract: In this study, nitrogen-doped carbon nanotube/Ag nanocomposites (denoted as N-C/Ag) have been synthesized in a urea solution using a hydrothermal method. The carbon nanotubes, $AgNO_3$ solution, urea and poly-dopamine (PDA) served as carbon, silver, nitrogen and carbon sources, respectively. The results show that the diameter of the carbon tubes was about 30 nm, and the Ag nanoparticles, with a diameter of ca. 10 nm, dispersed on the carbon tube surface. The Ag particle size decreased with a lower degree of crystallinity at a high temperature in the presence of urea. The friction and wear behavior of the oil acid (OA) modified N-C/Ag (OAN-C/Ag) as an additive in liquid paraffin (LP) were studied using a four-ball friction and wear tester. The results have shown that the coefficients of friction (COFs) and wear scar diameters (WSDs) of steel balls lubricated with LP-OAN-C/Ag decreased by 27.3% and 25.3%, respectively, relative to pure LP. Tribofilms containing Ag, carbon and nitride were formed on the worn steel ball surfaces. Details, the carbon, Fe_2O_3, azides and nitride, Ag and alloy and other compounds on the wear scars may improve tribological properties. The synergistic effect of carbon, Ag and urea plays a critical role during sliding.

Keywords: N-doped carbon nanotube/Ag nanocomposites; polydopamine; hydrothermal; wear; friction

Citation: Jia, S.; Zhao, J.; Hao, G.; Feng, J.; Zhang, C.; Wang, Z.; Jia, Z.; Bai, Y. Preparation and Tribological Behavior of Nitrogen-Doped Carbon Nanotube/Ag Nanocomposites as Lubricant Additives. *Lubricants* **2023**, *11*, 443. https://doi.org/10.3390/lubricants11100443

Received: 6 September 2023
Revised: 5 October 2023
Accepted: 10 October 2023
Published: 13 October 2023

Copyright: © 2023 by the authors. Licensee MDPI, Basel, Switzerland. This article is an open access article distributed under the terms and conditions of the Creative Commons Attribution (CC BY) license (https://creativecommons.org/licenses/by/4.0/).

1. Introduction

In recent years, nanomaterials have been widely studied in many fields [1–4]. Carbon nanomaterials, including fullerenes, carbon nanotubes (CNTs) and graphene, possess excellent mechanical properties, thermal properties and corrosion resistance [5]. In common with graphene, CNTs are characterized by a special sp²-hybridized structure, which has been the subject of considerable attention in interfacial science and tribology. However, CNTs exhibit shortcomings, notably aggregation and poor dispersion in matrices, which have limited the range of applications. Functionalization and modification of CNTs and associated composites can improve dispersion in base oil [6,7]. Hao et al. [6] synthesized a brushlike polystyrene with a method of reversible addition–fragmentation chain transfer, which was modified on the surfaces of hydroxylated CNTs. The results showed that the dispersity of the CNT-polystyrene in base oil was outstandingly enhanced. The tribological properties of the LP containing CNT–polystyrene were better than those of pure LP, which were attributed to the synergistic effect of CNT and polystyrene and the improved dispersibility in base oil. Jesús ACC et al. [7] investigated the tribological properties of single-walled and muti-walled carbon nanotubes being treated with carboxylic acid and strong acid, respectively, as additives of oil or water. The results showed that coefficients of friction and mass losses of twin-disk pairs lubricated with oil or water containing additives were smaller than those lubricated with pure oil or water, respectively. The carbon tribofilms were formed on the worn surfaces of disks with lubricants of oil or water containing carbon tubes and played an important role in reducing friction and wear resistance. Diana-Luciana C et al. [1] synthesized single-wall CNTs at a temperature of 850 °C and a pressure of 6 atm. The tribological properties of the CNTs as additives of mineral oil for different pairs were executed. The results show that mineral oil containing CNTs possesses a smaller coefficient of friction than pure base oil and oil with zinc dialkyldithiophosphates, respectively. The synergistic effect of CNTs and zinc dialkyldithiophosphates occurred due to the modification of zinc dialkyldithiophosphates on the surfaces of CNTs.

Furthermore, methods to improve the tribological properties of CNTs and composites containing CNTs have been reported in much of the literature. Chen et al. [8] investigated the tribological properties of a CNTs/ZnS hybrid in epoxy coatings and found that the hybrid can enhance friction reduction and anti-wear capability. Researchers have proposed that the micro-fluid characteristics of CNTs and synergistic effects in composites can endow lubricant coatings with excellent tribological properties. Ramaprabhu [9] synthesized Fe-carbon tube composites at 800 °C in a nitrogen atmosphere with the precursors of ferric chloride hexahydrate and melamine, respectively. The composites were directly added to different base oils (gearbox oil, vegetable oil and engine oil, respectively) to evaluate their tribological properties. The results showed that the engine oil containing the additives possessed better tribological properties than the other two base oils containing Fe-carbon tube composites, which might be attributed to the formation of stable tribofilms during rotation in engine oil containing additives. The Fe-carbon tube composites impeded the formation of iron oxides and residues on the worn surfaces and entered into the furrow to form tribofilms. Saeed Zeinali Heris [3] compared the physical and chemical properties of titanium dioxide nanoparticles (TiO_2) and muli-walled carbon nanotubes (MW-CNT) as additives to oil. To obtain stable lubricants, TiO_2 and MW-CNT were modified with oleic acid and triton 100, respectively. The depths of wear pins under the lubrication of base oil containing TiO_2 and MW-CNT (0.1 wt.%) were about 40 and 30 μm, respectively, which is about one-fourth of the depth of pure base oil.

A number of papers have revealed that the use of elements (such as S, P and N) as doped additives can result in enhanced tribological properties due to the formation of tribo-chemical films and/or strong absorption of the additives on worn surfaces [10,11]. Pyrolysis and hydrothermal reactions are common procedures in doping nitrogen and/or sulphur on carbon [10,12]. Sun et al. [10] investigated the tribological properties of nitrogen hybridized carbon quantum dots and MoS_2 as additives of the lubricant glycerol. Nitrogen hybridized carbon quantum dots were synthesized via the solvothermal method using polydopamine as the carbon and nitrogen source. A chemical reaction between the nitrogen hybridized carbon quantum dots and the metal disks was executed during rotation because of the nitrogen and oxygen on the carbon quantum dots. On the other hand, the atoms of sulfur in MoS_2 reacted with metal surfaces during sliding to protect the worn surfaces from severe wear. Liu et al. [11] synthesized nitrogen/sulfur-doped porous carbon nanospheres at high temperatures using melamine and thiourea as a nitrogen source and sulfur source, respectively. The excellent tribological properties of the 500 SN-containing composites were attributed to the protective tribofilms being formed between the N/S-doped carbon nanosphere and substrates. In detail, the tribofilm containing iron oxides, iron sulfide and carbonitride played an important role in anti-wear and reduced friction.

Soft metals (such as silver and copper) attract much attention because of their excellent embeddability and deposition ability during running to form tribofilms. Song [13] synthesized Cu/PDA/CNTs composites using $NaH_2PO_2 \cdot H_2O$ as a reducing agent at 80 °C. The results showed that the Cu/PDA/CNTs markedly reduced the coefficients of friction and wear track widths, which were attributed to the tribofilms and self-lubricating properties of the composites. However, fabricating CNT composites in an eco-friendly manner remains a challenge. Jia et al. [14] synthesized silver/polydopamine (Ag/PDA) composites in a facile environment using Ag^+ and dopamine as the Ag source and reductant, respectively. The results show that the COFs and WSDs of steel balls lubricated with poly-alpha-olefin

(PAO) containing oleic acid-modified Ag/PDA nanocomposites were lower than those lubricated with pure PAO. A tribofilm containing iron oxides, silver oxides and PDA was formed during rotation, which protected the steel balls from violent wear. Our team [15] also synthesized hexagonal boron nitride/copper nanocomposites through the method of ultrasonic exfoliation and in situ reduction. Steel balls lubricated with LP containing the OA-modified BN/Cu nanocomposites possess lower COF and WSD than pure LP. On the one hand, the tribofilm containing BN, oxides of boron, metal and carbon was formed on the worn surfaces to avoid violent wear. On the other hand, increasing the reduction time can enlarge the interplanar spaces and decrease the thickness of BN, which might decrease the van der Waals forces between the layers and provide good friction reductions. Waralorn Limbut et al. [2] synthesized a phosphor-doped carbon nanotube composited with silver nanoparticles using $Na_4P_2O_7$ and $AgNO_3$ as the phosphor source and silver source, respectively. N,N-dimethylformamide and $NaBH_4$ were used as a solvent and a reducing agent, respectively. Synthesizing composites in a facile way is still a challenge.

In this study, carbon nanotube/Ag nanocomposites have been synthesized with dopamine (DA) as a reducing agent at room temperature. The N-doped C/Ag was obtained using a hydrothermal method using urea as the nitrogen source. The OA-modified N-CNT/Ag were dispersed in paraffin in order to investigate their friction and wear properties. A possible wear mechanism has also been studied.

2. Experimental Procedures

2.1. Materials and Preparation

$AgNO_3$, DA hydrochloride, urea, hydroxylated multi-walled carbon nanotubes (HO-MW-CNTs) and other agents were purchased from Aladdin Biochemical Technology Company (Shanghai, China) and used as supplied. LP, with a viscosity of 17.2 mm^2/s (25 °C), was employed as the base oil, and AISI-52100 steel balls (Ø 12.7 mm) were used in testing. The $AgNO_3$ solution and Tris-buffer solution (pH = 8.5) were prepared following procedures reported previously [14]. A sample (8 g) of urea powder and 0.04 g CNTs were added to a 25 mL Tris-buffer solution and stirred for 30 min. Then, 4 mL $AgNO_3$ solution and 0.4 g DA were added to the above solution and stirred for 24 h at room temperature. The solution was transferred to the hydrothermal reactor at a temperature of 140 °C for 12 h to generate the target product (denoted as N_8-$C_{0.04}$/Ag_4). The composite product was modified with OA (denoted as OAN_8-$C_{0.04}$/Ag_4) as described previously [15]. Tests to investigate tribological behavior were conducted using an MRS-10A four-ball tester (392 N, 1450 rpm, 30 min).

2.2. Characterization

The microstructure of the synthesized N-C/Ag was evaluated using a high-resolution transmission electron microscope (JEM-2100, HRTEM), a field-emission scanning electron microscope (SIGMA 500/VP, FE-SEM) equipped with energy-dispersive X-ray analysis (SEM-EDXA, Kevex Sigma, St. Louis, MO, USA), a Horiba LabRam HR evolution spectrometer (with 532-nm laser) and a Bruker D8 Advanced X-ray Diffractometer. X-ray photoelectron spectroscopy (XPS) was conducted using an ESCALAB Xi+ X-ray photoelectron spectrometer to determine chemical bonding and surface structure. The FT-IR spectra were obtained using a Bruck IFs66v spectrometer.

3. Results and Discussion

3.1. Characterization

The FE-SEM and EDXA analyses of the N-C/Ag have shown that the tubes, with diameters of ca. 50 nm, were covered with a cracked coating (Figure 1a) that must represent the PDA/Ag composite. The EDXA measurements have demonstrated the presence of C, Ag, O and N, confirming the inclusion of silver on the carbon tubes (Figure 1b). The TEM analysis has established that the diameter of the carbon tubes was ca. 30 nm, and the Ag nanoparticles exhibited a diameter of ca. 10 nm dispersed on the carbon tube surface (Figure 1c). The high-resolution images of the composites illustrate the interlayer spacing (0.63 nm) of the carbon nanotubes and the lattice spacing (0.218 nm) of the (111) Ag plane [16]. Elemental mapping of the composites has revealed the presence of Ag, N and O covering the tubes, confirming that the PDA/Ag composites form the observed rough coating.

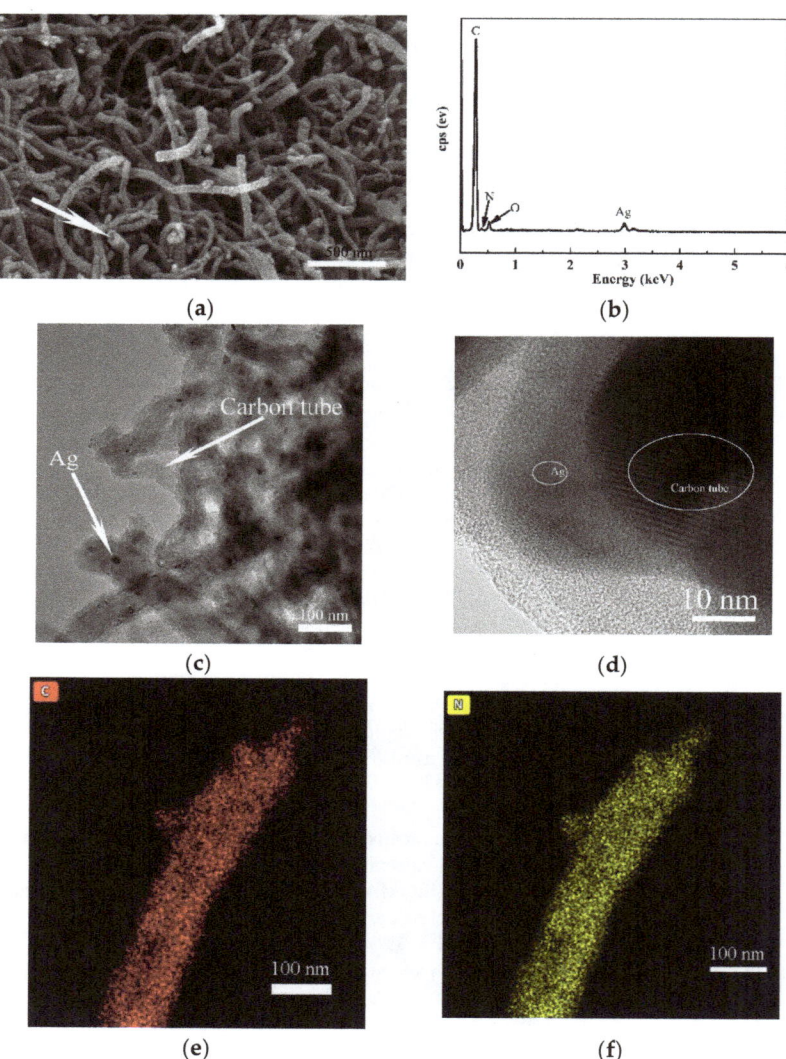

Figure 1. Cont.

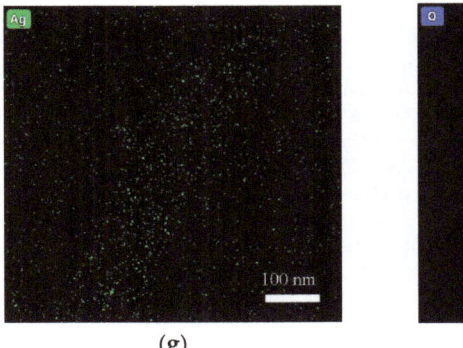
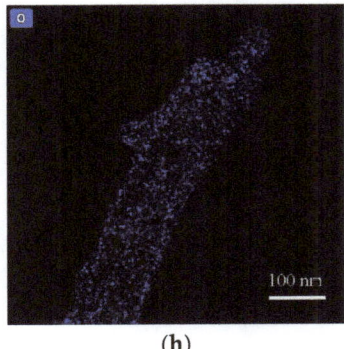

Figure 1. FE-SEM (**a**) and EDXA spectra (**b**), TEM (**c**,**d**) and corresponding elemental mapping (**e**–**h**) of the N-C/Ag.

The FT-IR spectrum of the OA-modified N-C/Ag is presented in Figure 2. The broad peak stretching from 3150 to 3350 cm^{-1} can be attributed to the asymmetry stretching vibration of aromatic −OH and −NH$_2$ due to OA, poly-dopamine and hydroxylated carbon nanotubes [14,17]. The peaks at ca. 2893 cm^{-1} and 2827 cm^{-1} are ascribed to the −CH$_3$ and −CH$_2$ groups, respectively. The peak at ca. 2255 cm^{-1} may be due to the cumulene bond of PDA [14]. The broad peak from 1694 to 1280 cm^{-1} is assigned to C=O and C=C [14,17], and the broad peak from 1167 to 1160 cm^{-1} is attributed to PDA quinoid groups [18].

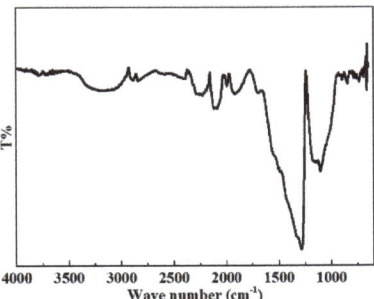

Figure 2. FT-IR spectrum of the N-C/Ag modified with OA.

3.2. Friction and Wear Behaviors

The WSD–concentration and COF–concentration curves for steel balls lubricated with LP-OAN$_8$-C$_{0.04}$/Ag$_4$ at different temperatures are presented in Figure 3a,b. The steel balls lubricated with LP-OAN$_8$-C$_{0.04}$/Ag$_4$ exhibited lower COFs and WSDs than those recorded using pure LP. Increasing the reaction temperature to 140 °C, the LP-OAN$_8$-C$_{0.04}$/Ag$_4$ decreased the COFs and WSDs from 0.195 and 0.745 mm for LP to 0.131 and 0.55 mm, respectively. A further increase to 160 °C resulted in equivalent WSDs to those achieved with the samples synthesized at room temperature (RT). In contrast, the COFs of lubricants containing additives synthesized at 160 °C with concentrations greater than 0.2 wt.% are smaller than those obtained for the room temperature sample. Accordingly, the hydrothermal reaction temperature was fixed at 140 °C for subsequent tests. The effects of varying the amounts of urea on the WSD–concentration and COF–concentration curves are shown in Figure 3c,d. It is clear that the balls lubricated with LP-OAN$_8$-C$_{0.04}$/Ag$_4$ exhibited lower WSDs than those recorded for other samples. The COFs of the balls lubricated with LP-OAN-C$_{0.04}$/Ag$_4$ with different amounts of urea are smaller than those obtained with pure LP. However, there are no obvious differences in COFs of the balls lubricated using samples with different amounts of urea. The amount of urea used was fixed at 8 g. The WSD–concentration and COF–concentration curves for LP-N_8-C/Ag_4 with different amounts of carbon nanotubes were determined and are shown in Figure 4e,f. At a concentration less than 0.5 wt.%, the WSDs of the balls lubricated with LP-OAN_8-C/Ag_4 containing nanotubes are lower than those without nanotubes. The steel ball WSDs for lubrication using LP-N_8-C/Ag_4s (0.2 wt.%) with 0.04 g and 0.08 g carbon tubes are ca. 0.56 mm, which is lower than those for LP-N_8-C/Ag_4 without carbon tubes (0.61 mm) and pure LP (0.75 mm). In the case of the COF–concentration curves for LP-N_8-C/Ag_4 with different amounts of carbon tubes, the steel ball COFs were much smaller than those measured for the other lubricants. At a concentration of 0.2 wt.%, the average COF associated with lubrication using LP containing N_8-$C_{0.08}$/Ag_4 was 0.125, lower than those of LP containing N_8-C_0/Ag_4 (0.135) and N_8-$C_{0.04}$/Ag_4 (0.155). The amount of carbon nanotubes was then fixed at 0.08 g.

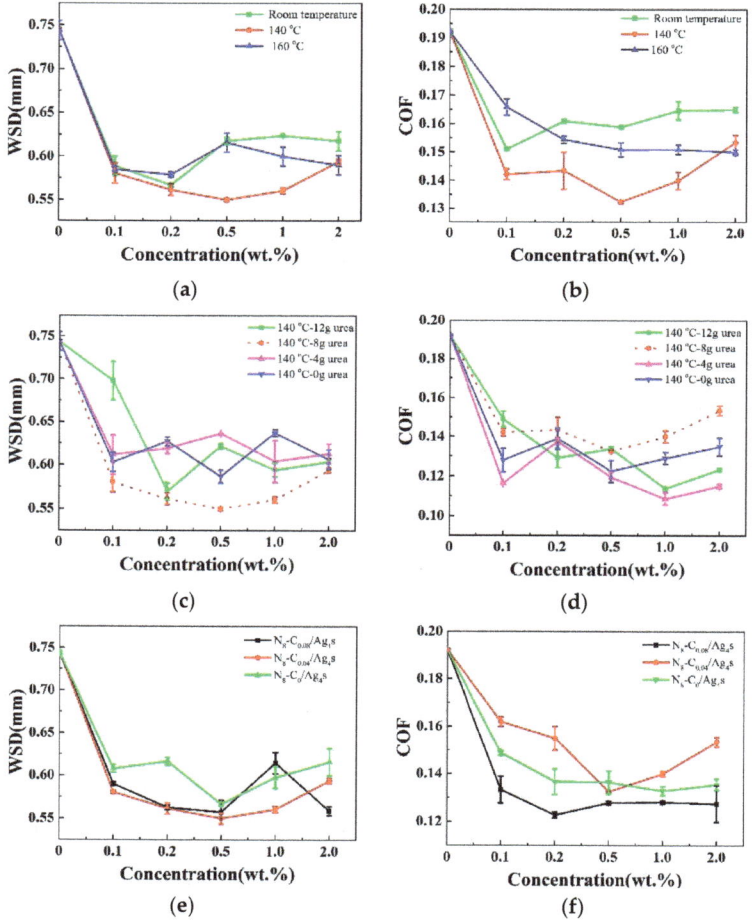

Figure 3. WSD–concentration and COF–concentration curves of LP-OAN_8-$C_{0.04}$/Ag_4 for different synthesis temperatures (**a**,**b**), different amounts of urea (**c**,**d**) at 140 °C and different amounts of CNTs (**e**,**f**).

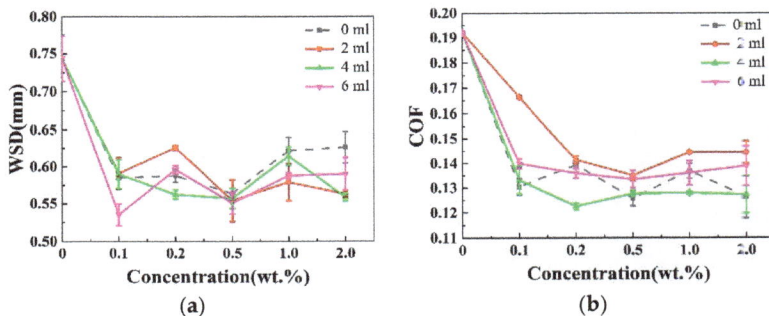

Figure 4. WSD–concentration (**a**) and COF–concentration (**b**) curves of LP-OAN$_8$-C$_{0.08}$/Ag with different volumes of AgNO$_3$ solution at 140 °C.

The effect of varying AgNO$_3$ solution volume on the WSD–concentration and COF–concentration curves for LP-OAN$_8$-C$_{0.08}$/Ag at 140 °C is presented in Figure 4. It can be seen that the WSDs of the steel balls with the three lubricants are lower than those with LP. For instance, at a concentration of 0.5 wt.%, the WSDs associated with the three lubricants are ca 0.56 mm, which is approximately three-quarters of the WSD with LP lubrication. The three WSD–concentration curves overlap with an increase in additive concentration. The COF–concentration curves reveal that the LP-OAN$_8$-C$_{0.08}$/Ag$_4$ possesses better friction-reducing capability than the other samples. At a concentration of 0.2 wt.%, the COF of steel balls lubricated with LP-OAN$_8$-C$_{0.08}$/Ag$_4$ is ca. 0.122, which is lower than those obtained with LP-OAN$_8$-C$_{0.08}$/Ag$_2$ (0.142) and LP-OAN$_8$-C$_{0.08}$/Ag$_6$ (0.138). The volume of the AgNO$_3$ solution was fixed at 4 mL. Use of LP-OAN$_8$-C$_{0.08}$/Ag$_4$ resulted in smaller COFs and WSDs than the samples without an AgNO$_3$ solution at almost all concentrations. At a concentration of 0.2 wt.%, the COFs and WSDs of steel balls lubricated with LP-OAN$_8$-C$_{0.08}$/Ag$_0$ are 0.14 and 0.58 mm, respectively, which are larger than those obtained with LP containing OAN$_8$-C$_{0.08}$/Ag$_4$ (0.12 and 0.55 mm, respectively).

3.3. Discussion

Taking an overview of the results generated, tribological properties can be enhanced through the combined use of a suitable temperature and the amount of carbon nanotubes, AgNO$_3$ solution and urea. This suggests a synergistic effect involving these process variables. In order to account for this effect, an additional investigation has been conducted.

Optical images of worn surfaces lubricated with LP and LP-OAN$_8$-C$_{0.08}$/Ag$_4$ are shown in Figure 5a,b, respectively. It is immediately evident that the worn surfaces lubricated with LP-OAN-C/Ag are smaller and smoother than those with LP. The element maps and EDXA spectrum of the worn surface lubricated with LP-OAN$_8$-C$_{0.08}$/Ag$_4$ show the presence of Fe, C, Ag, O and N, suggesting the formation of a tribofilm containing these elements during rotation. The urea-treated C/Ag nanoparticles generated a tribofilm containing a high percentage of C, O and N (Table 1).

Table 1. Percentage elemental content of the worn surfaces with lubrication using LP containing C$_{0.08}$/Ag$_4$ with/without urea.

	Fe (at.%)	C (at.%)	O (at.%)	N (at.%)	Ag (at.%)
N$_8$-C$_{0.08}$/Ag$_4$	72.26	22.23	4.12	1.05	0.34
N$_0$-C$_{0.08}$/Ag$_4$	86.68	11.15	1.21	0.19	0.77

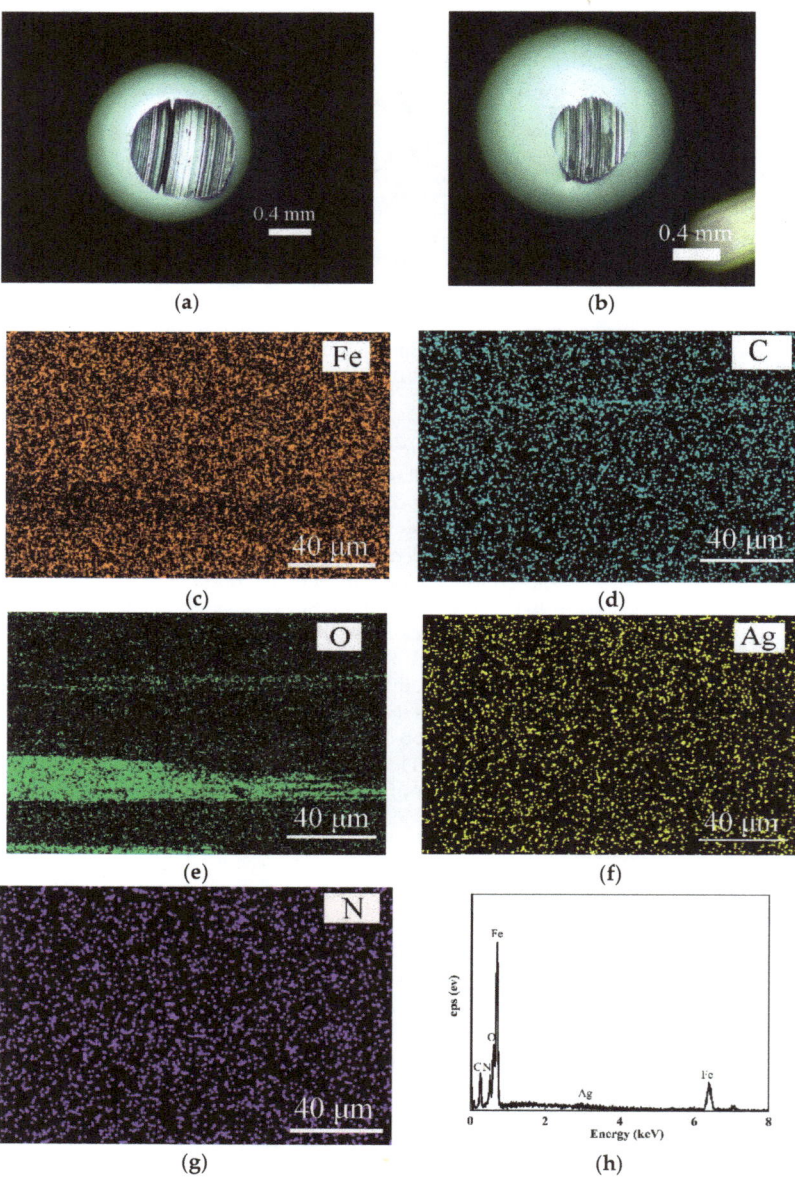

Figure 5. Optical images of worn surfaces lubricated with LP (**a**) and LP-OAN$_8$-C$_{0.08}$/Ag$_4$ (**b**). Elemental maps and EDXA spectrum of the worn surfaces lubricated with LP-OAN$_8$-C$_{0.08}$/Ag$_4$ (**c**–**h**).

The results of the XPS analysis of the worn steel ball surfaces are presented in Figure 6 for lubrication using LP-OAC/Ag with and without urea. The C1s signal was fitted to three peaks with binding energies of 284.75 eV, 286.14 eV and 288.18 eV, which are attributed to carbon, C-N and C-O, respectively [15,19,20]. The O1s signal was deconvoluted into three peaks at 529.92 eV, 531.27 eV and 533.54 eV, which are assigned to metal oxides, carbonates and nitrates, respectively [19]. Oxidation of Fe, C and N during rotation may have resulted in the formation of Fe_2O_3, carbonates and nitrates. The N1s signal was deconvoluted into five peaks at 397.36 eV, 398.49 eV, 399.60 eV, 402.12 eV and 405.92 eV, which are attributed to nitride, azide, organic matrix/cyanides, ammonium salt and nitrites, respectively. Nitrogen may have reacted with the available metal, nitrogen, oxygen and other elements to generate these surface species [15,20]. Comparing the N1s signal for the worn surfaces using lubricants containing additives with/without urea, the peaks due to nitrites and ammonium salt were less intense with the inclusion of urea. This suggests that the nitride associated with the tribofilm improved the tribological properties of the steel balls lubricated with LP-OAN-C/Ag [21,22]. The Fe2p peak at 710.92 eV is assigned to Fe_2O_3 [15,19]. The weak Ag3d signal was deconvoluted into three peaks at 367.53 eV, 368.13 eV and 368.72 eV, which are attributed to oxides, Ag and alloys, respectively [20,23]. The results indicate that a tribofilm containing C, O, N, Fe and Ag was formed on the worn surfaces of the steel balls. The presence of carbon, Fe_2O_3, azides and nitride, Ag and alloy and other compounds associated with the wear scars may improve tribological properties.

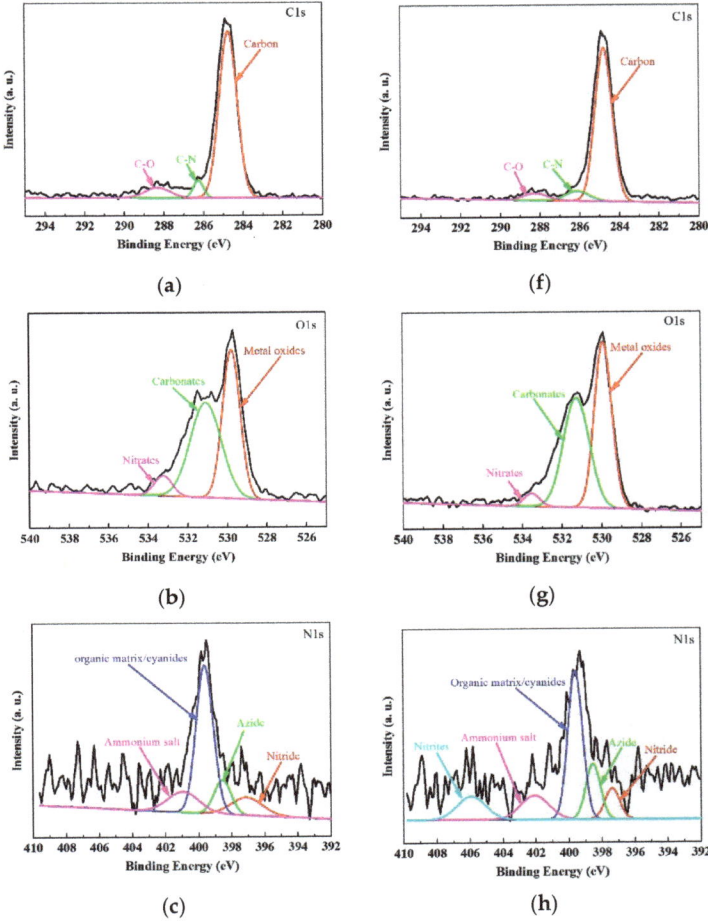

Figure 6. *Cont.*

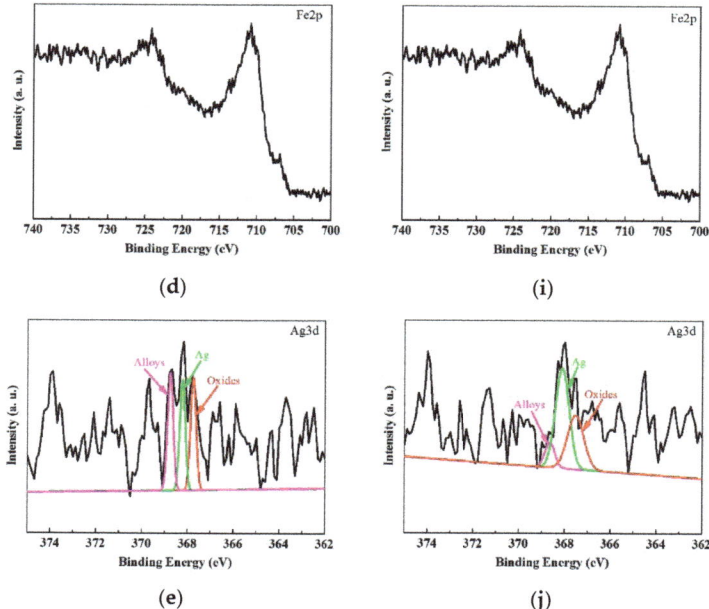

Figure 6. XPS analysis of the worn surfaces lubricated with LP-OAN$_8$-C$_{0.08}$/Ag$_4$ (**a–e**) and LP-OAN$_0$-C$_{0.08}$/Ag$_4$ (**f–j**).

In order to establish the relationship between temperature/urea and tribological properties, XRD analysis was conducted for samples with different temperatures and amounts of urea, and the results are presented in Figure 7. The peaks with 2θ values of 26.10°, 38.27°, 44.20°, 64.48°, 77.31° and 81.48° are assigned to (002), (111), (200), (220), (311) and (222) crystalline planes, respectively [8,14]. The (002) peak for carbon nanotubes was stable with increasing reaction temperature and/or added urea. The intensity of the (111) and (200) peaks for Ag was reduced and shifted to lower 2θ values as the reaction temperature was increased from ambient to 140 °C, and the peak intensities were further reduced with the addition of urea at high temperature. This indicates that Ag particle size decreased with a lower degree of crystallinity at high temperatures in the presence of urea [15]. Previous reports have suggested that "soft" Ag nanoparticles are deposited on worn surfaces [24]. Our results suggest that a smaller Ag particle size and reduced crystallinity may be beneficial to friction reduction and anti-wear properties.

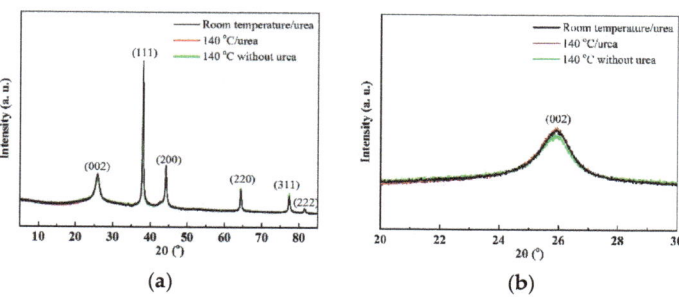

Figure 7. *Cont.*

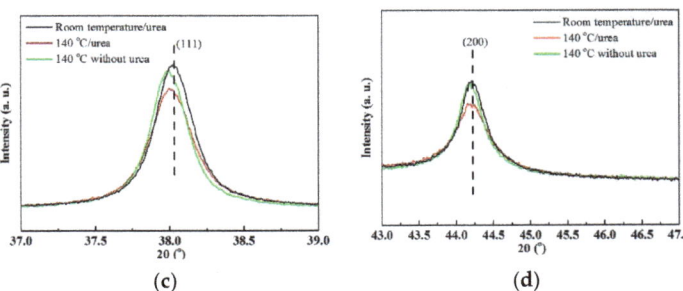

Figure 7. XRD patterns of samples with different reaction temperatures and urea inclusion (**a**) and magnified peaks at different 2θ values (**b–d**).

Structural analysis of the composites was also conducted to establish the tribological mechanism. The XPS spectra with/without urea at room temperature and 140 °C are presented in Figure 8. The C1s signals for the three samples were deconvoluted into peaks ascribed to C-C at 284.71 eV, C=N/C=O at 286.68 eV and carboxyl at 289.08 eV [14,25]. The carboxyl peak intensity was reduced with increasing reaction temperature, which may be attributed to a thermal-induced polymerization of dopamine [26]. The O1s peaks were deconvoluted into Ag-O (529.92 eV), C-O (531.26 eV) and nitrates (533.13 eV) [14,25]. The nitrogen component in PDA/urea may interact with the CNT and Ag nanoparticles to generate nitrates. The N1s peaks were deconvoluted into four peaks at 397.97 eV, 400.17 eV, 401.59 eV and 402.82 eV, attributed to nitride, organic matrix, ammonium salt and azide species, respectively [25,27]. An increase in reaction temperature from ambient to 140 °C resulted in the formation of nitrides (C_xN_y), as shown in Figure 8c,j,k [25]. Moreover, cyanide and nitride formation can be linked to the inclusion of urea during hydrothermal treatment. The Ag3d signal was deconvoluted into two peaks at 367.71 eV and 368.56 eV, which are ascribed to silver oxides and Ag metal, respectively [14]. Raman analysis was also conducted, and the results are presented in Figure 9. The intensity ratios (I_D/I_G) of the composites with urea at room temperature, without urea at 140 °C and with urea at 140 °C were 0.798, 0.905 and 0.957, respectively. The increased values of I_D/I_G may suggest that new carbon and/or N-disrupted carbon were formed as a result of the solvothermal method of doping [2]. Formation of C_xN_y at high temperature and pressure can improve the tribological properties of the carbon nanotube/Ag nanocomposites, where urea promotes the formation of cyanides and nitrides.

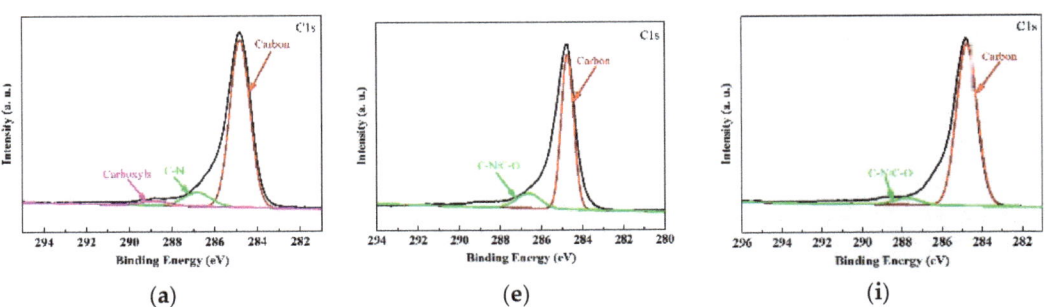

Figure 8. *Cont.*

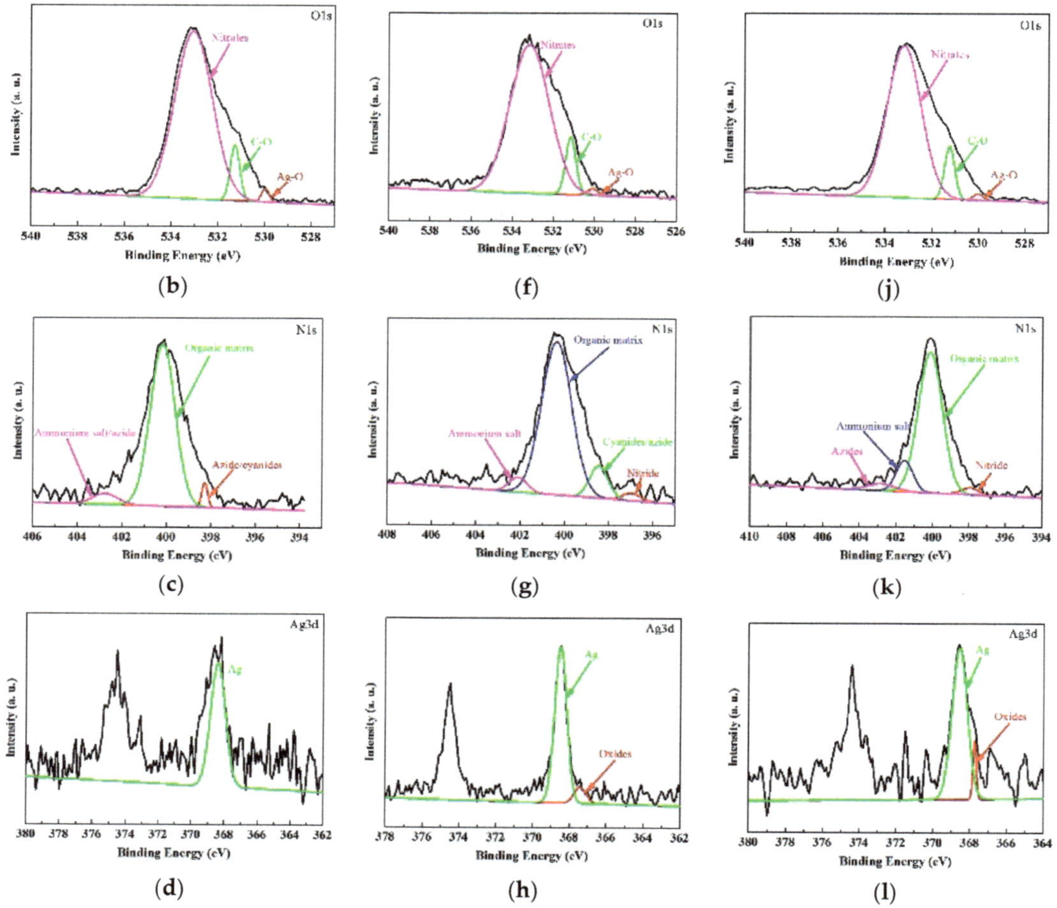

Figure 8. XPS spectra of samples with urea at room temperature (**a–d**), with urea at 140 °C (**e–h**) and without urea at 140 °C (**i–l**).

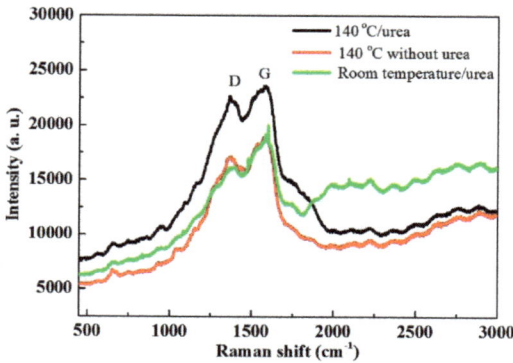

Figure 9. Raman spectra of samples with urea at room temperature, with urea at 140 °C and without urea at 140 °C.

4. Conclusions

Carbon nanotube/Ag nanocomposites have been synthesized in a urea solution using a hydrothermal method. The analysis reveals that the diameter of the carbon tubes was about 30 nm, and Ag nanoparticles with a diameter of ca. 10 nm covered the carbon tube surface. The XRD analysis shows that the (002) peak for carbon nanotubes was stable, and the intensity of the (111) and (200) peaks for Ag was reduced and shifted to lower 2θ values as the temperature was increased with the addition of urea, which means that the Ag particle size decreased with a lower degree of crystallinity at a high temperature in the presence of urea. Steel balls under the lubrication of LP-OAN$_8$-C$_{0.04}$/Ag$_4$ exhibited lower COFs and WSDs than those recorded using pure LP. Increasing the reaction temperature from room temperature to 140 °C, the LP-OAN$_8$-C$_{0.04}$/Ag$_4$ decreased the WSDs from 0.745 mm for LP to 0.55 mm. It is obvious that the steel balls under the lubrication of LP-OAN$_8$-C$_{0.04}$/Ag$_4$ exhibited lower WSDs than those recorded for other samples. The COFs of the balls lubricated with LP-OAN-C$_{0.04}$/Ag$_4$ with different amounts of urea are also smaller than those obtained with pure LP.

A synergism has been established between the preparation variables relating to the quantity of carbon nanotubes, the volume of a AgNO$_3$ solution and urea at a reaction temperature of 140 °C. A tribofilm containing Ag, carbon, nitride and other compounds was formed on the worn surfaces during sliding. Ag particle size decreased with a lower degree of crystallinity at a high temperature in the presence of urea; the smaller Ag particle size and reduced crystallinity may be beneficial for friction reduction and anti-wear properties. At a high temperature, the nitrogen component in PDA/urea may interact with the CNT and Ag nanoparticles to generate nitrates. The nitride (C_xN_y), organic matrix, ammonium salt and azide species were formed at high temperature. Moreover, urea promoted the formation of N-doped carbon and/or carbon, which contributes to the improvement in tribological properties.

Author Contributions: S.J. and J.Z.: Experiment, Data curation, Formal analysis, Investigation, Methodology, Writing—original draft. G.H. and J.F.: Data curation, Investigation, Methodology. C.Z.: Visualization, Investigation. Z.W.: Investigation, Supervision, Software. Z.J. and Y.B.: Conceptualization, Investigation, Funding acquisition, Project administration, Supervision, Writing—review and editing. All authors have read and agreed to the published version of the manuscript.

Funding: This research was funded by Shan Dong Province Nature Science Foundation (Grant ZR2020ME133).

Data Availability Statement: Not applicable.

Acknowledgments: The authors would like to express their gratitude to EditSprings (https://www.editsprings.cn (accessed on 9 October 2023)) for the expert linguistic services provided. The authors thank the support of Project of Shan Dong Province Nature Science Foundation (Grant ZR2020ME133).

Conflicts of Interest: The authors declare no conflict of interest.

References

1. Diana-Luciana, C.; Corina, A.; Cristian, P.; Razvan, R. The efficiency of Co-based single-wall carbon nanotubes (SWNTs) as an AW/EP additive for mineral base oils. *Wear* **2012**, *290–291*, 133–139. [CrossRef]
2. Kiattisak, P.; Jenjira, S.; Asamee, S.; Kasrin, S.; Kritsada, S.; Cheng, H.P.; Sangay, W.; Proespichaya, K.; Panote, T.; Warakorn, L. New electrode material integrates silver nanoprisms with phosphorus-doped carbon nanotubes for forensic detection of nitrite. *Electrochim. Acta* **2022**, *436*, 141439. [CrossRef]
3. Hadi, P.; Saeed, Z.H.; Yaghob, M. Comparison between multi-walled carbon nanotubes and titanium dioxide nanoparticles as additives on performance of turbine meter oil nano lubricant. *Sci. Rep.* **2021**, *11*, 11064. [CrossRef]
4. Manish, C.; Gehan, A.J.A. Thin films of fullerene-like MoS$_2$ nanoparticles with ultra-low friction and wear. *Nature* **2000**, *407*, 164–167. [CrossRef]
5. Zhai, W.; Srikanth, N.; Kong, L.B.; Zhou, K. Carbon nanomaterials in tribology. *Carbon* **2017**, *119*, 150–171. [CrossRef]
6. Pei, X.; Liu, W.; Hao, J. Functionalization of multiwalled carbon nanotube via surface reversible addition fragmentation chain transfer polymerization and as lubricant additives. *J. Polym. Sci. A Polym. Chem.* **2008**, *46*, 3014–3023. [CrossRef]

7. Jesús, A.C.C.; Paula, A.C.; Lina, M.H.P.; Javier, L.R.; Alejandro, T. Tribological properties of carbon nanotubes as lubricant additive in oil and water for a wheel-rail system. *J. Mater. Res. Technol.* **2016**, *5*, 68–76. [CrossRef]
8. Xiang, L.; Beibei, C.; Yuhan, J.; Xiaofang, L.; Jin, Y.; Changsheng, L.; Fengyuan, Y. Enhanced tribological properties of epoxy-based lubricating coatings using carbon nanotubes-ZnS hybrid. *Surf. Coat. Technol.* **2018**, *344*, 154–162. [CrossRef]
9. Nisha, R.; Mamta, S.L.; Muthusamy, K.; Sundara, R. Tribological study of iron infused carbon tubes additive in gearbox, engine, and vegetable-based lubricants. *Tribol. Int.* **2022**, *171*, 107538. [CrossRef]
10. He, J.; Sun, J.; Choi, J.; Wang, C.; Su, D. Synthesis of N-doped carbon quantum dots as lubricant additive to enhance the tribological behavior of MoS_2 nanofluid. *Friction* **2023**, *11*, 441–459. [CrossRef]
11. Wang, Y.; Zhang, T.; Qiu, Y.; Guo, R.; Xu, F.; Liu, S.; Ye, Q.; Zhou, F. Nitrogen-doped porous carbon nanospheres derived from hyper-cross linked polystyrene as lubricant additives for friction and wear reduction. *Tribol. Int.* **2022**, *169*, 107458. [CrossRef]
12. Wang, B.; Hu, E.; Tu, Z.; David, K.D.; Hu, K.; Hu, X.; Yang, W.; Guo, J.; Cai, W.; Qian, W.; et al. Characterization and tribological properties of rice husk carbon nanoparticles Co-doped with sulfur and nitrogen. *Appl. Surf. Sci.* **2018**, *462*, 944–954. [CrossRef]
13. Wang, Z.; Ren, R.; Song, H.; Jia, X. Improved tribological properties of the synthesized copper/carbon nanotube nanocomposites for rapeseed oil-based additives. *Appl. Surf. Sci.* **2018**, *428*, 630–639. [CrossRef]
14. Jia, Z.; Wang, Z.; Liu, C.; Zhao, L.; Ni, J.; Li, Y.; Shao, X.; Wang, C. The synthesis and tribological properties of Ag/polydopamine nanocomposites as additives in poly-alpha-olefin. *Tribol. Int.* **2017**, *114*, 282–289. [CrossRef]
15. Zang, C.; Yang, M.; Liu, E.; Qian, Q.; Zhao, J.; Zhen, J.; Zhang, R.; Jia, Z.; Han, W. Synthesis, characterization and tribological behaviors of hexagonal boron nitride/copper nanocomposites as lubricant additives. *Tribol. Int.* **2022**, *165*, 107312. [CrossRef]
16. Hemant, J.; Ambuj, M.; Kabiraj, D. The effects of metal concentration and annealing temperature on the optical properties of silver nanocomposite. *Mater. Today Proc.* **2022**, *57*, 234–238. [CrossRef]
17. Dash, M.P.; Tripathy, M.; Sasmal, A.; Mohanty, G.C.; Nayak, P.L. Poly (anthranilic acid)/multi-walled carbon nanotube composites: Spectral, morphological, and electrical properties. *J. Mater. Sci.* **2010**, *45*, 3858–3865. [CrossRef]
18. Amin, I.; Gholamali, F. VRH investigation of polyaniline-multiwalled carbon nanotube nanocomposite network. *Bull. Mater. Sci.* **2015**, *38*, 831–835. [CrossRef]
19. Jaiswal, V.; Kalyani Umrao, S.; Rastogi, R.B.; Kumar, R.; Srivastava, A. Synthesis, Characterization, and Tribological evaluation of TiO_2-reinforced boron and nitrogen co-doped reduced graphene oxide based hybrid nanomaterials as efficient antiwear lubricant additives. *ACS Appl. Mater. Interfaces* **2016**, *8*, 11698–11710. [CrossRef]
20. Cai, J.; Huang, J.; Wang, S.; Iocozzia, J.; Sun, J.; Sun, Y.; Yang, Y.; Lai, Y.; Lin, Z. Crafting mussel-inspired metal nanoparticle-decorated ultrathin graphitic carbon nitride for the degradation of chemical pollutants and production of chemical resources. *Adv. Mater.* **2019**, *31*, 1806314. [CrossRef]
21. Sinval, P.S.; Alexandre, M.A.; Ernane, R.S.; Marcelo, A.C. Surface modification of AISI H13 steel by die-sinking electrical discharge machining and TiAlN coating: A promising hybrid technique to improve wear resistance. *Wear* **2020**, *462–463*, 203509. [CrossRef]
22. Yang, X.; Meng, Y.; Tian, Y. Controllable friction and wear of nitrided steel under the lubrication of $[DMIm]PF_6$/PC solution via electrochemical potential. *Wear* **2016**, *360–361*, 104–113. [CrossRef]
23. Yang, M.; Jia, Z.; Pang, X.; Shao, X.; Zhen, J.; Zhang, R.; Ni, J.; Jiang, J.; Yu, B. The Synthesis and tribological properties of carbonized polydopamine/Ag composite films. *J. Mater. Eng. Perform.* **2019**, *28*, 7213–7226. [CrossRef]
24. Meng, Y.; Su, F.; Chen, Y. Supercritical fluid synthesis and tribological applications of silver nanoparticle-decorated graphene in engine oil nanofluid. *Sci. Rep.* **2016**, *6*, 31246. [CrossRef]
25. Wu, X.; Li, Z.; Tian, Y.; Lu, Y.; Jiang, H.; Liu, L.; Gai, L. Grinding to produce polydopamine-modified polypyrrole nanotubes with enhanced performance for sodium-ion capacitor. *Electrochim. Acta* **2022**, *434*, 141338. [CrossRef]
26. Wang, H.; Liu, Z.; Zhang, X.; Lv, C.; Yuan, R.; Zhu, Y.; Mu, L.; Zhu, J. Durable self-healing superhydrophobic coating with biomimic "chloroplast" analogous structure. *Adv. Mater. Interfaces* **2016**, *3*, 1600040. [CrossRef]
27. Huang, Q.; Liu, M.; Wan, Q.; Jiang, R.; Mao, L.; Zeng, G.; Huang, H.; Deng, F.; Zhang, X.; Wei, Y. Preparation of polymeric silica composites through polydopamine-mediated surface initiated ATRP for highly efficient removal of environmental pollutants. *Mater. Chem. Phys.* **2017**, *193*, 501–511. [CrossRef]

Disclaimer/Publisher's Note: The statements, opinions and data contained in all publications are solely those of the individual author(s) and contributor(s) and not of MDPI and/or the editor(s). MDPI and/or the editor(s) disclaim responsibility for any injury to people or property resulting from any ideas, methods, instructions or products referred to in the content.

Article

Tribological Behavior and Wear Protection Ability of Graphene Additives in Synthetic Hydrocarbon Base Stocks

Ge Du [1,2], Hongmei Yang [2,*], Xiuli Sun [2] and Yong Tang [2]

1. School of Materials and Chemistry, University of Shanghai for Science and Technology, Shanghai 200093, China; duge@sioc.ac.cn
2. State Key Laboratory of Organometallic Chemistry, Shanghai Institute of Organic Chemistry, Chinese Academy of Sciences, Shanghai 200032, China
* Correspondence: yanghm@sioc.ac.cn

Abstract: Graphene carbon materials show good tribological properties due to their unique layered structures. In this work, the tribological properties of graphene (GN) and fluorinated graphene (FGN) were studied in two kinds of synthetic hydrocarbon base stocks at different working conditions. Firstly, the structures of GN and FGN were characterized comparatively using FT-IR, Raman, XRD, and TGA. Secondly, the tribological properties of GN and FGN as the lubrication additives both in PAO6 and CTL6 were studied on a four-ball tester. Finally, the surfaces of friction counterparts, before and after tribological tests, were analyzed to disclose the lubrication mechanism using UV, micro-Raman, and EDS. The results show that GN and FGN can be stably dispersed in the selected synthetic hydrocarbon base stocks with 1 wt.% T161 as the dispersant, and the optimal addition of graphene additive is 100 ppm, which shows better friction reducing and anti-wear properties. GN and FGN also show better tribological performance at a higher load (not less than 392 N), and their compatibility with PAO6 is better. The worn surface analysis shows that the graphene additive participates in the lubrication film formation during friction by frictional chemical reaction with friction counterparts, which could improve the stability and tribological performance, resulting in an increased application temperature of synthetic hydrocarbon base stock by at least 10 °C.

Keywords: tribological behavior; wear protection; graphene; fluorinated graphene; synthetic hydrocarbon

Citation: Du, G.; Yang, H.; Sun, X.; Tang, Y. Tribological Behavior and Wear Protection Ability of Graphene Additives in Synthetic Hydrocarbon Base Stocks. *Lubricants* 2023, 11, 200. https://doi.org/10.3390/lubricants11050200

Received: 27 March 2023
Revised: 24 April 2023
Accepted: 28 April 2023
Published: 29 April 2023

Copyright: © 2023 by the authors. Licensee MDPI, Basel, Switzerland. This article is an open access article distributed under the terms and conditions of the Creative Commons Attribution (CC BY) license (https://creativecommons.org/licenses/by/4.0/).

1. Introduction

Global industrial synthetic and semisynthetic use continues to grow, driven by modern equipment, advanced technology, and government regulations. Synthetics are now required in many applications due to severe operating conditions or corrosive environments. Synthetic lubricants—formulated products consisting of synthetic base stocks plus additives—have physical and chemical properties that are generally superior to those of conventional mineral oil-based lubricants, including good thermal and oxidative stability [1], low-temperature fluidity [2], low volatility, high viscosity indexes, and fire resistance. As a result, synlubes are the lubricants of choice in applications with especially demanding performance requirements [3]. According to his Markit, the global market for synthetic base oils was 1.3 million metric tons in 2020, and polyalphaolefins (PAOs) dominated the market with a share of 47% (volume), followed by esters (28%), and polyalkylene glycols (PAG, 13%). Therefore, the compatibility of synthetic base stocks, especially synthetic hydrocarbons and additives, has become the focus of researchers in recent years [4,5].

In the past few decades, various new two-dimensional materials, including MXenes [6,7], MoS_2 [8,9], WS_2 [10], as well as graphene [11], have become a hot topic in the field of friction [12]. Among them, graphene has been widely considered by researchers due to its low surface energy and unique tribological properties [13–15] caused by easy-sliding multi-layer structures [16]. Omrani et al. [17] modified graphene using oleic acid and found that 0.02 wt.% oleic acid-modified graphene could be well-dispersed in PAO9 and improve the friction-reducing and anti-wear properties effectively. Yin et al. [18] prepared a boronized functional layer on the surface of bearing steel through an electrochemical boronization process, which improved the adsorption of GO on the bearing steel. By adding 1 wt.% GO nanosheets, an ultra-low friction state with a coefficient of friction (COF) of 0.03 was achieved. Ismail et al. [19] achieved a high oil-soluble graphene oxide that was modified through the copper-catalyzed cycloaddition, which reacted with an alkyne, and applied the prepared modified graphene oxide as an additive; the results showed that the modified graphene oxide could significantly improve the friction reducing and anti-wear performance of the base oil when the addition was 0.01 wt.%. At present, the dispersion stability of nanoparticle additives as liquid additives is particularly important. Shi et al. [20] used molecular dynamics (MD) simulation and experiments to study the dispersion behavior and mechanism of the dispersant polyisobutylene succinimide polyamine (PIB) concentration on soot aggregates under different pressures and temperatures and established static, limiting, and shear models.

Fluorinated graphene (FGN) is a derivative of graphene wherein its hydrogen atoms are partially or completely replaced by fluorine atoms [21]. Therefore, FGN not only inherits the sheet structure of graphene but also introduces fluorine atoms with a larger radius, which can weaken the interactions between interlayers and enhance the lubricating performance due to an easier shear capability [11]. Researchers found that FGN exhibited better tribological performance compared to pristine graphene at the macroscale [22]. To achieve better tribological properties, Fan et al. [23] prepared fluorinated graphene containing different F concentrations through direct fluorination with F_2, which was added to liquid paraffin oil as an additive. The results show that fluorination with a C/F ratio of 1.0 could reduce friction and wear by 50.4% and 90.9%, respectively. They also [24] prepared fluorinated graphene oxide with a relatively low F content under mild temperature conditions, which could lead to a 47% and 31% lower wear rate compared to that with the lubrication of pure water and GO water suspension when used as the additive, respectively. With the assistance of microwave heating, Ma et al. [25] fulfiled the substitution of FGN with NaOH at a quite mild reaction conditions to give a high fluorine content HOFG with excellent water dispersibility. The study on the tribological behaviors of HOFGs showed that the extreme-pressure and wear-resistance properties of pure water could be improved to over 240% and up to 30%, respectively.

However, studies on the tribological properties of graphene additives at present mostly focus on a single oil or water system, and the performance of graphene additives in different base stocks with the same viscosity grade has not been explored. Meanwhile, the conditions of tribological tests are mostly carried out at room temperature, and the adaptability of graphene additives in different working conditions is not studied in detail. Therefore, two synthetic hydrocarbon base stocks, that is, PAO6 and CTL6, with the same viscosity grade were selected, and the high molecular weight polysuccinimide T161 was used as a dispersant to study the tribological performance of GN and FGN under different test conditions. The study of graphene additives in various synthetic hydrocarbon base stocks under different tribological conditions is helpful in establishing the compatibility law of graphene additives and synthetic base stocks and disclosing its lubrication mechanism in synthetic hydrocarbon base stocks, which will provide theoretical guidance for the subsequent formulation of research based on graphene additives.

2. Materials and Methods

2.1. Materials

Graphene (GN) and fluorinated graphene (FGN) were purchased from Hubei Zhuoxi Fluorochemical Co., Ltd. (Yingcheng, China), and high molecular weight poly-(isobutylene succinimide) (T161, ashless dispersant) was purchased from Wuxi South Petroleum Additives Co., Ltd. (Wuxi, China). PAO6 (Durasyn 166, Ineos, London, UK), as well as CTL6 (Coal to oil, Lu'an Chemical Group, Changzhi, China), were commercially obtained and used as the synthetic hydrocarbon base stock.

2.2. Preparation of Oils

In general, 0~5 wt.% of the dispersant T161 was first added to the selected synthetic hydrocarbon base stock (PAO6 or CTL6) and stirred at 60 °C for 1 h to give the corresponding base oil. Further, the graphene additive (GN or FGN) was subsequently added to the prepared base oil with an addition of 0~200 ppm in mass percentage, and the mixture was ultrasonically dispersed for 10 min in a 20 °C water bath.

2.3. Characterization

The contents of C and H in GN and FGN were determined by an automatic element analyzer—Vario Micro Cube (Elementar Analysensysteme GmbH, Langenselbold, Germany)—according to the general rules of analytical methods for the element analyzer JY/T 017-2020. Additionally, the F content in FGN was determined by the oxygen bottle combustion-chemical titration method using a 25 mL Titrette Automatic Liquid Microtitrator (BRAND GMBH + CO KG, Wertheim, Germany) and a micro-analysis balance with a minimum fraction value of 1 ug, according to the general rules in the Chinese Pharmacopoeia (2020 edition)—oxygen bottle combustion method 0703. The detailed steps are as follows: FGN was accurately weighed and wrapped into the filter paper, which was decomposed using the oxygen bottle method, and 10 mL of deionized water was applied as the absorption solution.

The micro-morphologies of GN and FGN were tested using scanning electron microscopy with an accelerating voltage of 5~10 kV (SEM, JEOL JSM-6390LV) The specific surface area was achieved by the Brunauer-Emmett-Teller (BET) method, and the measurement was performed according to the physical adsorption of N_2 via a fully automatic surface area and porosity analyzer (ASAP 2020 HD88, Micromeritics, Norcross, GA, USA).

Fourier transform infrared spectroscopy (FT-IR, Thermo Fisher Nicolet iN 10) was recorded by scanning from 500 to 4000 cm^{-1} in the Attenuated Total Reflectance (ATR) configuration. Raman spectroscopy (Thermo DXR) was detected with an excitation wavelength of 532 nm. X-ray diffraction (XRD, PANalytical XPert PRO) was carried out with Cu Kα radiation (λ = 0.154 nm), and the diffraction data were recorded as 2θ from 5~60° at a speed of 4° min^{-1}. Thermo gravimetric analysis (TGA, TA Q500) was performed under an N_2 atmosphere with a flow rate of 60 mL/min and a heating rate of 10 °C/min from 25 to 950 °C. Ultraviolet-visible spectroscopy (UV-Vis, Shimadzu UV-2700) was applied to analyze the chemical structure and composition changes of oils between, before, and after tribological tests. An energy dispersive spectrometer (EDS, OXFORD x-act) was used to analyze the components of the tribofilm after tribological tests.

2.4. Tribological Tests

The tribological behaviors of GN and FGN in the prepared oils were evaluated on a Tenkey MS-10A four-ball tester (Xiamen TenKey Automation Co., Ltd., Xiamen, China). The tribological tests were operated at 1200 rpm under different loads ranging from 96 to 490 N, the test duration was 30 min at varied temperatures (54~100 °C), the balls used were made of GCr15 bearing steel with a diameter of 12.7 mm, and the upper limit of COF was set to be 0.4. All tribological tests were performed at least three times to ensure repeatability.

3. Results and Discussion

3.1. Characterization of Graphene Additives

The content analyses of C, H, and F in GN and FGN were carried out by elemental analysis, as shown in Figure 1, and the O content in GN was obtained by subtraction of the C and H content from 100%. For GN, the element content of C, O, and H was 94.13%, 5.46%, and 0.41%, respectively, and the atomic number ratio of C, O, and H was estimated to be 19.13:0.83:1, indicating that GN contains a small amount of oxygen, which is likely to have been reduced by graphene oxide. The element content of C, F, and H was 36.29%, 63.36%, and 0.35% for FGN, and the atomic number ratio of C, F, and H was estimated to be 8.64:9.53:1; that is, the F/C ratio in FGN was as high as 1.1:1, indicating that FGN has a high degree of fluorination.

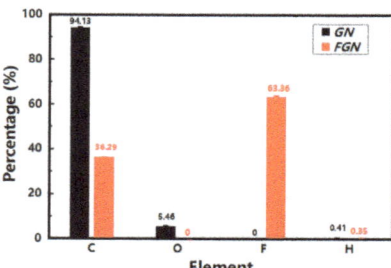

Figure 1. The elemental analysis of GN and FGN.

The micro-morphologies of GN and FGN were conducted by SEM, and the results are displayed in Figure 2. Both GN and FGN showed coiled-layered structures; however, the layer spacing in GN was smaller than that in FGN, which can be seen in the circled portion in Figure 2a,c. At the same time, the lamellar folds in FGN were weaker than that in GN (see Figure 2b,d); this may be because the introduction of fluorine weakened the π-π stacking between graphene layers.

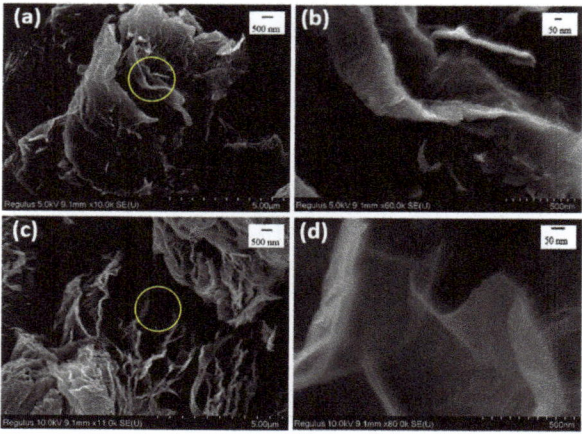

Figure 2. SEM images: (a) 5 μm and (b) 500 nm of GN; (c) 5 μm and (d) 500 nm of FGN.

The specific surface area (SSA) and pore-size characterization of GN and FGN were further performed by N_2 adsorption-desorption experiments, and the results are shown in Figure 3. Both GN and FGN are type IV isotherms according to the classification of the International Union of Pure and Applied Chemistry (IUPAC) (Figure 3a), and the H3 type hysteresis loops were obvious at high relative pressure, indicating the existence of mesoporous pores [26]. Meanwhile, the BET SSA of GN was higher than that of FGN, which was 526.1 m^2/g and 313.2 m^2/g, respectively, indicating that GN has more pores than FGN. The pore size in GN was bigger than that in FGN through the pore-size distribution calculated by DFT in Figure 3b. The fluorination process could cause the collapse of the porous structure, and the introduction of the F atom could significantly increase the weight per unit volume, so the formation of a C-F bond could result in the reduction of SSA and pore size [27].

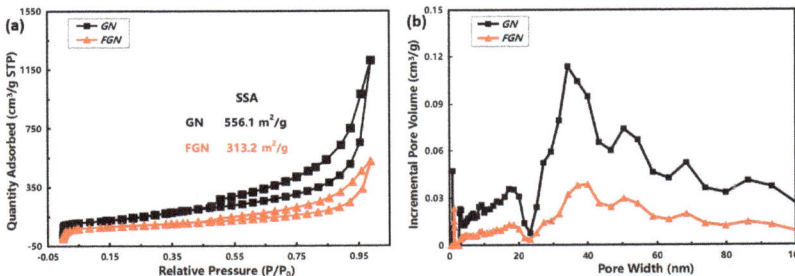

Figure 3. N$_2$ adsorption-desorption analysis of GN and FGN: (**a**) isotherm curves; (**b**) pore size distribution curves.

The FT-IR spectra of GN and FGN are shown in Figure 4a. The strong absorption peaks at 1213 and 1319 cm^{-1} in FGN are assigned to the stretching vibrations of covalent C-F bonds on the nanosheet and at the edge of the nanosheet for high F/C ratio, respectively [28,29], which is consistent with the result of the elemental analysis in Figure 1. The wide band at 3400 cm^{-1} in GN and FGN can be assigned to the O-H stretching vibration due to moisture, and the hydrogen bonding between F and H$_2$O causes the O-H stretching vibration to be more significant in FGN [30]. Raman is widely used to characterize carbon materials, where the D peak at ~1340 cm^{-1} represents a defect of the carbon skeleton caused by Csp^3, and the G peak at ~1580 cm^{-1} represents the carbon skeleton structure of Csp^2, corresponding to the stretching vibration of C=C [31]. The calculated I$_D$/I$_G$ can refer to the content ratio of Csp^3/Csp^2, which could estimate the degree of the defects and distinguish the disordered or ordered structure carbons [32]. The Raman results of GN and FGN are shown in Figure 4b. The I$_D$/I$_G$ of FGN (1.25) is significantly higher than that of GN (1.01), indicating that the defects are increased after fluorination in FGN. In addition, the G peak of FGN is downshifted, indicating that the defect type for FGN is mainly a point defect [33]. The layered structures of GN and FGN were further investigated by XRD, as shown in Figure 4c. FGN shows an obvious broad peak at 2θ = 12.3°, which is attributed to the diffraction peak of the crystal surface of the hexagonal crystal system (001) with high fluoride content, indicating that the regularity of FGN is high. The broad diffraction peak associated with the (001) reflection also indicates a high exfoliation degree for FGN. Based on Bragg's law, the reduced value of 2θ indicates a larger interplanar distance for FGN [34]; the interplanar distance is calculated to be 0.72 nm, which is two times that of GN (0.35 nm). FGN has two more diffraction peaks located at 2θ = 26.2° and 42.7°, which originated from the (002) and (100) crystalline planes of graphite. The peak at 26.2° corresponds to the (002) reflection of graphite, indicating the existence of Csp^3 due to the formation of C-F bonds in the FGN interlayers. The peak at 42.7° is attributed to the (100) reflection and is associated with the disorder induced by the fluorination in fully fluorinated graphene [35]. These data clearly illustrate that fluorination would give rise to a large interplanar distance due to the electrostatic repulsive force originating from the fluorine atom [36], which means an easier shear capability and the enhanced lubricant performance of FGN [11].

The thermal stabilities of GN and FGN are displayed using TGA, as shown in Figure 4d and Table 1. The initial weight loss temperature (T$_i$) and the maximum weight loss temperature (T$_{max}$) of GN were 547 °C and 721 °C, respectively, which is much higher than that of FGN (T$_i$ and T$_{max}$ was 347 °C and 551 °C, respectively). Meanwhile, the weight loss at T$_{max}$ of GN and FGN were 10.5% and 68.5%, respectively. For FGN, it appears that a sharp weight loss from 350 to 600 °C was caused by the defluorination process [37]; therefore, GN's thermal stability is much better.

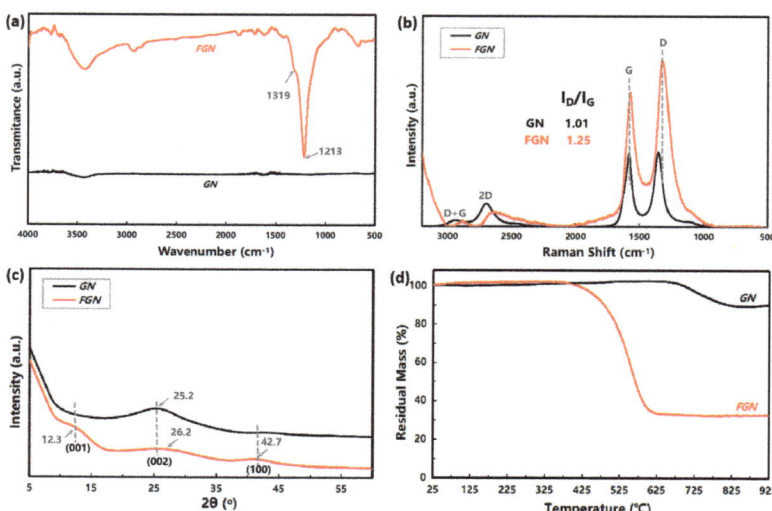

Figure 4. Characterizations of GN and FGN: (**a**) FT-IR; (**b**) Raman; (**c**) XRD; and (**d**) TGA.

Table 1. TGA analysis of GN and FGN.

	Initial Weight Loss Temperature (T_i), °C	Maximum Weight Loss Temperature (T_{max}), °C	Weight Loss at T_{max}, %
GN	547	721	10.5
FGN	347	551	68.5

3.2. Stability Monitoring of the Prepared Oils

According to the study reported by Kong et al. [4], high molecular weight poly-(isobutylene succinimide) could improve the dispersibility of graphene in PAO4 effectively. T161 is one high molecular weight poly-(isobutylene succinimide), and the images of the GN or FGN dispersion oils with or without T161 are shown in Figure 5. The settlement of GN could be observed in the dispersion oils, both in PAO6 and CTL6 without T161, after 1 day, and the settlement was almost complete after 7 days (in Figure 5a,e). Meanwhile, the dispersion oils with 1 wt.% T161, shown in Figure 5c,g, kept good dispersion after 7 days; that is, T161 can effectively improve the dispersion stability of GN in the synthetic hydrocarbon base stocks. For FGN, whether the oil was with or without T161, no significant FGN settlement was observed over time (Figure 5b,d,f,h). Combined with the characterization results of Raman and XRD, the introduction of F atoms in FGN could increase its interlayer spacing, making it easier to be exfoliated and stably dispersed in oils. However, lacking recognition of the gray-white FGN in dispersion oils may also cause settlement that was not observed.

3.3. Tribological Performance

3.3.1. Tribological Performance of Graphene Additives at Different Additions

In order to select a suitable addition of graphene additive, GN dispersion oils at various concentrations of 0 ppm, 50 ppm, 100 ppm, 150 ppm, and 200 ppm in mass percentage were prepared separately using the base oil PAO6 containing 1 wt.% T161. The tribological behaviors of the prepared oils were compared with the base stock of PAO6, and the results are shown in Figure 6. When the test temperature was 85 °C, and the load was 392 N, PAO6 and 1%T161-PAO6 appeared as spikes in friction profiles (Figure 6b), and the wear scars on the steel balls were irregular (Figure 6d). The spikes disappeared when adding different concentrations of GN, which would provide effective protection for friction counterparts. When the GN concentration was 100 ppm, the friction profile was more stable, and the average COF (ave. COF) and wear scar diameter (ave. WSD) were smaller (Figure 6c,d); that is, lubrication performance was optimal at this time. Therefore, all subsequent additions of graphene additives in this study used 100 ppm.

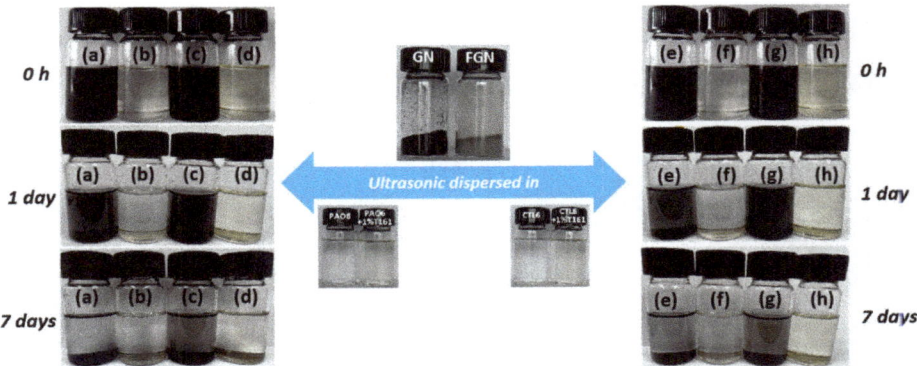

Figure 5. Images of dispersion oils with 100 ppm of graphene additives over time: (**a**) GN, (**b**) FGN, (**c**) GN-1%T161, and (**d**) FGN-1%T161 in PAO6; (**e**) GN, (**f**) FGN, (**g**) GN-1%T161, and (**h**) FGN-1%T161 in CTL6.

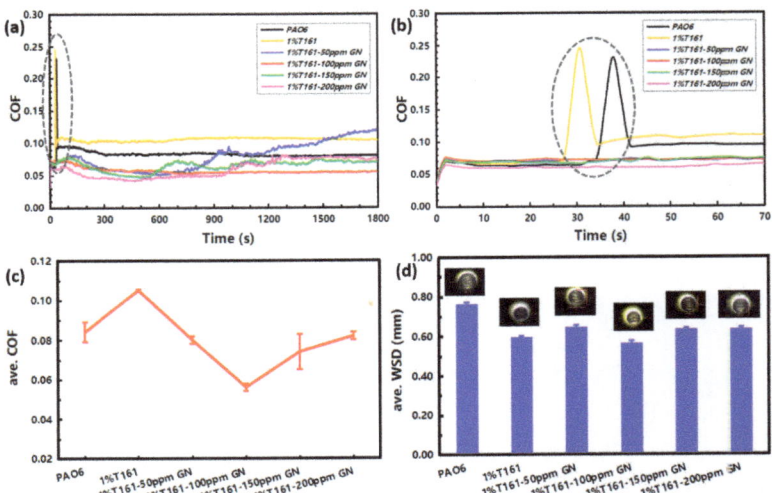

Figure 6. Tribological behaviors of dispersion oils with different additions of GN in PAO6: (**a**) the full (0~1800 s) and (**b**) locally amplified (0~70 s) friction profiles; (**c**) ave. COF; (**d**) ave. WSD and wear scars. All tests were run at 85 °C with a load of 392 N.

3.3.2. Tribological Performance of Graphene Additives under Different Test Loads

The tribological performance of graphene additives with increasing loads is displayed in Figure 7, from which we can see that all the dispersion oils in PAO6, as well as CTL6, cannot fulfill the tribological tests at 490 N. This means that their COFs exceed the set upper limit of 0.4. Apart from 1%T161-100 ppm FGN in CTL6, the COF of the base stock, 1%T161, 1%T161-100 ppm GN, and 1%T161-100 ppm FGN, shows a trend of first reducing and then increasing with an increasing load (Figure 7a,b), which is in line with the Stribeck curve [38]; that is, the thickness of the oil film decreases with the increasing load, and the transition from mixed lubrication to boundary lubrication occurs. According to the wear mechanism reported by J. F. Archard in 1953 [39], the wear is proportional to the positive pressure, so the higher the load, the more serious the wear (see Figure 7c,d). In PAO6, T161 increased the friction, but it also stabilized the oil film against wear (Figure 7a,c). When the test load was higher, such as at 392 N, the addition of GN and FGN could effectively reduce friction and wear. However, when further increasing the load to 490 N, none of the samples could operate stably. In CTL6, GN and FGN showed good friction-reducing performance at low test loads (<392 N) and stabilized the oil film at high loads (490 N), fulfilling the friction test that base stock cannot (Figure 7b,d).

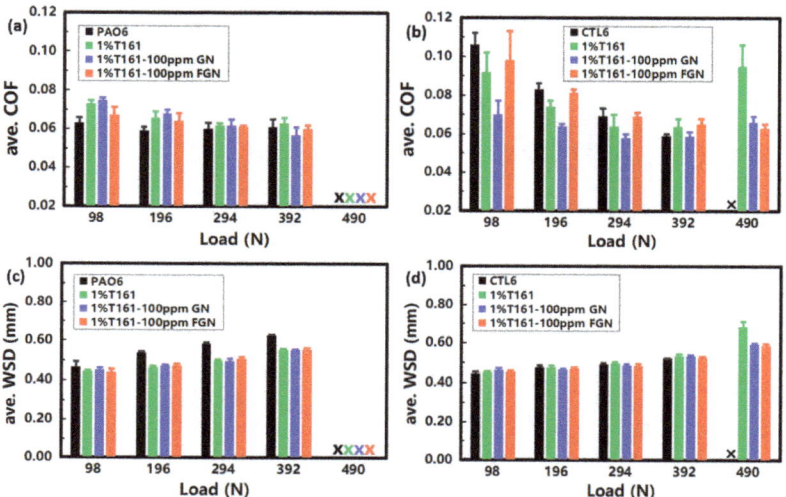

Figure 7. Tribological performance of dispersion oils when the test load varied from 98~490 N: (a) ave. COF and (c) ave. WSD in PAO6; (b) ave. COF and (d) ave. WSD in CTL6. The tests were run at 54 °C, where "x" represents that the tribological test cannot be successfully completed.

3.3.3. Tribological Performance of Graphene Additives at Different Test Temperatures

When the test temperature rises, the viscosity of base stocks will decrease to different degrees, and the thickness and stability of the oil films are the key factors in determining lubrication performance. In PAO6, the addition of T161 at 75~80 °C could stabilize the oil film and reduce the COF and WSD (Figure 8a,c); however, when the temperature rose to 85 °C or higher, T161 also increased friction (see Figure 8a). Meanwhile, GN or FGN could stabilize the oil film in time, which could relieve friction and wear effectively. In CTL6, the application of T161 at 75~85 °C could stabilize the oil film (Figure 8b,d), but it increased friction at 90 °C or higher temperatures (Figure 8b). At this time, GN or FGN could play an important role in friction reduction and anti-wear, and FGN showed better tribological performance. Overall, GN or FGN can increase the application temperature of PAO6 and CTL6 by at least 10 °C at 1200 rpm with a load of 392 N.

3.3.4. Tribological Performances of Graphene Additives in Different Base Stocks

When studying the influence of test load and temperature on the tribological performance of graphene additives, it was found that the performance in different base stocks was quite different, and the further friction-reducing and anti-wear properties in PAO6 and CTL6 were compared, as shown in Figure 9. From the figure, it can be seen that CTL6 shows a lower ave. COF and WSD compared to PAO6, meaning CTL6 has better friction-reducing and anti-wear properties. In PAO6, the ave. COF from low to high is 1%T161-100 ppm FGN~1%T161-100 ppm GN < 1%T161 < PAO6; namely, FGN and GN have a comparable friction-reducing property in PAO6. However, the order of ave. COF in CTL6 is 1%T161-100 ppm FGN < 1%T161-100 ppm GN~1%T161~CTL6; that is, the friction-reducing performance of FGN is better than that of GN in CTL6. In PAO6, ave. WSD from small to large is 1%T161-100 ppm GN~1%T161-100 ppm FGN < 1%T161 < PAO6; namely, both GN and FGN have comparable anti-wear performance in PAO6. However, the order of ave. WSD in CTL6 is 1%T161-100 ppm FGN~1%T161-100 ppm GN < 1%T161~CTL6; that is, the anti-wear performance of FGN and GN in CTL6 is also comparable. Overall, FGN shows a better tribological performance in both PAO6 and CTL6, combined with the XRD characterization in Section 3.1, which may be attributed to the easier shear capability after fluorination.

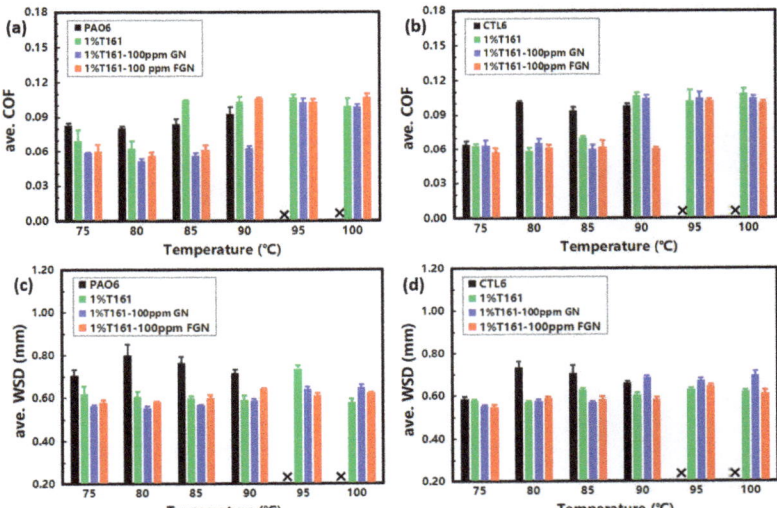

Figure 8. Tribological performance of dispersion oils when the test temperature varied from 75~100 °C: (**a**) ave. COF and (**c**) ave. WSD in PAO6; (**b**) ave. COF and (**d**) ave. WSD in CTL6. The tests were run with a load of 392 N, where "x" represents that the tribological test cannot be successfully completed.

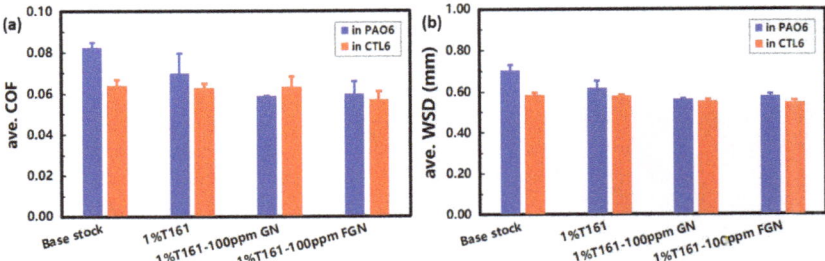

Figure 9. Comparison of (**a**) friction reducing and (**b**) anti-wear performance of graphene additives in PAO6 and CTL6. The tests were run at 75 °C with a load of 392 N.

3.3.5. The Lubrication Mechanism

To better understand the lubricating mechanism of the graphene additive in synthetic base stock, UV-Vis, micro-Raman, and EDS were applied to analyze the friction counterparts of 1%T161-100 ppm GN in PAO6 before and after the tribological test, which was run at 85 °C with a load of 392 N, and the results are displayed in Figures 10 and 11.

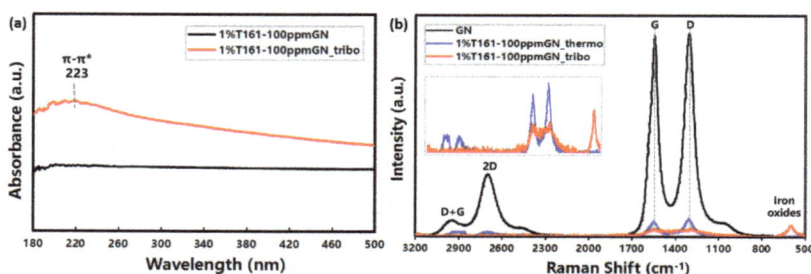

Figure 10. (**a**) UV-Vis spectra and (**b**) micro-Raman analysis of the friction counterparts.

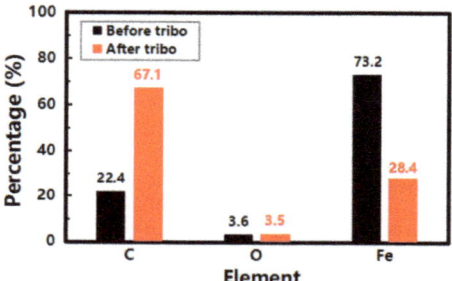

Figure 11. EDS analysis of the friction counterpart before and after the tribological tests.

Compared to the UV spectra before and after the tribological test in Figure 10a, there is a wide peak at 223 nm that can be assigned to the absorption of the π-π* transition on the wear scar, which is probably caused by the π-π* transition of C=C in the GN skeleton, indicating that the tribofilm contains the GN skeleton. The lubricating film during friction may be an adsorption film formed by simple physical adsorption or a reaction film generated by the frictional chemical reaction. In order to distinguish the composition of the lubricating film, a steel ball was soaked the oil and tested in an 85 °C oven for 0.5 h to form the thermal adsorption film. The results are shown in Figure 10b; 1%T161-100 ppmGN_thermo and 1%T161-100 ppmGN_tribo represent the thermal adsorption and the tribofilms after the tribological test, respectively. The weak Raman signal intensity of both thermal adsorption and tribofilms compared to GN is due to the small addition of GN (only 100 ppm). The Raman response of the thermal adsorption film is nearly the same as GN; that is, the D peak at ~1340 cm^{-1}, G peak at ~1580 cm^{-1}, 2D peak at ~2720 cm^{-1}, and D + G peak at ~2940 cm^{-1}. However, besides the D peak, G peak, and D + G peak, the peak that appeared at ~600 cm^{-1} can be assigned to iron oxides, indicating that the lubrication film during friction is formed by the frictional chemical reaction between GN and friction counterparts, which is different from the thermal adsorption film.

In Figure 11, the surface of the friction counterpart before tribological tests is composed of Fe (73.2%) and C (22.4%). However, the content of Fe and C became 28.4% and 67.1% after tribological tests, respectively. The C content increased more than three times, and the Fe content decreased by nearly 90%, while the O content was essentially unchanged. This means that the tribofilm is mainly composed of C. Combining the analysis of the UV and micro-Raman results, GN is involved in the formation of tribofilms, which improves the tribological performance of the base stock.

4. Conclusions

The study of graphene additives in various synthetic hydrocarbon base stocks at different tribological conditions is helpful in establishing compatibility law. In this work, the tribological properties of GN and FGN were studied in PAO6 and CTL6 under different working conditions. The results show that GN and FGN can be stably dispersed in the selected synthetic hydrocarbon base stocks with 1 wt.% T161 as the dispersant, and the optimal addition of the graphene additive is 100 ppm, which shows better friction-reducing and anti-wear properties. GN and FGN show better tribological performance at a higher load (not less than 392 N), and their compatibility with PAO6 is better. The worn surface analysis discloses that the graphene additive participates in the formation of lubrication film during friction by a frictional chemical reaction with friction counterparts, which could improve the stability and tribological performance, resulting in an increased application temperature of synthetic hydrocarbon base stock by at least 10 °C. These results provide theoretical guidance for the formulation of research based on graphene additives.

Author Contributions: Conceptualization, H.Y., X.S. and Y.T.; methodology, H.Y. and X.S.; validation, H.Y.; formal analysis, H.Y. and G.D.; investigation, G.D.; data curation, H.Y. and G.D.; writing—original draft preparation, H.Y. and G.D.; writing—review and editing, H.Y. and G.D.; supervision, H.Y. All authors have read and agreed to the published version of the manuscript.

Funding: This research was funded by the Ling Chuang Research Project of China National Nuclear Corporation that was granted by CAS Key Laboratory of Nuclear Materials and Safety Assessment of the Chinese Academy of Sciences.

Conflicts of Interest: The authors declare that they have no known competing financial interests or personal relationships that could influence the work reported in this paper.

References

1. Pournorouz, Z.; Mostafavi, A.; Pinto, A.; Bokka, A.; Jeon, J.; Shin, D. Enhanced thermophysical properties via PAO superstructure. *Nanoscale Res. Lett.* **2017**, *12*, 29. [CrossRef] [PubMed]
2. Nifant'ev, I.; Bagrov, V.; Vinogradov, A.; Vinogradov, A.; Ilyin, S.; Sevostyanova, N.; Batashev, S.; Ivchenko, P. Methylenealkane-Based Low-Viscosity Ester Oils: Synthesis and Outlook. *Lubricants* **2020**, *8*, 50. [CrossRef]
3. Dong, S.Q.; Mi, P.K.; Xu, S.; Zhang, J.; Zhao, R.D. Preparation and Characterization of Single-Component Poly-α-olefin Oil Base Stocks. *Energy Fuels* **2019**, *33*, 9796–9804. [CrossRef]
4. Kong, S.; Wang, J.; Hu, W.; Li, J. Effects of Thickness and Particle Size on Tribological Properties of Graphene as Lubricant Additive. *Tribol. Lett.* **2020**, *68*, 112. [CrossRef]
5. Yu, Q.; Zhang, C.; Dong, R.; Shi, Y.; Wang, Y.; Bai, Y.; Zhang, J.; Cai, M.; Zhou, F.; Novel, N. P-containing oil-soluble ionic liquids with excellent tribological and anticorrosion performance. *Tribol. Int.* **2018**, *132*, 118–129. [CrossRef]
6. Pogorielov, M.; Smyrnova, K.; Kyrylenko, S.; Gogotsi, O.; Zahorodna, V.; Pogrebnjak, A. MXenes—A new class of two-dimensional materials: Structure, properties and potential applications. *Nanomaterials* **2021**, *11*, 3412. [CrossRef]
7. Rakhadilov, B.K.; Maksakova, O.V.; Buitkenov, D.B.; Kylyshkanov, M.K.; Pogrebnjak, A.D.; Antypenko, V.P.; Konoplianchenko, Y.V. Structural-phase and tribo-corrosion properties of composite Ti3SiC2/TiC MAX-phase coatings: An experimental approach to strengthening by thermal annealing. *Appl. Phys. A* **2022**, *128*, 145. [CrossRef]
8. Liu, T.; Qin, J.; Wang, J.; Li, J. On the Tribological Properties of RGO–MoS2 Composites Surface Modified by Oleic Acid. *Tribol. Lett.* **2022**, *70*, 14. [CrossRef]
9. Xu, Z.; Lou, W.; Wu, X.; Wang, X.; Hao, J. Investigating the tribological behavior of PEGylated MoS2 nanocomposites as additives in polyalkylene glycol at elevated temperature. *RSC Adv.* **2017**, *7*, 53346–53354. [CrossRef]
10. Ratoi, M.; Niste, V.B.; Zekonyte, J. WS2 nanoparticles-potential replacement for ZDDP and friction modifier additives. *RSC Adv.* **2014**, *4*, 21238–21245. [CrossRef]
11. Berman, D.; Erdemir, A.; Sumant, A.V. Graphene: A new emerging lubricant. *Mater. Today* **2014**, *17*, 31–42. [CrossRef]
12. Omrani, E.; Menezes, P.L.; Rohatgi, P.K. Effect of Micro-and Nano-Sized Carbonous Solid Lubricants as Oil Additives in Nanofluid on Tribological Properties. *Lubricants* **2019**, *7*, 25. [CrossRef]
13. Pape, F.; Poll, G. Investigations on Graphene Platelets as Dry Lubricant and as Grease Additive for Sliding Contacts and Rolling Bearing Application. *Lubricants* **2019**, *8*, 3. [CrossRef]
14. Wu, Q.; Li, H.; Wu, L.; Bo, Z.; Wang, C.; Cheng, L.; Wang, C.; Peng, C.; Li, C.; Hu, X.; et al. Synergistic Lubrication and Antioxidation Efficacies of Graphene Oxide and Fullerenol as Biological Lubricant Additives for Artificial Joints. *Lubricants* **2022**, *11*, 11. [CrossRef]

15. Liu, Y.; Yu, S.; Shi, Q.; Ge, X.; Wang, W. Graphene-family lubricant additives: Recent developments and future perspectives. *Lubricants* **2022**, *10*, 215. [CrossRef]
16. Kim, K.-S.; Lee, H.-J.; Lee, C.; Lee, S.-K.; Jang, H.; Ahn, J.-H.; Kim, J.-H.; Lee, H.-J. Chemical Vapor Deposition-Grown Graphene: The Thinnest Solid Lubricant. *ACS Nano* **2011**, *5*, 5107–5114. [CrossRef]
17. Zhang, W.; Zhou, M.; Zhu, H.; Yu, T.; Wang, K.; Wei, J.; Ji, F.; Li, X.; Li, Z.; Zhang, P.; et al. Tribological properties of oleic acid-modified graphene as lubricant oil additives. *J. Phys. D Appl. Phys.* **2011**, *44*, 225303. [CrossRef]
18. Yin, S.; Wu, H.; Yi, X.; Huang, Z.; Ye, C.; Li, P.; Zhang, Y.; Shi, J.; Hua, K.; Wang, H. Enhanced graphene oxide adhesion on steel surface through boronizing functionalization treatment: Toward the robust ultralow friction. *Carbon* **2023**, *206*, 201–210. [CrossRef]
19. Ismail, N.A.; Bagheri, S. Highly oil-dispersed functionalized reduced graphene oxide nanosheets as lube oil friction modifier. *Mater. Sci. Eng. B* **2017**, *222*, 34–42. [CrossRef]
20. Shi, J.; Yi, X.; Wang, J.; Jin, G.; Lu, Y.; Wu, H.; Fan, X. Carbonaceous soot dispersion characteristic and mechanism in lubricant with effect of dispersants by molecular dynamics simulation and experimental studies. *Carbon* **2022**, *200*, 253–263. [CrossRef]
21. Nair, R.R.; Ren, W.; Jalil, R.; Riaz, I.; Kravets, V.G.; Britnell, L.; Blake, P.; Schedin, F.; Mayorov, A.S.; Yuan, S.; et al. Fluorographene: A two-dimensional counterpart of Teflon. *Small* **2010**, *6*, 2877–2884. [CrossRef]
22. Matsumura, K.; Chiashi, S.; Maruyama, S.; Choi, J. Macroscale tribological properties of fluorinated graphene. *Appl. Surf. Sci.* **2018**, *432*, 190–195. [CrossRef]
23. Fan, K.; Chen, X.; Wang, X.; Liu, X.; Liu, Y.; Lai, W.; Liu, X. Toward excellent tribological performance as oil-based lubricant additive: Particular tribological behavior of fluorinated graphene. *ACS Appl. Mater. Interfaces* **2018**, *10*, 28828–28838. [CrossRef]
24. Fan, K.; Liu, J.; Wang, X.; Liu, Y.; Lai, W.; Gao, S.; Qin, J.; Liu, X. Towards enhanced tribological performance as water-based lubricant additive: Selective fluorination of graphene oxide at mild temperature. *J. Colloid Interface Sci.* **2018**, *531*, 138–147. [CrossRef] [PubMed]
25. Ma, L.; Li, Z.; Jia, W.; Hou, K.; Wang, J.; Yang, S. Microwave-assisted synthesis of hydroxyl modified fluorinated graphene with high fluorine content and its high load-bearing capacity as water lubricant additive for ceramic/steel contact. *Colloids Surf. A Physicochem. Eng. Asp.* **2021**, *610*, 125931. [CrossRef]
26. Meloni, D.; Monaci, R.; Solinas, V.; Auroux, A.; Dumitriu, E. Characterization of the active sites in mixed oxides derived from LDH precursors by physico-chemical and catalytic techniques. *Appl. Catal. A Gen.* **2008**, *350*, 86–95. [CrossRef]
27. Wang, X.; Wang, W.; Xu, D.; Liu, Y.; Lai, W.; Liu, X. Activation effect of porous structure on fluorination of graphene based materials with large specific surface area at mild condition. *Carbon* **2017**, *124*, 288–295. [CrossRef]
28. Boopathi, S.; Narayanan, T.N.; Kumar, S.S. Improved heterogeneous electron transfer kinetics of fluorinated graphene derivatives. *Nanoscale* **2014**, *6*, 10140–10146. [CrossRef] [PubMed]
29. Lu, A.-H.; Hao, G.-P.; Sun, Q. Can Carbon Spheres Be Created through the Stober Method? *Angew. Chem.-Int. Ed.* **2011**, *50*, 9023–9025. [CrossRef]
30. Bon, S.B.; Valentini, L.; Verdejo, R.; Garcia Fierro, J.L.; Peponi, L.; Lopez-Manchado, M.A.; Kenny, J.M. Plasma Fluorination of Chemically Derived Graphene Sheets and Subsequent Modification with Butylamine. *Chem. Mater.* **2009**, *21*, 3433–3438. [CrossRef]
31. Tuinstra, F.; Koenig, J.L. Raman Spectrum of Graphite. *J. Chem. Phys.* **1970**, *53*, 1126–1130. [CrossRef]
32. Yang, H.; Li, J.; Zeng, X. Correlation between Molecular Structure and Interfacial Properties of Edge or Basal Plane Modified Graphene Oxide. *ACS Appl. Nano Mater.* **2019**, *1*, 2763–2773. [CrossRef]
33. Fan, K.; Fu, J.; Liu, X.; Liu, Y.; Lai, W.; Liu, X.; Wang, X. Dependence of the fluorination intercalation of graphene toward high-quality fluorinated graphene formation. *Chem. Sci.* **2019**, *10*, 5546–5555. [CrossRef]
34. Ci, X.; Zhao, W.; Luo, J.; Wu, Y.; Ge, T.; Xue, Q.; Gao, X.; Fang, Z. How the fluorographene replaced graphene as nanoadditive for improving tribological performances of GTL-8 based lubricant oil. *Friction* **2020**, *9*, 488–501. [CrossRef]
35. Quan, Y.; Liu, Q.; Li, K.; Zhang, H.; Yang, Y.; Zhang, J. Simultaneous fluorination and purification of natural block coaly graphite into fluorinated graphene with tunable fluorination degree. *Mater. Today Commun.* **2022**, *32*, 104130. [CrossRef]
36. Zhang, X.; Yu, L.; Wu, X.; Hu, W. Experimental Sensing and Density Functional Theory Study of H2S and SOF2 Adsorption on Au-Modified Graphene. *Adv. Sci.* **2015**, *2*, 1500101. [CrossRef] [PubMed]
37. Sun, C.; Feng, Y.; Li, Y.; Qin, C.; Zhang, Q.; Feng, W. Solvothermally exfoliated fluorographene for high-performance lithium primary batteries. *Nanoscale* **2014**, *6*, 2634–2641. [CrossRef]
38. Lu, X.; Khonsari, M.; Gelinck, E. The Stribeck curve: Experimental results and theoretical prediction. *J. Tribol.* **2006**, *128*, 789–794. [CrossRef]
39. Archard, J.F. Contact and Rubbing of Flat Surfaces. *J. Appl. Phys.* **1953**, *24*, 981–988. [CrossRef]

Disclaimer/Publisher's Note: The statements, opinions and data contained in all publications are solely those of the individual author(s) and contributor(s) and not of MDPI and/or the editor(s). MDPI and/or the editor(s) disclaim responsibility for any injury to people or property resulting from any ideas, methods, instructions or products referred to in the content.

Article

Enhanced Thermally Conductive Silicone Grease by Modified Boron Nitride

Yumeng Wang [1], Ning Shi [2], Min Liu [3], Sheng Han [1,*] and Jincan Yan [1,3,*]

[1] School of Chemical and Environmental Engineering, Shanghai Institute of Technology, Shanghai 201418, China
[2] School of Chemistry and Chemical Engineering, Shanghai University of Engineering Science, Shanghai 201620, China
[3] School of Chemistry and Chemical Engineering, Huizhou University, Huizhou 516007, China
* Correspondence: hansheng654321@sina.com (S.H.); jcyan@sit.edu.cn (J.Y.)

Abstract: In this work, a chemical modification method was used to prepare silicone grease with high thermal conductivity. We report two preparation methods for thermal conductive fillers, which are hydroxylated boron nitride-grafted carboxylic silicone oil (h-BN-OH@CS) and amino boron nitride-grafted carboxylic silicone oil (h-BN-NH$_2$@CS). When h-BN-OH@CS and h-BN-NH$_2$@CS were filled with 30 wt% in the base grease, the thermal conductivity was 1.324 W m^{-1} K^{-1} and 0.982 W m^{-1} K^{-1}, which is 6.04 and 4.48 times that of the base grease (0.219 W m^{-1} K^{-1}), respectively. The interfacial thermal resistance is reduced from 11.699 °C W^{-1} to 1.889 °C W^{-1} and 2.514 °C W^{-1}, respectively. Inorganic filler h-BN and organic filler carboxylic silicone oil were chemically grafted to improve the compatibility between h-BN and the base grease. The covalent bond between functionalized h-BN and carboxylic silicone oil is stronger than the van der Waals force, which can reduce the viscosity of the silicone grease.

Keywords: silicone grease; boron nitride; thermal conductivity; interfacial thermal resistance; chemical modification

---

Citation: Wang, Y.; Shi, N.; Liu, M.; Han, S.; Yan, J. Enhanced Thermally Conductive Silicone Grease by Modified Boron Nitride. *Lubricants* **2023**, *11*, 198. https://doi.org/10.3390/lubricants11050198

Received: 24 March 2023
Revised: 24 April 2023
Accepted: 26 April 2023
Published: 29 April 2023

Copyright: © 2023 by the authors. Licensee MDPI, Basel, Switzerland. This article is an open access article distributed under the terms and conditions of the Creative Commons Attribution (CC BY) license (https://creativecommons.org/licenses/by/4.0/).

1. Introduction

Filling thermal interface materials (TIMs) between devices and heat sinks can reduce the interface thermal resistance (IR) [1]. TIMs include silicone grease, silicone pad, thermally conductive phase change materials, and so on [2]. Lubricating grease has been widely used as a kind of TIM due to its easy processing and lightweight and excellent flexibility [3]. It is mainly composed of the base silicone grease and the thermally conductive (TC) filler [4]. The base silicone grease itself has a low thermal and high heat resistance which limits the application [5]. Therefore, the introduction of TC nano-fillers helps to reduce the IR [6–8]. Commonly used fillers for silicone greases include metal nanoparticles [9] and carbon black materials [10], which can also be electrically conductive. However, the applications in integrated circuits and high-power devices are hindered by the introduction of thermally and electrically conductive fillers. Therefore, it is necessary to reduce the interface resistance and form thermally conductive pathways in the base silicone grease [11].

Thermally conductive insulating fillers are mainly made of ceramic materials such as alumina oxide, silicon oxide, and so on [12]. The h-BN has high thermal conductivity, oxidation resistance, and electrical insulation [13]. Researchers have found that h-BN can effectively increase the TC, and the thermal conductivity mainly depends on the filler content, i.e., h-BN [14,15]. However, the high content of thermally conductive fillers tends to agglomerate and reduce the strength of composite materials and correspondingly increase the cost of manufacturing processes [16]. Therefore, in order to maximize the functionality of the thermally conductive filler, different methods of modification of h-BN need to be chosen [17]. It was found that h-BN modified with physical and chemical methods provides better thermal-conductivity enhancement than unmodified and agglomerated h-BN [18]. The h-BN plates were embedded in polycarbonate materials by Sun using a hot-press alignment method, and the ordered arrangement of the BN plates provided a pathway for phonon transport. When the h-BN content reaches 18.5 vol%, the composite was able to achieve a maximum TC of 3.1 W m^{-1} K^{-1} along the orientation direction [19]. Luo prepared vertically oriented, densely packed h-BN/epoxy resin composites by a vacuum filtration and slicing method. The thermal conductivity of the composites was 9.0 W m^{-1} K^{-1} when 44.0 vol% of h-BN was added [20]. Ahn modified the BN/polyvinyl butyral (PVB) composites with ellagic acid hydrate to give a higher thermal conductivity than PVB with pristine BN [21]. The BN surface is functionalized with octadecylamine and hyperbranched aromatic polyamides by the Yu group, and the TC of the epoxy resin was significantly improved [22]. Qian improved the thermal conductivity of BN nanosheet hybrids by copper phthalocyanine-grafted BN, and the TC was up to 0.63 W m^{-1} K^{-1} when the addition of the filler was 50 wt% [23]. Han polymerized PS-COOH on the surface of BN-OH. The TC of the composite material containing 12 wt% BN-OH was 1.131 W m^{-1} K^{-1}, better than pure PS (0.186 W m^{-1} K^{-1}) and BN/PS (0.312 W m^{-1} K^{-1}) [24]. When 2–3 wt% of the hydroxyl- and amino-functionalized BN nanotubes were added to epoxy resins, Young's modulus, strain at break, and thermal conductivity were improved [25]. Currently, h-BN is widely used in polymers, but its application in greases needs to be further developed.

In this work, hydroxyl- and amino-functionalized h-BN were bonded with carboxyl silicone oil (CS) by a chemical modification method. The thermal stability of silicone grease with modified h-BN was analyzed using thermogravimetric curves. The effects of modified h-BN-OH and h-BN-NH$_2$ on the viscosity and thermal conductivity of silicone greases were compared. The results showed that the TC was improved by the addition of the functionalized h-BN. The mechanism was investigated, and it is suggested that the formed ester and amide bonds can bridge adjacent thermally conductive materials to form thermally conductive pathways.

2. Experimental Section
2.1. Materials

SiO$_2$ (ca. 15 ± 5 nm) was purchased from Rohn Reagent Co., Ltd. (Shanghai, China). Dimethyl silicone oil, CS (molecular weight ca. 2000) was purchased from Anhui Aiyota Silicone Oil Co., Ltd. (Bengbu, China). h-BN (purity > 99.9%) with a particle size of 1–2 μm was purchased from Aladdin Co., Ltd. (Shanghai, China). The chemical reagents were purchased from commercial sources and, prior to usage, were not purified.

2.2. The Preparation of Functionalized BN
2.2.1. The Preparation of h-BN-OH and h-BN-NH$_2$

We dispersed h-BN at 5 mg mL^{-1} in a NaOH solution, then heated the h-BN dispersion to 120 °C and stirred it magnetically at a speed of 350 r·min^{-1} for 48 h. After cooling down, the product was filtered, and the sample was washed with deionized water (DIW) until neutral. It was then vacuum dried at 60 °C for 18 h to obtain h-BN-OH.

The h-BN-NH$_2$ was prepared by a one-step method. The h-BN powder (1.0 g) and urea (4.0 g) were added into the isopropanol solution (25 mL), and the solution was premixed and ball milled for 48 h at a rate of 300 r·min^{-1}. After the ball mill operation, the mixture was removed into a beaker and left to stand, and the upper suspension was injected into an extraction flask. The washed sample was vacuum dried at 60 °C for 24 h to obtain h-BN-NH$_2$.

2.2.2. The Preparation of h-BN-OH@CS

The preparation of h-BN-OH@CS is shown in Figure 1. Toluene (40 mL) was used as the solvent, h-BN-OH (2 g) and CS (18 g) were added to a container, and we adjusted the pH of the reaction system to 3 with concentrated sulphuric acid. The oil bath was heated to 90 °C, and the pre-reaction was carried out for 1–1.5 h. The p-toluenesulfonic acid was added to the reaction flask as a catalyst and heated to 120 °C for 72 h. The crude product after the reaction is washed with a 5 wt% NaOH solution. The upper solution was washed with DIW to neutral and rotary evaporated to remove the solvent and dried for 24 h to obtain h-BN-OH@CS (10 wt%).

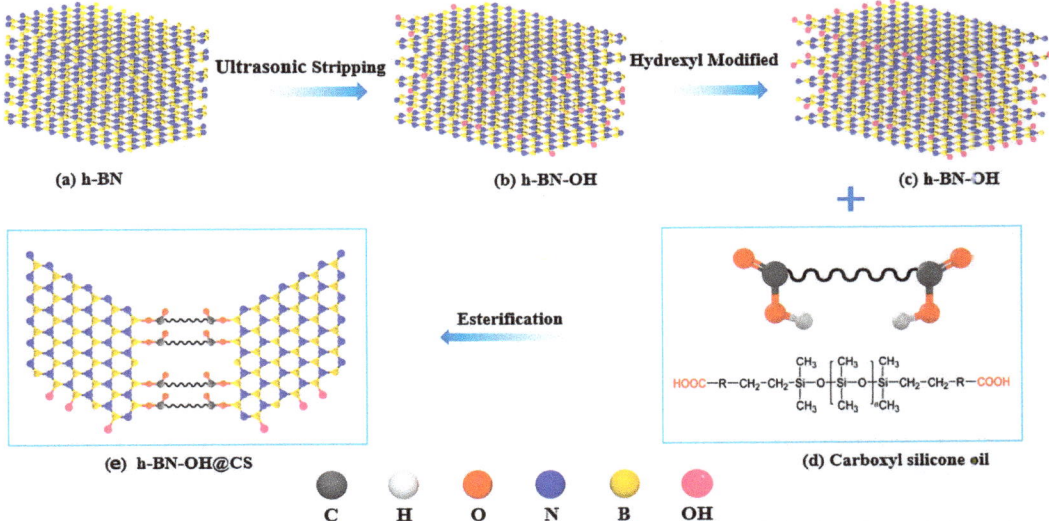

Figure 1. The preparation of h-BN-OH@CS.

2.2.3. The Preparation of h-BN-NH$_2$@CS

The preparation of h-BN-NH$_2$@CS was similar to that of h-BN-OH@CS. N, N-dimethylformamide (40 mL) was used as a solvent, CS (9 g) and h-BN-NH$_2$ (1 g) were added to the beaker, and dicyclohexylcarbodiimide (0.5 g) was used as a catalyst. After the amidation process, h-BN-NH$_2$ (10 wt%) was grafted with CS to give h-BN-NH$_2$@CS.

2.2.4. Preparation of Thermally Conductive Silicone Grease

The filler (30 wt%) was filled with the base silicone grease at room temperature, and the mixed components were stirred in a three-roller mill at 150 r min^{-1} for 20 min. Then the grease was degassed by a vacuum-drying oven at room temperature under 0.01 MPa for 2 h.

2.3. Characterization

Fourier-transform infrared spectra (FTIR) were collected by an infrared spectrometer (Spectrum ONE, PerkinElmer, Waltham, MA, USA) in the range 500–4000 cm^{-1}. Microscopic scanning electron micrographs (SEM) of thermally conductive fillers were analyzed by a scanning electron microscope (Gemini 300, Zeiss, Tubingen, Germany). The thermally conductive fillers were observed using transmission electron microscopy (TEM) (eol2100f, JEOL, Tokyo, Japan) at 200 kV. Thermo gravimetric analysis (TGA) (HCT-4, Hengjiu, China) reached 800 °C at a heating rate of 10 °C min^{-1} in a nitrogen atmosphere. Thermal resistance and thermal conductivity were carried out with a heat flow method thermal conductivity tester (DRL-III, Xiangtan Xiangyi, Xiangtan, China) at an average temperature of 50 °C, pressure 50 N, sample area 615 mm^2, and deviation ±5%. The Rheometer (RheoStress 6000, HAAKE, Germany) was used to test the viscosity of the sample. The apparent viscosity of the sample was obtained at a shear rate of 1.25 s^{-1}. The modified fillers were characterized using an X-ray photoelectron spectrometer (XPS) (Thermo Scientific K-Alpha, Waltham, MA, USA) at a chamber pressure of less than 2.0 × 10^{-7} mbar. UV-Visible absorption spectroscopy (UV-Vis) (Evolution 350, Thermo Fisher, Waltham, MA, USA) was used to measure the absorbance of the samples. The X-ray diffraction instrument (XRD) (Ultima IV, Rigaku, Tokyo, Japan) with a scanning angle range of 5–60° and a scanning speed of 2° min^{-1}.

3. Results and Discussions

3.1. FTIR Analysis

The FTIR analysis of h-BN, h-BN-OH, and h-BN-NH$_2$ is shown in Figure 2a. Due to defects on the h-BN surface, a broad peak of hydroxyl groups at 3500 cm^{-1}. In the spectrum of h-BN-NH$_2$, two peaks at 1100 cm^{-1} and 1650 cm^{-1} can be attributed to the bending vibrations of the N-H bond. The peak at 3208–3433 cm^{-1} in h-BN-NH$_2$ indicates that the NH$_2$ group was grafted onto the h-BN surface after the amination process [26]. The FTIR analysis of h-BN-OH@CS and h-BN-NH$_2$@CS is shown in Figure 2b. The peaks at 1737 cm^{-1} and 1712 cm^{-1} are attributed to the C=O double bond stretching vibrations in the carboxyl group, and the peak at 986 cm^{-1} is attributed to the C-O single bond stretching vibrations. The grafted sample shows a methyl and methylene structure at 2966 cm^{-1} and a broad peak representing B-OH and -OH at 3000–3600 cm^{-1}. The stretching vibration peak at 1733 cm^{-1} is attributed to the ester group formed by the esterification reaction. The presence of the ester group peaks demonstrated that h-BN-NH$_2$ and h-BN-OH were grafted onto the CS [27]. As is shown in FTIR curves of h-BN-NH$_2$@CS, at peaks of 3500–3100 cm^{-1}, 1680–1630 cm^{-1}, 1655–1590 cm^{-1}, and 1420–1400 cm^{-1}, the vibrations correspond to N-H, C=O, N-H, and C-N stretching vibration, respectively [28].

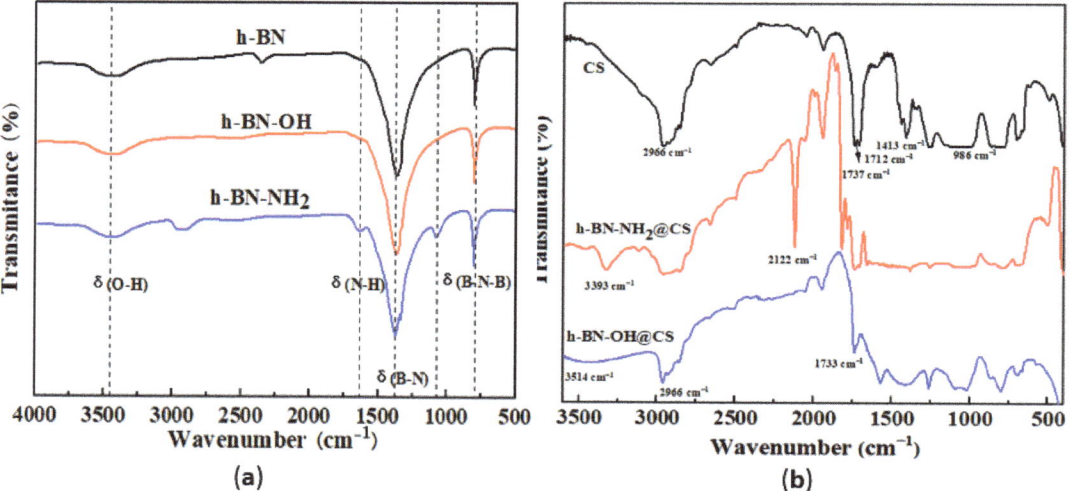

Figure 2. FTIR analyses of (**a**) h-BN, h-BN-OH, and h-BN-NH$_2$; (**b**) CS, h-BN-OH@CS, and h-BN-NH$_2$@CS.

3.2. SEM Analysis

The SEM-EDS analyses of h-BN, h-BN-OH, and h-BN-NH$_2$ are shown in Figure 3. It can be seen in Figure 3a,b that a small amount of –OH can be found on the surface of unmodified h-BN. The EDS mapping images of N and O elements in h-BN are shown in Figure 3c,d, where the percentages of B, N, and O atoms are 51.48%, 46.41%, and 2.12%, respectively. Figure 3e,f show the SEM analysis of h-BN exfoliated by NaOH. In the presence of NaOH, more -OH can be found on the surface of h-BN. The EDS mapping images of N and O elements in h-BN-OH are shown in Figure 3g,h, where the percentages of B, N, and O atoms are 50.63%, 42.66%, and 6.71%, respectively. Figure 3i,j show the

SEM analysis of h-BN-NH$_2$ treated by the ball milling method. The EDS images of N and O elements in h-BN-NH$_2$ are shown in Figure 3k,l, where the percentages of B, N, and O atoms are 50.45%, 47.37%, and 2.18%, respectively. Therefore, the surface of the ball-ground h-BN was loaded with a certain quantity of hydroxyl or amino groups.

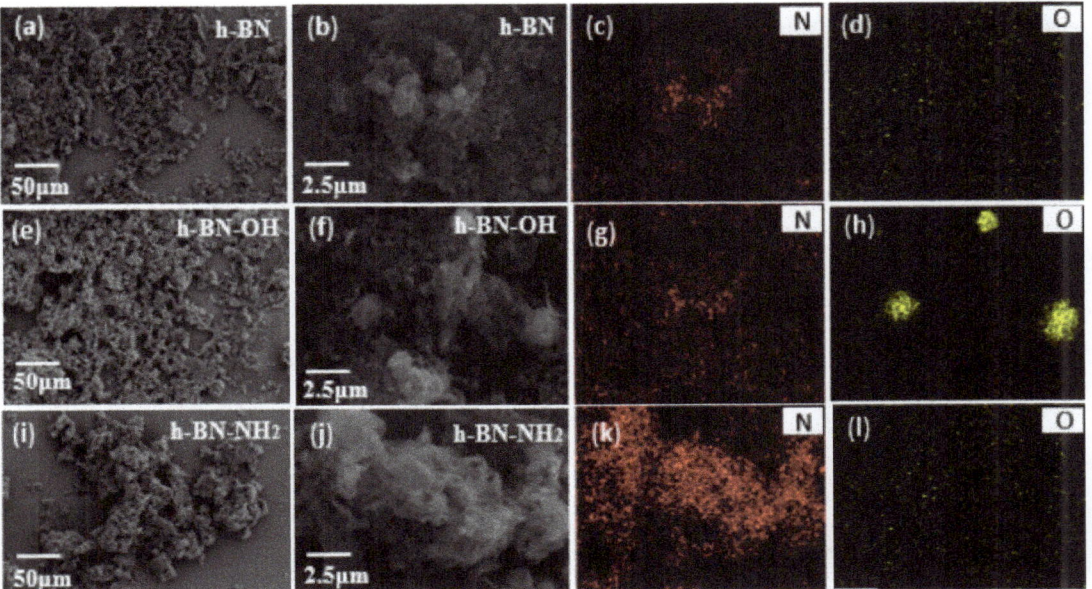

Figure 3. SEM images of (**a**,**b**) h-BN (**e**,**f**) h-BN-OH; (**i**,**j**) h-BN-NH$_2$ and EDS mapping images of the corresponding distribution of (**c**,**g**,**k**) N and (**d**,**h**,**l**) O element.

3.3. XRD Analysis

The XRD patterns of h-BN, h-BN-OH, and h-BN-NH$_2$ samples are shown in Figure 4. According to the PDF card, functionalized h-BN exhibits a decrease in the (002) peak intensity compared to pristine h-BN [29]. Additionally, the (002) peak of functionalized h-BN shifts towards lower angles, with an increase in interlayer spacing for h-BN-OH and h-BN-NH$_2$. Chemical modification and ball milling disrupt the van der Waals forces between h-BN layers, while the functional groups loaded onto h-BN increase the interlayer distance, achieving the goal of exfoliation.

3.4. TEM Analysis

The TEM images of h-BN, h-BN-OH, and h-BN-NH$_2$ are shown in Figure 5. The crystal plane spacing of unmodified h-BN is 0.341 nm (Figure 5a,b). After immersion in NaOH solution, the h-BN surface is loaded with hydroxyl groups, the crystal plane spacing of h-BN-OH is 0.344 nm (Figure 5c,d), and the crystal plane spacing of h-BN-CH is slightly greater than that of h-BN [30]. The TEM images of h-BN-NH$_2$ are shown in Figure 5e,f; h-BN after being ball-milled, h-BN surface loaded with -NH$_2$ after ball-milling by urea, and the crystal plane spacing of h-BN-NH$_2$ is 0.214 nm [31]. The relative thickness of h-BN can be visually estimated through its transparency. In Figure 5a, the h-BN shows low transparency and tends to stack with relatively thick layers. However, h-BN-OH and h-BN-NH$_2$ on the substrate tend to be transparent (Figure 5c,e), indicating the few-layer structure of functionalized h-BN [32].

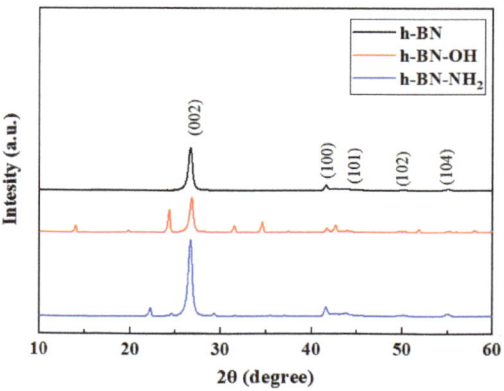

Figure 4. XRD patterns of h-BN, h-BN-OH, and h-BN-NH$_2$.

3.5. XPS Analysis

The XPS spectra of h-BN, h-BN-OH, and h-BN-NH$_2$ are shown in Figure 6. In Figure 6a, the peaks of B 1s, N 1s, C 1s, and O 1s appear at 191 eV, 398.2 eV, 285 eV, and 532 eV, respectively. The presence of carbon elements in the XPS spectra arises from adsorbed CO_2 on the h-BN surface [33]. In the B 1s and N1s spectrum (Figure 6b,c), two peaks at 190.5 and 398.2 eV correspond to the B-N and N-B bonds. As shown in Figure 6d, the peak at 192.0 eV in the B 1s spectrum is attributed to the B-O bond, indicating hydroxylation at the B site [34]. In the O 1s spectrum (Figure 6e), the main peak at 532.6 eV is caused by -OH, and the smaller peak at 533.5 eV is possibly due to the O-B bond. In N 1s spectrum (Figure 6f), the peak at 398.2 eV can be ascribed as the N-B bond in the N1s core-level spectrum, and the result demonstrates that -OH was grafted on the h-BN surface. Figure 6g,h show the B 1s and N 1s spectra of h-BN-NH$_2$, respectively. The peak at 190.5 eV in the B 1s spectrum corresponds to the B-N bond in h-BN, and the peak at 191.5 eV is possibly due to the B-O bond formed during ball milling. In the N 1s spectrum, the two peaks at 398.9 eV and 398.2 eV are attributed to the N-H and N-B bonds, respectively, indicating the loading of -NH$_2$ on the h-BN surface [35].

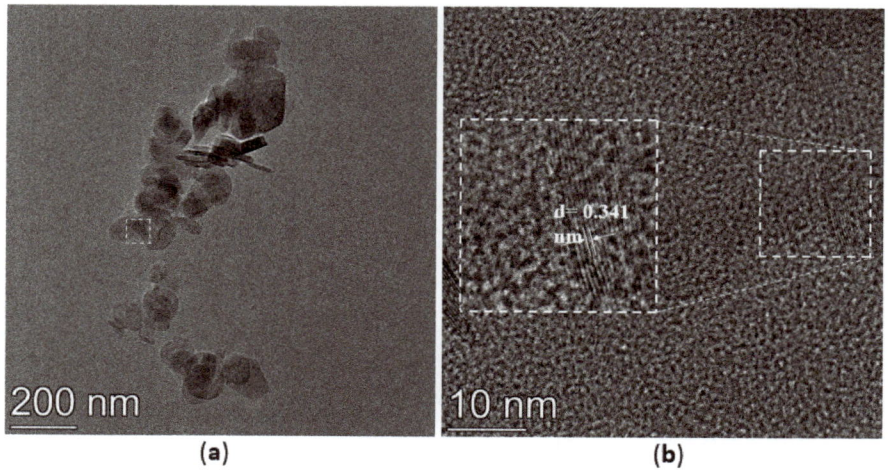

Figure 5. *Cont.*

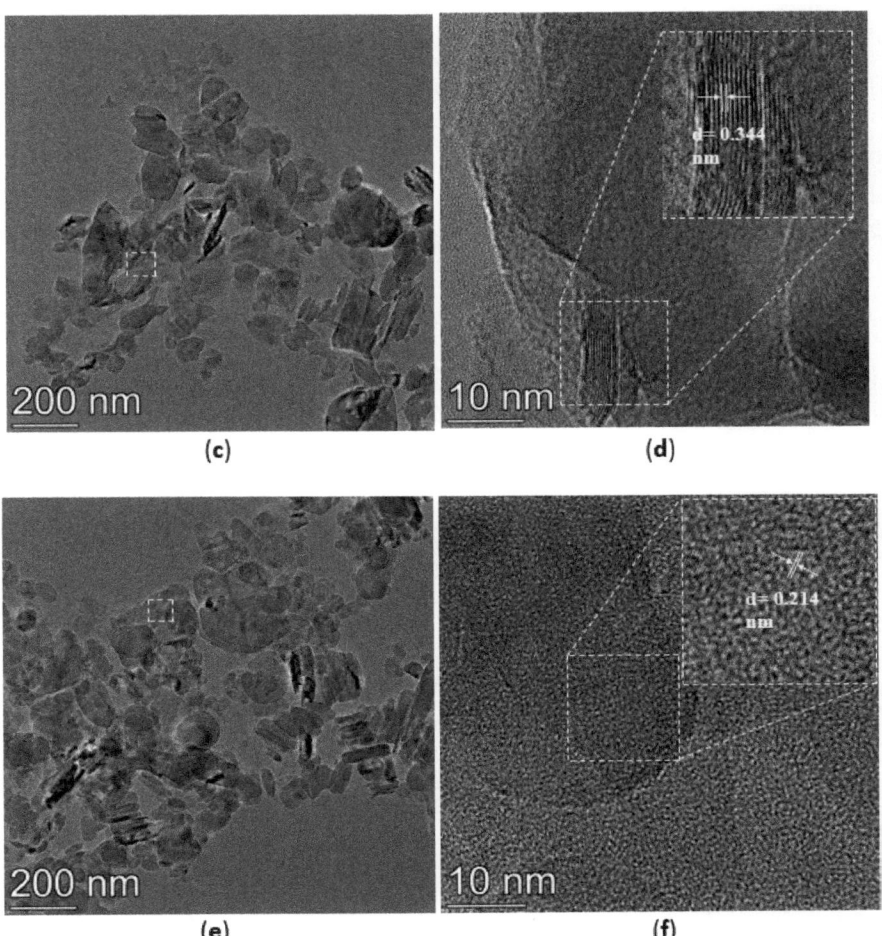

Figure 5. TEM images of (a,b) h-BN, (c,d) h-BN-OH, and (e,f) h-BN-NH$_2$.

3.6. UV-Visible Analysis

The UV-Vis spectra of h-BN, h-BN-OH, and h-BN-NH$_2$ samples are shown in Figure 7. The three samples were dispersed in water to form a dispersion solution of 1 mg mL^{-1}, which was then left to settle at room temperature for more than 48 h and centrifuged three times (1500 rpm, 15 min each time). After centrifugation, the solution was left to stand for 24 h to allow the insoluble substances to settle. By studying the UV-Visible spectra, we investigated the difference between the UV-Visible spectra of h-BN-OH, h-BN-NH$_2$, and h-BN. The h-BN had poor dispersibility in water, but functionalized h-BN was found to improve its solubility in water through UV-Visible spectroscopy [36]. The absorbance values at a wavelength of 300 nm for h-BN, h-BN-OH, and h-BN-NH$_2$ were 0.458, 0.701, and 0.759, respectively. Thus, it can be concluded that under the same preparation and testing conditions, the dispersion solubility of h-BN-OH and h-BN-NH$_2$ in water was higher than that of h-BN [37].

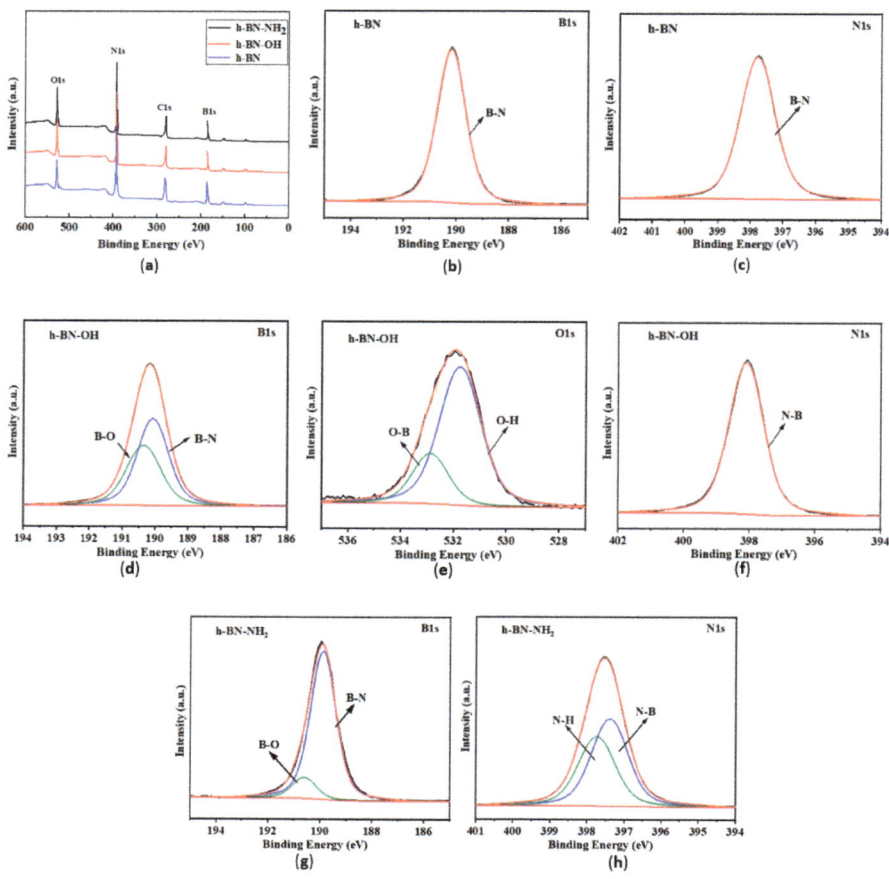

Figure 6. (a) XPS analysis of h-BN, h-BN-OH, and h-BN-NH$_2$; (b) B 1s and (c) N 1s spectrum of h-BN-OH; (d) B 1s (e) O 1s and (f) N 1s spectrum of h-BN-OH; (g) B 1s and (h) N 1s spectrum of h-BN-NH$_2$.

3.7. Thermo Gravimetric Analysis

The thermogravimetric analysis of h-BN-OH, h-BN-NH$_2$, and grafted products is shown in Figure 8. The TG curve for h-BN-OH shows a slight decrease in the 0–800 °C range, with a total weight loss of approximately 5 wt%, which can be traced back to the grafting of –OH on the h-BN surface because generally pure h-BN is highly thermal and stable below 800 °C [38]. The base silicone decomposes rapidly at 300 °C, and the decomposition rate slows down to a temperature of 600 °C. When h-BN-OH (20 wt%) was grafted with the CS, no weight loss occurred between 130 and 200 °C for h-BN-OH@CS, and no evaporation of carboxyl groups was present in the sample. As shown in the TG curve, there is a mass loss in the range of 25–50 °C and 180–200 °C for h-BN-NH$_2$, with a total weight loss of approximately 20 wt%, which was attributed to the grafting of the –NH$_2$ on the h-BN surface. h-BN-NH$_2$ (20 wt%) grafted with CS resulted in a smaller mass loss in the 0–200 °C range for h-BN-NH$_2$@CS. The thermal stability of the silicone is improved by the grafted hBN-NH$_2$. Most significantly, the presence of h-BN-NH$_2$ significantly improved the thermal stability of the CS and retarded the thermal decomposition process [39]. In summary, the h-BN-grafted thermally conductive filler can effectively enhance the thermal stability of the base silicone grease. h-BN-OH/h-BN-NH$_2$ has good compatibility with CS and forms closed chains in the matrix, limiting the thermal movement of the CS segments [40].

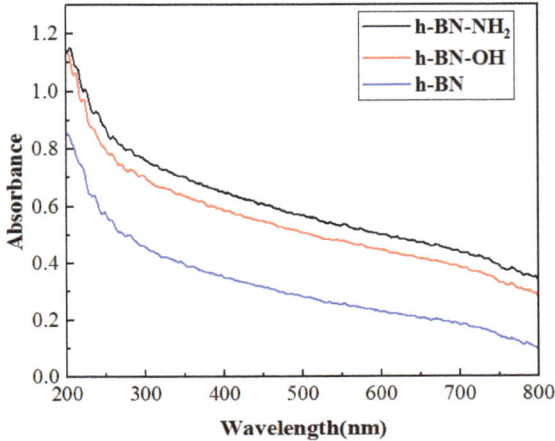

Figure 7. UV-Vis spectra of h-BN, h-BN-OH, and h-BN-NH$_2$.

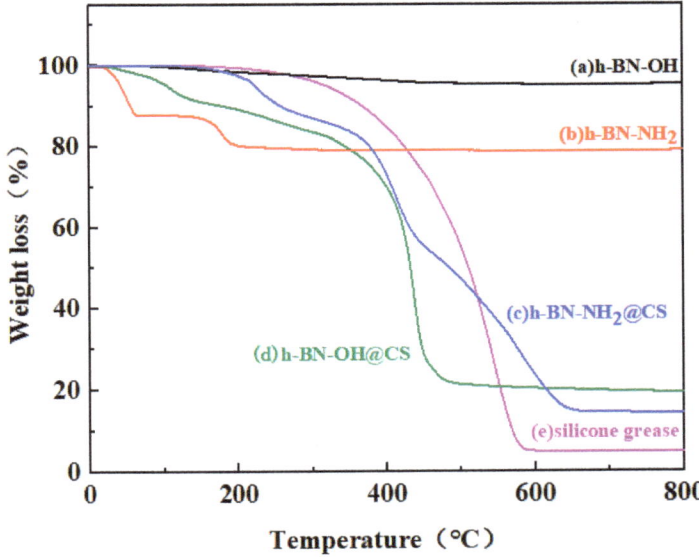

Figure 8. TG curves of (a) h-BN-OH, (b) h-BN-NH$_2$, (c) h-BN-NH$_2$@CS, (d) h-BN-OH@CS, and (e) silicone grease.

3.8. Viscosity Analysis

The variation of the viscosity of the silicone grease with the amount of filler is shown in Figure 9. The viscosity of base silicone grease is 21,220 mPa s, h-BN (10–60 wt%) was filled with the CS (h-BN@CS), and the viscosity is 27,980–73,460 mPa s when the amount of 30 wt% h-BN @CS is in the base silicone grease. h-BN-OH (10–60 wt%) was grafted with the CS (h-BN-OH@CS), and the viscosity is 25,180–63,400 mPa s when the amount of 30 wt% h-BN-OH @CS is in the base silicone grease. Compared to the unmodified h-BN, the h-BN modification reduced the viscosity of the silicone grease by 10.01–13.7%.

A total of 10–60 wt% of h-BN-NH$_2$ was grafted with a CS to form a thermally conductive filler h-BN-NH$_2$@CS, and the viscosity was 24,180–62,180 mPa s when the amount of 30 wt% h-BN-NH$_2$@CS was in the base silicone grease. This indicates that the viscosity of the silicone grease is reduced by the modified h-BN, and h-BN-NH$_2$@CS had a higher viscosity reduction than h-BN-OH@CS. It can be due to the fact that the modified grafting improved the IR of the silicone grease and increased the thermal conductivity of the base silicone grease. h-BN or modified h-BN grafting products filled into the base silicone grease showed a significant increase in viscosity, which can be due to the presence of the filler significantly changing the rheology of the base silicone grease [41]. When loaded with thermally conductive fillers, there are mutual frictional and chemical forces between the h-BN particles that can cause rheological distortions in the base silicone grease, thus increasing the viscosity of the base silicone grease [42]. The viscosity of the silicone grease is reduced by the filling of the modified h-BN. This is mainly due to the covalent bonds formed between the modified h-BN and the CS, and the forces of the covalent bonds are larger than the van der Waals gravitational forces, making it easier for the h-BN particles to be dispersed evenly in the base silicone grease, after which a reduction in IR can be obtained [43].

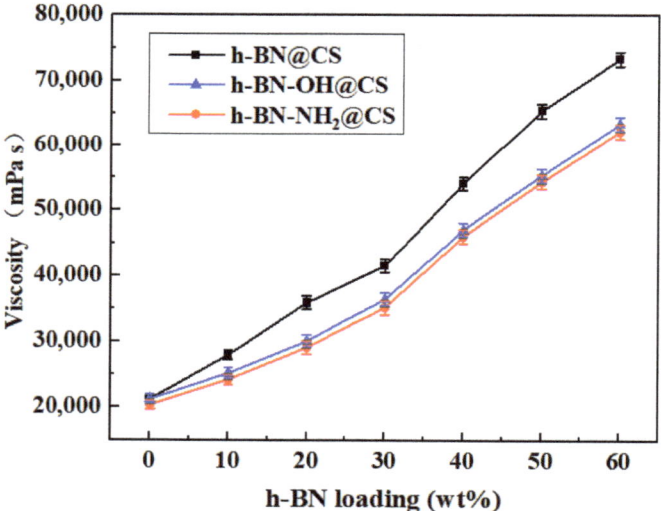

Figure 9. The viscosity of thermally conductive silicone grease varies with the h-BN@CS, h-BN-OH@CS, and h-BN-NH$_2$@CS loadings.

3.9. Thermal Conductivity and Interface Thermal Resistance

The TC and IR of h-BN@CS, h-BN-OH@CS, and h-BN-NH$_2$@CS are shown in Figure 10. The TC of the base silicone grease was 0.219 W m^{-1} K^{-1}, and 10–60 wt% of h-BN-OH was grafted with CS to form a thermally conductive filler. After the modification, the TC of the silicone grease was increased from 0.219 W m^{-1} K^{-1} to 1.324 W m^{-1} K^{-1}, and the IR was reduced from 11.699 °C W^{-1} to 1.889 °C W^{-1}. When 10–60 wt% of h-BN-NH$_2$ was grafted with a CS to form a thermally conductive filler that is filled in silicone grease, the TC of the silicone grease was increased from 0.219 W m^{-1} K^{-1} to 0.982 W m^{-1} K^{-1}, and the IR was reduced from 11.699 °C W^{-1} to 2.514 °C W^{-1}. The TC of the modified grafted silicone grease was improved, and the IR was reduced [44]. Notably, when 60 wt% of h-BN-OH was grafted with CS, the TC of the nanocomposite reached 1.324 W m^{-1} K^{-1}, which is 6.05 times that of the pure silicone grease, and the IR was reduced from 11.699 °C W^{-1} to 1.889 °C W^{-1}. This can be attributed to the strong chemical bonding between the modified thermally conductive filler, and h-BN can be evenly dispersed in silicone grease [45]. When the content of modified thermally conductive fillers in the silicone grease increases, the thermal conductivity increases [46]. However, as the amount of h-BN filler increases, the van der Waals forces between the h-BN and the matrix material become greater [47]. Being filled directly with h-BN in thermally conductive silicone, the resulting silicone is unstable. The ester and amide bonds are able to create thermally conductive channels in the silicone grease, which is beneficial to heat transfer [48].

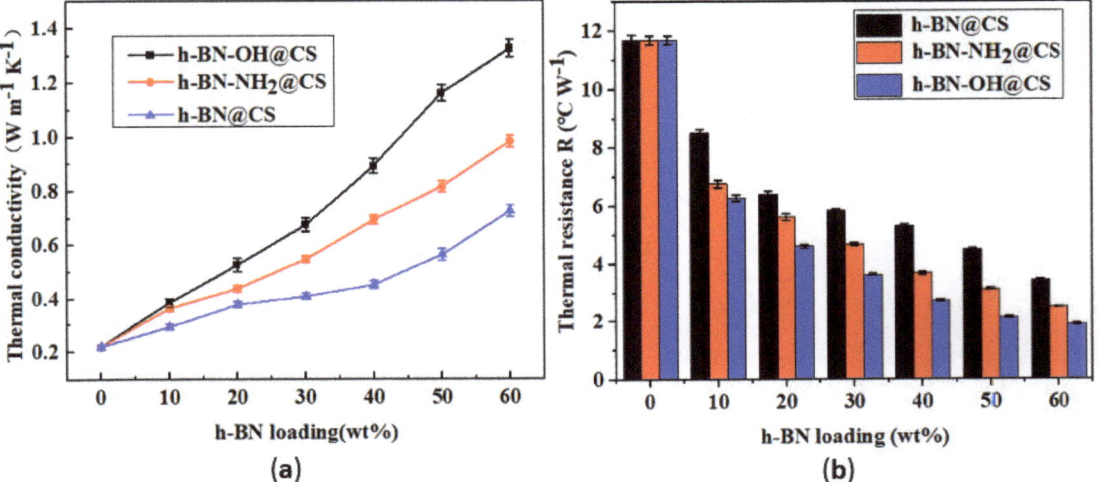

Figure 10. (a) TC of silicone grease variations with h-BN@CS, h-BN-OH@CS, and h-BN-NH$_2$@CS loadings; (b) IR of silicone grease with h-BN@CS, h-BN-OH@CS, and h-BN-NH$_2$@CS loadings.

The thermal conductivity of filled-high TC silicone grease is mainly influenced by hydrodynamics, mutual friction between filler particles, and mutual attraction between molecules [49]. Because of the hydrodynamic interaction between h-BN and silicone oil molecules, the inorganic filler h-BN interferes with the silicone grease, which leads to the distortion of the flow lines. Spherical fillers lead to less distortion of the flow lines than flake fillers. Therefore, the grafting of h-BN results in a thermally conductive filler with fewer angles and smoother edges. Furthermore, TC is also influenced by the viscosity of the thermally conductive silicone grease and the forces between the thermally conductive filler and the silicone grease molecules [50]. It was found that the formation of microscopic routes that transport heat contributes to the filling of the materials [51]. At low content, the thermally conductive filler spreads in the silicone grease, and the TC is not significantly enhanced owing to the influence of the interfacial layer among the filler that is thermally conductive and due to the heat transfer being weak [52]. When the quantity of filling is raised to a specified limit, the thermally conductive fillers commence making contact, forming a thermally conductive pathway or thermally conductive network chain. The heat flow is transferred along the path with the lowest thermal resistance, resulting in a substantial increase in the thermal conductivity of the silicone grease.

In summary, h-BN-OH and h-BN-NH$_2$ at 10–60 wt% were chemically grafted with CS, and h-BN-OH@CS and h-BN-NH$_2$@CS were used as thermally conductive fillers. When they were filled with 30 wt% into the base silicone, the TC of the silicone grease is 0.386–1.324 W m^{-1} K^{-1} and 0.364–0.982 W m^{-1} K^{-1}, respectively, which is higher than the base silicone grease (0.219 W m^{-1} K^{-1}). The IR of the base silicone grease drops from 11.699 °C W^{-1} to 1.889 °C W^{-1} and 2.514 °C W^{-1}. By analyzing the viscosity of the lubricant, it was demonstrated that there is an interaction between the modified h-BN and the base silicone grease molecules. The dispersion of the filler can be effectively improved, and thermal conductivity pathways can be formed. Then the IR was reduced, and the TC of the silicone grease was improved.

4. Conclusions

In this work, two types of thermally conductive fillers were prepared by grafting carboxylic acid silicone oil onto 10–60 wt% hydroxyl and amino-functionalized h-BN. The effect of h-BN modification on the TC of the base grease was studied, and the thermal mechanism of the thermally conductive fillers on the grease was explored. After thermal stability testing, the modified thermally conductive fillers effectively enhanced the thermal stability of base grease. The h-BN-OH@CS has better thermal stability than h-BN-NH$_2$@CS. The addition of h-BN or modified boron nitride graft products significantly increased the viscosity of the silicone grease. When the modified thermal conductive filler was added to the silicone grease, the viscosity was smaller than when unmodified h-BN particles were directly added to the grease. The viscosity-reducing effect of the h-BN-NH$_2$@CS was more significant than that of the h-BN-OH@CS. The modified thermal conductive fillers, h-BN-NH$_2$@CS and h-BN-OH@CS, effectively improved the thermal conductivity of the grease and reduced the IR. When there is less filling, the filling separates in the grease, which is not conducive to heat transfer. As the content of the modified thermally conductive filler increased, the thermal conductivity increased. However, with increasing h-BN content, the formed grease became unstable and did not easily form thermal transfer paths. By forming ester bonds and amide bonds, h-BN could form thermal conductivity paths in the silicone grease, thereby promoting heat transfer.

Author Contributions: Conceptualization, Y.W., J.Y. and S.H.; methodology, Y.W., J.Y. and S.H.; data curation, Y.W.; writing—original draft preparation, Y.W.; writing—review and editing, Y.W., M.L., J.Y. and N.S.; project administration, S.H. and J.Y.; funding acquisition, S.H. and J.Y. All authors have read and agreed to the published version of the manuscript.

Funding: The authors are grateful for the financial support of the National Natural Science Foundation of China (Nos. 22075183, 22008155, 22278269, and 21975161), Industrial Collaborative Innovation Project of Shanghai (Grant Nos. 2021-cyxt1-kj37 and XTCX-KJ-2022-70), Natural Science Foundation Project of Shanghai (Grant No. 22ZR1426100), Leading Talents Program of Shanghai (Grant No. 053), Foundation of Science and Technology Commission of Shanghai Municipality (Grant No. 22010503900), Foundation of Department of Education of Guangdong Province (Grant No. 2018KTSCX219), Foundation of Guangdong Basic and Applied Research (Grant No. 2020A1515011102).

Data Availability Statement: The data that support the findings of this study are available from the corresponding authors upon reasonable request.

Conflicts of Interest: The authors declare that they have no known competing financial interests or personal relationships that could have appeared to influence the work reported in this paper.

References

1. Liu, C.; Wu, W.; Wang, Y.; Liu, X.R.; Chen, Q.M.; Xia, S.X. Silver Nanoparticle-Enhanced Three-Dimensional Boron Nitride/Reduced Graphene Oxide Skeletons for Improving Thermal Conductivity of Polymer Composites. *ACS Appl. Polym. Mater.* **2021**, *3*, 3334–3343. [CrossRef]
2. Jiang, Z.-H.; Xue, C.-H.; Guo, X.-J.; Liu, B.-Y.; Wang, H.-D.; Fan, T.-T.; Jia, S.-T.; Deng, F.-Q. Thermally Conductive, Superhydrophobic, and Flexible Composite Membrane of Polyurethane and Boron Nitride Nanosheets by Ultrasonic Assembly for Thermal Management. *ACS Appl. Polym. Mater.* **2023**, *5*, 1264–1275. [CrossRef]
3. Liu, Z.Q.; Huang, J.H.; Cao, M.; Jiang, G.W.; Hu, J.; Chen, Q. Preparation of Binary Thermal Silicone Grease and Its Application in Battery Thermal Management. *Materials* **2020**, *13*, 4763. [CrossRef]
4. Swamy, M.C.K.; Satyanarayan. A Review of the Performance and Characterization of Conventional and Promising Thermal Interface Materials for Electronic Package Applications. *J. Electron. Mater.* **2019**, *48*, 7623–7634. [CrossRef]
5. Wu, T.T.; Hu, Y.X.; Liu, X.Q.; Wang, C.H. Effect Analysis on Thermal Management of Power Batteries Utilizing a Form-Stable Silicone Grease/Composite Phase Change Material. *ACS Appl. Energy Mater.* **2021**, *4*, 6233–6244. [CrossRef]

6. Zhang, R.C.; Huang, Z.R.; Huang, Z.H.; Zhong, M.L.; Zang, D.M.; Lu, A.; Lin, Y.F.; Millar, B.; Garet, G.; Turner, J.; et al. Uniaxially stretched polyethylene/boron nitride nanocomposite films with metal-like thermal conductivity. *Compos. Sci. Technol.* **2020**, *196*, 108154. [CrossRef]
7. Chen, L.; Xiao, C.; Tang, Y.L.; Zhang, X.; Zheng, K.; Tian, X.Y. Preparation and properties of boron nitride nanosheets/cellulose nanofiber shear-oriented films with high thermal conductivity. *Ceram. Int.* **2019**, *45*, 12965–12974. [CrossRef]
8. Yang, S.Y.; Huang, Y.F.; Lei, J.; Zhu, L.; Li, Z.M. Enhanced thermal conductivity of polyethylene/boron nitride multilayer sheets through annealing. *Compos. Part A Appl. Sci. Manuf.* **2018**, *107*, 135–143. [CrossRef]
9. Zheng, X.R.; Kim, S.; Park, C.W. Enhancement of thermal conductivity of carbon fiber-reinforced polymer composite with copper and boron nitride particles. *Compos. Part A Appl. Sci. Manuf.* **2019**, *121*, 449–456. [CrossRef]
10. Kusunose, T.; Sekino, T. Thermal conductivity of hot-pressed hexagonal boron nitride. *Scr. Mater.* **2016**, *124*, 138–141. [CrossRef]
11. Ren, J.W.; Li, Q.H.; Yan, L.; Jia, L.C.; Huang, X.L.; Zhao, L.H.; Ran, Q.C.; Fu, M.L. Enhanced thermal conductivity of epoxy composites by introducing graphene@boron nitride nanosheets hybrid nanoparticles. *Mater. Des.* **2020**, *191*, 108663. [CrossRef]
12. Chakraborty, P.; Xiong, G.P.; Cao, L.; Wang, Y. Lattice thermal transport in superhard hexagonal diamond and wurtzite boron nitride: A comparative study with cubic diamond and cubic boron nitride. *Carbon* **2018**, *139*, 85–93. [CrossRef]
13. Zheng, J.C.; Zhang, L.; Kretinin, A.V.; Morozov, S.V.; Wang, Y.B.; Wang, T.; Li, X.J.; Ren, F.; Zhang, J.Y.; Lu, C.Y.; et al. High thermal conductivity of hexagonal boron nitride laminates. *2D Mater.* **2016**, *3*, 011004. [CrossRef]
14. Lewis, J.S.; Barani, Z.; Magana, A.S.; Kargar, F.; Balandin, A.A. Thermal and electrical conductivity control in hybrid composites with graphene and boron nitride fillers. *Mater. Res. Express* **2019**, *6*, 8. [CrossRef]
15. Ou, X.H.; Chen, S.S.; Lu, X.M.; Lu, Q.H. Enhancement of thermal conductivity and dimensional stability of polyimide/boron nitride films through mechanochemistry. *Compos. Commun.* **2021**, *23*, 100549. [CrossRef]
16. Wang, S.; Xue, H.Q.; Araby, S.; Demiral, M.; Han, S.S.; Cui, C.; Zhang, R.; Meng, Q.S. Thermal conductivity and mechanical performance of hexagonal boron nitride nanosheets-based epoxy adhesives. *Nanotechnology* **2021**, *32*, 355707. [CrossRef]
17. Hutchinson, J.M.; Moradi, S. Thermal Conductivity and Cure Kinetics of Epoxy-Boron Nitride Composites—A Review. *Materials* **2020**, *13*, 3634. [CrossRef]
18. Zhong, B.; Zou, J.X.; An, L.L.; Ji, C.Y.; Huang, X.X.; Liu, W.; Yu, Y.L.; Wang, H.T.; Wen, G.W.; Zhao, K.; et al. The effects of the hexagonal boron nitride nanoflake properties on the thermal conductivity of hexagonal boron nitride nanoflake/silicone rubber composites. *Compos. Part A Appl. Sci. Manuf.* **2019**, *127*, 105629. [CrossRef]
19. Sun, N.; Sun, J.J.; Zeng, X.L.; Chen, P.; Qian, J.S.; Xia, R.; Sun, R. Hot-pressing induced orientation of boron nitride in polycarbonate composites with enhanced thermal conductivity. *Compos. Part A Appl. Sci. Manuf.* **2018**, *110*, 45–52. [CrossRef]
20. Yu, C.P.; Zhang, J.; Li, Z.; Tian, W.; Wang, L.J.; Luo, J.; Li, Q.L.; Fan, X.D.; Yao, Y.G. Enhanced through-plane thermal conductivity of boron nitride/epoxy composites. *Compos. Part A Appl. Sci. Manuf.* **2017**, *98*, 25–31. [CrossRef]
21. Ahn, H.J.; Cha, S.H.; Lee, W.S.; Kim, E.S. Effects of amphiphilic agent on thermal conductivity of boron nitride/poly(vinyl butyral) composites. *Thermochim. Acta* **2014**, *591*, 96–100. [CrossRef]
22. Yu, J.H.; Mo, H.L.; Jiang, P.K. Polymer/boron nitride nanosheet composite with high thermal conductivity and sufficient dielectric strength. *Polym. Adv. Technol.* **2015**, *26*, 514–520. [CrossRef]
23. Xiao, Q.; Zhan, C.H.; You, Y.; Tong, L.F.; Wei, R.B.; Liu, X.B. Preparation and thermal conductivity of copper phthalocyanine grafted boron nitride nanosheets. *Mater. Lett.* **2018**, *227*, 33–36. [CrossRef]
24. Han, W.F.; Chen, M.Y.; Song, W.; Ge, C.H.; Zhang, X.D. Construction of hexagonal boron nitride@polystyrene nanocomposite with high thermal conductivity for thermal management application. *Ceram. Int.* **2020**, *46*, 7595–7601. [CrossRef]
25. Guan, J.W.; Ashrafi, B.; Martinez-Rubi, Y.; Jakubinek, M.B.; Rahmat, M.; Kim, K.S.; Simard, B. Epoxy resin nanocomposites with hydroxyl (OH) and amino (NH2) functionalized boron nitride nanotubes. *Nanocomposites* **2018**, *4*, 10–17. [CrossRef]
26. Terao, T.; Bando, Y.; Mitome, M.; Zhi, C.; Tang, C.; Golberg, D. Thermal Conductivity Improvement of Polymer Films by Catechin-Modified Boron Nitride Nanotubes. *J. Phys. Chem. C* **2009**, *113*, 13605–13609. [CrossRef]
27. Jiang, Y.X.; Li, J.Q.; Leng, J.; Zhang, J. Extrusion-Based Additive Manufacturing Samples with Desirable Thermal Conductivities Prepared by Incorporating Hybrid Hexagonal Boron Nitride(h-BN) and Novel Process Strategy. *Macromol. Mater. Eng.* **2022**, *307*, 2100715. [CrossRef]
28. Yang, N.; Ji, H.F.; Jiang, X.X.; Qu, X.W.; Zhang, X.J.; Zhang, Y.; Liu, B.Y. Preparation of Boron Nitride Nanoplatelets via Amino Acid Assisted Ball Milling: Towards Thermal Conductivity Application. *Nanomaterials* **2020**, *10*, 1652. [CrossRef]
29. Kim, S.M.; Hsu, A.; Park, M.H.; Chae, S.H.; Yun, S.J.; Lee, J.S.; Cho, D.H.; Fang, W.J.; Lee, C.; Palacios, T.; et al. Synthesis of large-area multilayer hexagonal boron nitride for high material performance. *Nat. Commun.* **2015**, *6*, 8662. [CrossRef] [PubMed]
30. Hu, J.; Huang, Y.; Yao, Y.; Pan, G.; Sun, J.; Zeng, X.; Sun, R.; Xu, J.-B.; Song, B.; Wong, C.-P. Polymer Composite with Improved Thermal Conductivity by Constructing a Hierarchically Ordered Three-Dimensional Interconnected Network of BN. *ACS Appl. Mater. Interfaces* **2017**, *9*, 13544–13553. [CrossRef] [PubMed]
31. Chen, Y.M.; Gao, X.; Wang, J.L.; He, W.; Silberschmidt, V.V.; Wang, S.X.; Tao, Z.H.; Xu, H. Properties and application of polyimide-based composites by blending surface functionalized boron nitride nanoplates. *J. Appl. Polym. Sci.* **2015**, *132*, 41889. [CrossRef]
32. Ren, J.K.; Stagi, L.; Carbonaro, C.M.; Malfatti, L.; Casula, M.F.; Ricci, P.C.; Castillo, A.E.D.; Bonaccorso, F.; Calvillo, L.; Granozzi, G.; et al. Defect-assisted photoluminescence in hexagonal boron nitride nanosheets. *2D Mater.* **2020**, *7*, 045023. [CrossRef]

33. Burghaus, U. Surface chemistry of CO_2—Adsorption of carbon dioxide on clean surfaces at ultrahigh vacuum. *Prog. Surf. Sci.* **2014**, *89*, 161–217. [CrossRef]
34. Soong, Y.C.; Chiu, C.W. Multilayered graphene/boron nitride/thermoplastic polyurethane composite films with high thermal conductivity, stretchability, and washability for adjustable-cooling smart clothes. *J. Colloid Interface Sci.* **2021**, *599*, 611–619. [CrossRef]
35. Guo, F.H.; Zhao, J.; Li, F.X.; Kong, D.Y.; Guo, H.G.; Wang, X.; Hu, H.Q.; Zong, L.B.; Xu, J.T. Polar crystalline phases of PVDF induced by interaction with functionalized boron nitride nanosheets. *CrystEngComm* **2020**, *22*, 6207–6215. [CrossRef]
36. Sainsbury, T.; Satti, A.; May, P.; Wang, Z.; McGovern, I.; Gun'ko, Y.K.; Coleman, J. Oxygen Radical Functionalization of Boron Nitride Nanosheets. *J. Am. Chem. Soc.* **2012**, *134*, 18758–18771. [CrossRef] [PubMed]
37. Ren, J.; Stagi, L.; Innocenzi, P. Hydroxylated boron nitride materials: From structures to functional applications. *J. Mater. Sci.* **2021**, *56*, 4053–4079. [CrossRef]
38. Bhang, J.; Ma, H.; Yim, D.; Galli, G.; Seo, H. First-Principles Predictions of Out-of-Plane Group IV and V Dimers as High-Symmetry, High-Spin Defects in Hexagonal Boron Nitride. *Acs Appl. Mater. Interfaces* **2021**, *13*, 45768–45777. [CrossRef] [PubMed]
39. Ji, S.-Y.; Jung, H.-B.; Kim, M.-K.; Lim, J.-H.; Kim, J.-Y.; Ryu, J.; Jeong, D.-Y. Enhanced Energy Storage Performance of Polymer/Ceramic/Metal Composites by Increase of Thermal Conductivity and Coulomb-Blockade Effect. *ACS Appl. Mater. Interfaces* **2021**, *13*, 27343–27352. [CrossRef]
40. de los Reyes, C.A.; Hernández, K.; Martínez-Jiménez, C.; Walz Mitra, K.L.; Ginestra, C.; Smith McWilliams, A.D.; Pasquali, M.; Martí, A.A. Tunable Alkylation of White Graphene (Hexagonal Boron Nitride) Using Reductive Conditions. *J. Phys. Chem. C* **2019**, *123*, 19725–19733. [CrossRef]
41. Ren, L.L.; Zeng, X.L.; Sun, R.; Xu, J.B.; Wong, C.P. Spray-assisted assembled spherical boron nitride as fillers for polymers with enhanced thermally conductivity. *Chem. Eng. J.* **2019**, *370*, 166–175. [CrossRef]
42. Yazdan, A.; Wang, J.Z.; Nan, C.W.; Li, L.L. Rheological Behavior and Thermal Conductivities of Emulsion-Based Thermal Pastes. *J. Electron. Mater.* **2020**, *49*, 2100–2109. [CrossRef]
43. Terao, T.; Zhi, C.; Bando, Y.; Mitome, M.; Tang, C.; Golberg, D. Alignment of Boron Nitride Nanotubes in Polymeric Composite Films for Thermal Conductivity Improvement. *J. Phys. Chem. C* **2010**, *114*, 4340–4344. [CrossRef]
44. Huang, X.; Iizuka, T.; Jiang, P.; Ohki, Y.; Tanaka, T. Role of Interface on the Thermal Conductivity of Highly Filled Dielectric Epoxy/AlN Composites. *J. Phys. Chem. C* **2012**, *116*, 13629–13639. [CrossRef]
45. Jiang, Y.; Yilmaz, N.E.D.; Barker, K.P.; Baek, J.; Xia, Y.; Zheng, X.L. Enhancing Mechanical and Combustion Performance of Boron/Polymer Composites via Boron Particle Functionalization. *ACS Appl. Mater. Interfaces* **2021**, *13*, 28908–28915. [CrossRef]
46. Yang, W.; Wang, Y.F.; Li, Y.; Gao, C.; Tian, X.J.; Wu, N.; Geng, Z.S.; Che, S.; Yang, F.; Li, Y.F. Three-dimensional skeleton assembled by carbon nanotubes/boron nitride as filler in epoxy for thermal management materials with high thermal conductivity and electrical insulation. *Compos. Part B Eng.* **2021**, *224*, 109168. [CrossRef]
47. Xu, C.Y.; Miao, M.; Jiang, X.F.; Wang, X.B. Thermal conductive composites reinforced via advanced boron nitride nanomaterials. *Compos. Commun.* **2018**, *10*, 103–109. [CrossRef]
48. Xu, C.K.; Wei, C.M.; Li, Q.H.; Li, Z.H.; Zhang, Z.X.; Ren, J.W. Robust Biomimetic Nacreous Aramid Nanofiber Composite Films with Ultrahigh Thermal Conductivity by Introducing Graphene Oxide and Edge-Hydroxylated Boron Nitride Nanosheet. *Nanomaterials* **2021**, *11*, 2544. [CrossRef] [PubMed]
49. Yang, S.; Huang, Z.; Hu, Q.; Zhang, Y.; Wang, F.; Wang, H.; Shu, Y. Proportional Optimization Model of Multiscale Spherical BN for Enhancing Thermal Conductivity. *ACS Appl. Electron. Mater.* **2022**, *4*, 4659–4667. [CrossRef]
50. Li, C.N.; Cao, X.W.; Tong, Y.Z.; Yang, Z.T.; Gao, D.L.; Ru, Y.; He, G.J. Hybrid Filler with Nanoparticles Grown in Situ on the Surface for the Modification of Thermal Conductive and Insulating Silicone Rubber. *ACS Appl. Polym. Mater.* **2022**, *4*, 7152–7161. [CrossRef]
51. Christensen, G.; Lou, D.; Hong, H.P.; Peterson, G.P. Improved thermal conductivity of fluids and composites using boron nitride (BN) nanoparticles through hydrogen bonding. *Thermochim. Acta* **2021**, *700*, 178927. [CrossRef]
52. Wei, Q.G.; Yang, D. Formation of Thermally Conductive Network Accompanied by Reduction of Interface Resistance for Thermal Conductivity Enhancement of Silicone Rubber. *ACS Appl. Electron. Mater.* **2022**, *4*, 3503–3511. [CrossRef]

Disclaimer/Publisher's Note: The statements, opinions and data contained in all publications are solely those of the individual author(s) and contributor(s) and not of MDPI and/or the editor(s). MDPI and/or the editor(s) disclaim responsibility for any injury to people or property resulting from any ideas, methods, instructions or products referred to in the content.

Communication

Tribological Property of Al$_3$BC$_3$ Ceramic: A Lightweight Material

Jinjun Lu [1,2,*], Rong Qu [1], Fuyan Liu [2,3], Tao Wang [2,4], Qinglun Che [2,5], Yanan Qiao [1] and Ruiqing Yao [1,2,*]

1. Key Laboratory of Synthetic and Natural Functional Molecule of Ministry of Education, College of Chemistry and Materials Science, Northwest University, Xi'an 710127, China; 202110204@stumail.nwu.edu.cn (R.Q.); 18435996130@163.com (Y.Q.)
2. State Key Laboratory of Solid Lubrication, Lanzhou Institute of Chemical Physics, Chinese Academy of Sciences, Lanzhou 730000, China; liufy@czu.cn (F.L.); tao.wang@biam.ac.cn (T.W.); cheqinglun@163.com (Q.C.)
3. Changzhou Institute of Technology, Changzhou 213002, China
4. Beijing Institute of Aeronautical Materials, Beijing 100095, China
5. School of Mechanical and Automotive Engineering, Qingdao University of Technology, Qingdao 266520, China
* Correspondence: jjlu@nwu.edu.cn (J.L.); yaoruiqing@nwu.edu.cn (R.Y.)

Abstract: Lightweight materials with a density less than 3 g/cm^3 as potential tribo-materials for tribological applications (e.g., space tribology) are always desired. Al$_3$BC$_3$ ceramic, a kind of ternary material, is one of the lightweight materials. In this study, dense Al$_3$BC$_3$ ceramic is prepared via a reactive hot-pressing process in a vacuum furnace. Its tribological properties are investigated in two unlubricated conditions (one is at elevated temperature up to 700 °C in air, and another is in a vacuum chamber of back pressures from 10^5 Pa to 10^{-2} Pa at room temperature) and lubricated conditions (i.e., water and ethanol as low-viscosity fluids). At 400 °C and lower temperatures in air, as well as in vacuum, the tribological property of Al$_3$BC$_3$ ceramic is poor due to the fracture of grains and formation of a mechanically mixed layer. The beneficial influence of adsorbed gas species on reducing friction is very limited. Due to the formation of lubricious tribo-oxide at 600 °C and 700 °C, the friction coefficient is reduced from ca. 0.9 at room temperature and 400 °C to ca. 0.4. In the presence of low-viscosity fluids, a high friction coefficient and wear but a polished surface are observed in water, while a low friction coefficient and wear occur in ethanol. A lubricious carbide-derived carbon (CDC) coating on top of Al$_3$BC$_3$ ceramic through high-temperature chlorination can be fabricated and the wear resistance of CDC can be improved by adjusting the chlorination parameters. The above results suggest that Al$_3$BC$_3$ ceramic is a potential lubricating material for some tribological applications.

Keywords: Al$_3$BC$_3$ ceramic; lightweight; wear and tribology; liquid lubrication; self-lubrication

1. Introduction

There is ongoing demand for light-weight lubricating materials for many tribological applications (e.g., gyro for space tribology). Because of its low density (2.66 g/cm^3) and high elastic modulus (137 GPa), Al$_3$BC$_3$ ceramic is an attractive material for light-weight structural components [1–3]. From the viewpoint of tribology, lamellar Al$_3$BC$_3$ ceramic is also considered as an attractive candidate for tribological components under both unlubricated and lubricated conditions. For example, based on the ratio of hardness to Young's modulus (Hv/E = 0.068), Al$_3$BC$_3$ ceramic is predicted to have excellent wear resistance at room temperature [1]. In addition, the stiffness of Al$_3$BC$_3$ ceramic at 1600 °C is 79%, as high as that at room temperature, which is much higher than that of Ti$_3$SiC$_2$ [1]. This suggests that Al$_3$BC$_3$ ceramic is a candidate for the production of high-temperature wear-resistant materials [1–4].

Up until now, there have been no published reports on the fundamental tribological property of Al$_3$BC$_3$ ceramic under both unlubricated conditions (e.g., air bearings for gyro

Citation: Lu, J.; Qu, R.; Liu, F.; Wang, T.; Che, Q.; Qiao, Y.; Yao, R. Tribological Property of Al$_3$BC$_3$ Ceramic: A Lightweight Material. *Lubricants* **2023**, *11*, 492. https://doi.org/10.3390/lubricants11110492

Received: 20 October 2023
Revised: 12 November 2023
Accepted: 13 November 2023
Published: 14 November 2023

Copyright: © 2023 by the authors. Licensee MDPI, Basel, Switzerland. This article is an open access article distributed under the terms and conditions of the Creative Commons Attribution (CC BY) license (https://creativecommons.org/licenses/by/4.0/).

and lubricated conditions (e.g., bearings and surface finishing). As a result, the application of Al_3BC_3 ceramic is greatly hindered due to the lack of basic data of the tribological property. Currently, a tentative exploration of the tribological properties of Al_3BC_3 seems to be the first step.

The importance of the tribological properties in air at elevated temperatures and in vacuum is highly valued due to the physical, chemical, and mechanical properties of Al_3BC_3 ceramic. In addition, technical solutions to obtain low friction of the Al_3BC_3 ceramic by liquid lubrication or solid lubrication are no doubt necessary in some cases. It is well known that water and ethanol are environmentally friendly and efficient as a lubricant and a polishing agent as well. A carbide-derived carbon coating on top of Al_3BC_3 ceramic is also expected to be an effective method of self-lubrication. In this connection, the fundamental tribological property, as well as the liquid lubrication and self-lubrication of Al_3BC_3 ceramic are presented in this study.

2. Materials and Methods

2.1. Preparation of Al_3BC_3 Ceramic and Carbide-Derived Carbon Coating

Lamellar Al_3BC_3 ceramic is prepared by reactive hot-pressing at 1800 °C and 15 MPa starting from powders of Al, B_4C, and graphite in a vacuum furnace. The powders of Al, B_4C, and graphite (−200 mesh) with a mole proportion of 3.2:0.25:2.30 are used to prepare Al_3BC_3 ceramic with low impurities. Al_3BC_3 has a lamellar structure (see Figure 1). The Al_3BC_3 ceramic for the tribological tests is dense with a porosity less than 2%. Some of the physical and mechanical properties of Al_3BC_3 ceramic include a density of 2.56 g/cm^3, three-point bending strength of 175 MPa, microhardness (10 gf) of 12.1 GPa, and indentation fracture toughness of 2.1 MPa·m$^{1/2}$.

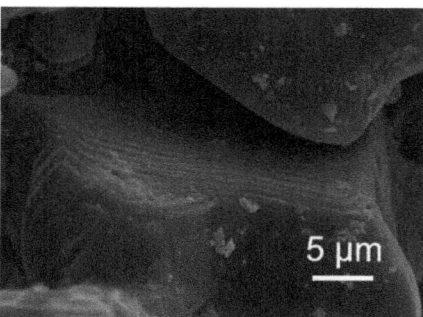

Figure 1. SEM micrograph of the edge of Al_3BC_3 grains.

Carbide-derived carbon (CDC) coating on top of Al_3BC_3 ceramic is prepared via a high-temperature chlorination process [5]. By varying the concentration of Cl_2 in a Cl_2 + Ar mixture and the duration of chlorination, two CDC coatings are prepared. Schedule A is prepared in 6.67% Cl_2 + Ar at 800 °C for 2 h and schedule B is a two-step process of 6.67% Cl_2 + Ar at 800 °C for 20 min and 3.33% Cl_2 + Ar at 800 °C for 1 hr. Dechlorination is not conducted and residual adsorbed Cl_2 is not removed. The thickness of the CDC coating is 20 to 30 μm. The advantage of CDC coating is its low internal stress [6]. The microstructure of the CDC coating on top of the Al_3BC_3 ceramic is different from that of the CDC coating on SiC and Ti_3SiC_2 (see Figure 2). A possible chemical reaction for high-temperature chlorination is as follows:

$$Al_3BC_3(s) + Cl_2(g) \rightarrow AlCl_3(g) + BCl_4(g) + C(s), \qquad (1)$$

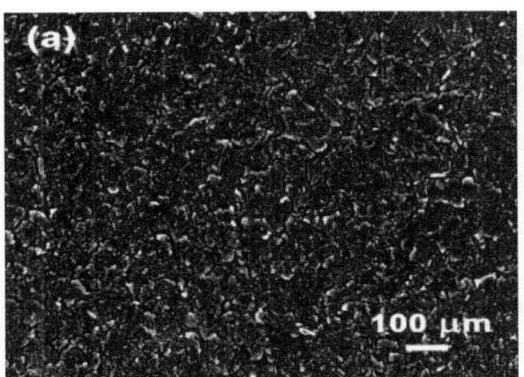 

Figure 2. (a) Microstructure of CDC coating under low magnification; (b) the white particles are AlCl$_3$ (EDS spectrum in (b)).

2.2. Tribological Tests

Two kinds of tribo-meters are employed to conduct tribological tests. A CSM THT tribo-meter with a ball-on-disk configuration is used for the evaluation of tribological properties (1) at high temperature, (2) in liquid lubrication, and (3) in the presence of a CDC coating. A home-made vacuum friction tester with a ball-on-disk configuration is used for tribological tests in vacuum. A G3 grade Si$_3$N$_4$ ball with a diameter of 3.175 mm is used for both tribo-meters. Fluctuation of the friction coefficient data of a given tribo-couple is frequently observed in this study, and it depends on the materials and assembly of the tribo-couple and sensitivity of the sensor to detect a friction force.

2.2.1. CSM THT Tribo-Meter with a Ball-on-Disk Configuration

A silicon nitride ball (3.175 mm in diameter) sliding on a bare Al$_3$BC$_3$ ceramic disk or CDC on Al$_3$BC$_3$ ceramic disk (25 mm in diameter and 3 mm in thickness) is used. For liquid lubrication, the disk is immersed in either flooded distilled water or flooded ethanol. Test conditions are 0.1 m/s for sliding speed, 5 N for normal load, temperature from 25 °C (for liquid lubrication and CDC coating) to 400 °C, 600 °C, and 700 °C. The wear volume is determined by a surface stylus profiler.

2.2.2. A Home-Made Tribo-Meter with a Ball-on-Disk Configuration

A silicon nitride ball (3.175 mm in diameter) sliding on a bare Al$_3$BC$_3$ ceramic disk is used. The test begins at a back pressure of 10^5 Pa for 10 min. After evacuation to a back pressure of 10^4 Pa, the test starts by using the same Si$_3$N$_4$ ball on the same wear track. The same process is repeated for back pressures of 10^3 Pa, 10^2 Pa, 10^1 Pa, 100 Pa, 10^{-1} Pa, and 10^{-2} Pa. The test conditions are 10 rpm rotation speed at a diameter of 12 mm, 2 N for normal load, and room temperature. The wear volume is determined by a surface stylus profiler.

3. Results

3.1. Tribological Property at Elevated Temperatures and in Vacuum

At room temperature, a stage of low friction coefficient (SLFC, ca. 0.2 in vacuum test, Figure 3a and ca. 0.4 in tests of high-temperature series, Figure 3b) before a stage of high friction coefficient (SHFC) is observed. In Figure 3a, such a SLFC occurs at back pressures of 10^5 Pa and 10^4 Pa, while no such SHFC occurs at back pressures of 10^3 Pa and lower. This implies that the SLFC is associated with an adsorption and desorption mechanism of gas species on the worn surface. In the vacuum test, a tribological test is begun at a back pressure of 10^5 Pa. After a ten-minute sliding test, the chamber is evacuated to a back pressure of 10^4 Pa and the re-adsorption of gas species on the sliding surface occurs. This enables a SLFC stage at 10^4 Pa. At back pressures of 10^3 Pa and lower, the time for the re-adsorption of gas species might be too long to fulfill multi-layer adsorption. Therefore, SHFC occurs at the very beginning of the sliding at back pressures of 10^3 Pa and lower.

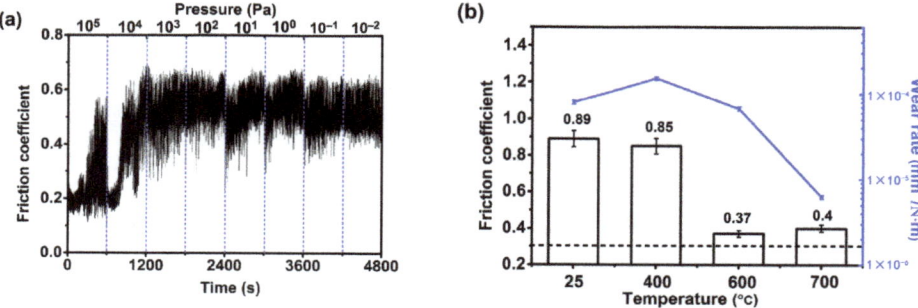

Figure 3. (a) Frictional trace of Al_3BC_3 ceramic in vacuum at room temperature; (b) average friction coefficient and wear rate at elevated temperatures in air. The black dash line is the average friction coefficient of PM304.

The SLFC is not observed due to the desorption of gas species at 400 °C together with a lack of lubricious tribo-oxide, Figures 3b and 4. The low friction coefficients at 600 °C and 700 °C are mainly attributed to tribo-oxides (B_2O_3). On one hand, the ionic potential of B_2O_3 is as high as 12, suggesting that B_2O_3 is a good high-temperature solid lubricant [7]. On the other hand, B_2O_3 is a glaze substance at 600 °C and 700 °C.

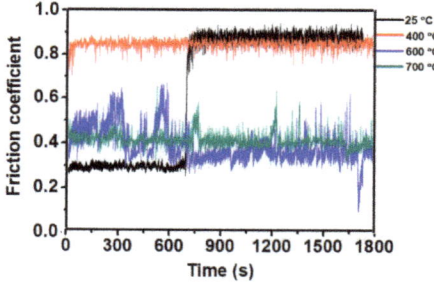

Figure 4. Frictional traces of Al_3BC_3 ceramic disk in sliding against a Si_3N_4 ball at elevated temperatures.

In summary, the tribological property of Al_3BC_3 ceramic in sliding against a Si_3N_4 ball is not plausible in vacuum at room temperature or in air at temperatures of 400 °C and lower. At temperatures of 600 °C and 700 °C, the tribological property of Al_3BC_3 ceramic is much better. Notably, the friction coefficients of Al_3BC_3 ceramic at 600 °C and 700 °C are comparable to that of a high temperature self-lubricating composite PM304 at 630 °C (Figure 3b, data from the authors' laboratory). The frictional traces at elevated temperatures in air can be found in Figure 4.

3.2. Lubrication in Water and Ethanol

It is clear that the high friction (Figure 5a) and severe wear (Figure 5b) of Al_3BC_3 ceramic and its counterpart material under an unlubricated condition can be reduced by using either flooded water or flooded ethanol as a lubricant. Specifically, Figure 5a suggests that under unlubricated conditions, a SLFC stage occurs (the same reason as the above-mentioned) and the friction coefficient increases gradually to a high friction coefficient of 0.6. It is obvious that the fracture of grains and formation of a discontinuous mechanically mixed layer (Figure 5c) are the two main characteristics of the worn surface of Al_3BC_3 ceramic under unlubricated sliding. The fracture of grains and formation of such a tribo-layer on Al_3BC_3 ceramic are commonly found in ceramics, including MAX phase materials [8,9].

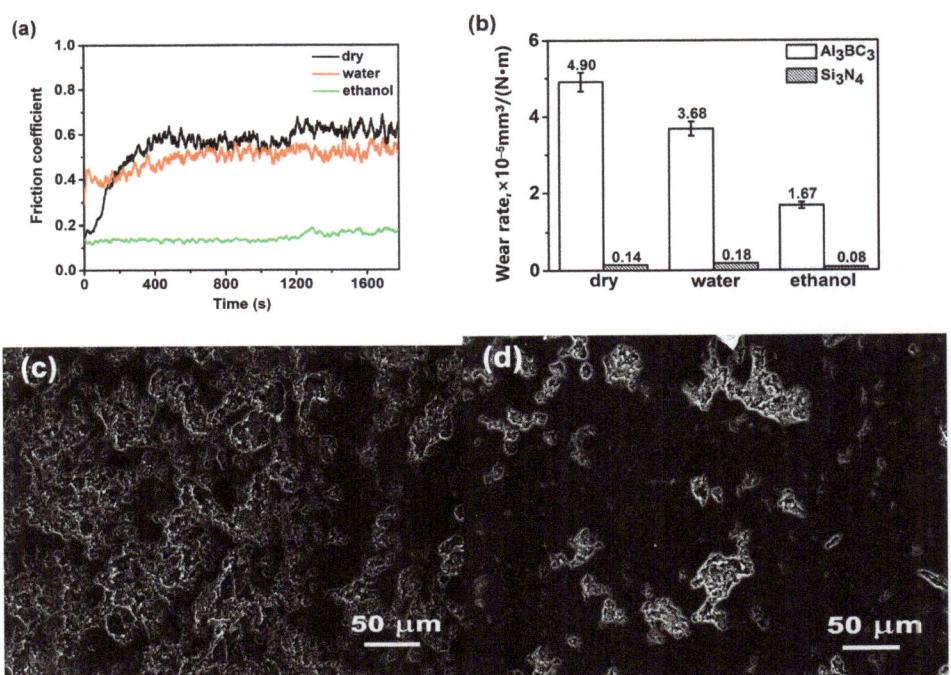

Figure 5. (a) Frictional traces of Al_3BC_3 ceramic under unlubricated and lubricated conditions; (b) wear rates of Al_3BC_3 ceramic and Si_3N_4 ceramic under unlubricated and lubricated conditions; (c) SEM micrograph of the worn surfaces of Al_3BC_3 ceramic under unlubricated conditions; (d) SEM micrograph of the worn surfaces of Al_3BC_3 ceramic in water.

Although no such SLFC and friction coefficient are stably as high as ca. 0.5 in distilled water, the worn surface is smooth as a result of a polishing effect. Water, at the least, can be a good polishing agent for the planarization of Al_3BC_3 ceramic. Additionally, it is reasonable that a well-polished surface enables the application of self-lubricating film (e.g., diamond-like carbon film and graphene film) in good quality.

Ethanol, as a good lubricant, significantly reduces the friction coefficient, which is observed for many ceramics [9]. Some reports emphasize the role of tribo-chemical products, while some focus on the role of the physical adsorption of ethanol on the sliding surface in the lubricating behavior of ethanol. It is an interesting but also challenging topic.

It is well known that water and ethanol have almost identical viscosities at room temperature. Therefore, the role of viscosity for the different tribological properties in water and in ethanol is insignificant. The water dissolution of friction-induced fresh asperities might be a good reason for the polished worn surface. A low and stable friction coefficient suggests that the robust physical adsorption of ethanol on the sliding surface rather than tribochemistry is the main reason. In case of friction dominated by tribochemistry, the friction coefficient will decrease with sliding duration.

3.3. Self-Lubrication by CDC Coating

As seen in Figure 6a, CDC coatings significantly reduce the friction coefficient to 0.3 or even lower. The problem of the CDC coating is its relatively high wear rate (see Figure 6a). Similar results are found for CDC coating on top of Ti_3SiC_2. The wear resistance of CDC can be improved by adjusting the chlorination parameters (see Figure 6a). As seen in Figure 6b, the two CDC coatings are composed of poly-crystalline graphite, similar to Ti_3SiC_2 [5]. In addition, schedule A produces a 'thicker' CDC coating while schedule B produces a 'thinner' CDC coating. That might be the reason for the better wear resistance by schedule B.

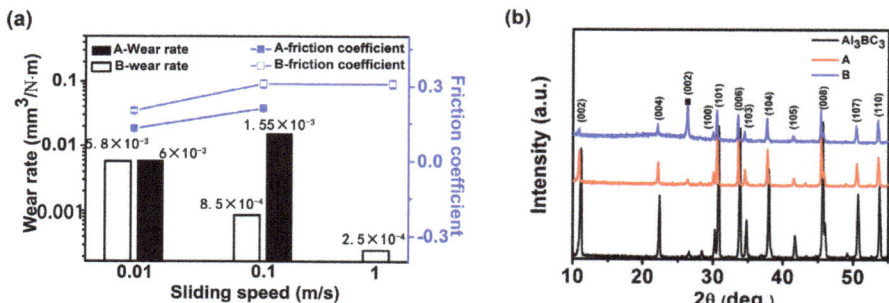

Figure 6. (a) Friction coefficient and wear rate of CDC coatings prepared using different chlorination schedules; (b) XRD patterns of Al_3BC_3 ceramic, CDC coating by schedule A, and CDC coating by schedule B. < indicates the (002) peak of graphite. Small amount of graphite in Al_3BC_3 ceramic.

The CDC coating is a highly porous coating of low density. The porosity of CDC is higher than 50% [5,6]. Therefore, it will not increase the density of the Al_3BC_3 ceramic. In addition, the internal stress of the CDC coating is much lower than diamond-like carbon film, which represents an advantage over the diamond-like carbon film. This means that the thickness of the CDC coating can be as high as several millimeters without risk of peeling [6].

4. Discussion

Al_3BC_3 ceramic behaves like brittle ceramics (i.e., fracture of grains rather than plastic deformation) under tribological loading. Al_3BC_3 ceramic is not a plausible lubricating material in vacuum at room temperature as well as at temperatures lower than 600 °C. Tribo-oxidation (glazed B_2O_3) in oxidizing atmosphere renders Al_3BC_3 ceramic with reduced friction and wear at 600 °C and 700 °C. Liquid lubrication by low-viscosity fluids and solid lubrication by carbon coating suggest that Al_3BC_3 ceramic can be used as a tribological component, either by liquid lubrication or by surface modification. The surface chemistry of Al_3BC_3 ceramic in various atmospheres and liquids is the key to understanding its tribological behavior. There are still many topics that deserve further investigation. For example, Al_3BC_3 ceramic with a low-friction coating (e.g., diamond-like carbon film) can be a candidate material for fabricating gyro for space tribology.

Author Contributions: Investigation, F.L., T.W. and Q.C.; data curation, F.L., T.W., Q.C. and Y.Q.; methodology, R.Q.; investigation, R.Q., F.L., T.W. and Q.C.; writing—original draft preparation, J.L.; writing—review and editing, J.L. and R.Y.; supervision, J.L.; funding acquisition, J.L. All authors have read and agreed to the published version of the manuscript.

Funding: This work is supported by Natural Science Foundation of China (51775434), Hundred Talent of Shaanxi Province.

Data Availability Statement: The data that support the findings of this study are available from the corresponding author upon reasonable request.

Conflicts of Interest: The authors declare no conflict of interest.

References

1. Li, F.Z.; Zhou, Y.C.; He, L.F.; Liu, B.; Wang, J.Y. Synthesis, microstructure, and mechanical properties of Al_3BC_3. *J. Am. Ceram. Soc.* **2008**, *91*, 2343–2348. [CrossRef]
2. Lee, S.H.; Tanaka, H. Thermal stability of Al_3BC_3. *J. Am. Ceram. Soc.* **2009**, *92*, 2172–2174. [CrossRef]
3. Wang, J.Y.; Zhou, Y.C.; Liao, T.; Lin, Z.J. First-principles prediction of low shear-strain resistance of Al_3BC_3: A metal borocarbide containing short linear BC2 units. *Appl. Phys. Lett.* **2006**, *89*, 021917. [CrossRef]
4. Dong, B.; Deng, C.J.; Di, J.H.; Ding, J.; Zhu, Q.Y.; Liu, H.; Zhu, H.X.; Yu, C. Oxidation behavior of Al_3BC_3 powders at 800–1400 °C in ambient air. *J. Mater. Res. Technol.* **2023**, *23*, 670–679. [CrossRef]
5. Sui, J.; Lu, J. Formulated self-lubricating carbon coatings on carbide ceramics. *Wear* **2011**, *271*, 1974–1979. [CrossRef]
6. Nikitin, A.; Gogotsi, Y. *Nanostructured Carbide-Derived Carbon: Encyclopedia of Nanoscience and Nanotechnology*, 1st ed.; American Scientific Publisher: Valencia CA, USA, 2004; pp. 1–22.
7. Erdemir, A. A crystal chemical approach to the formulation of self-lubricating nanocomposite coatings. *Surf. Coat. Technol.* **2005**, *200*, 1792–1796. [CrossRef]
8. Gupta, S.; Barsoum, M.W. On the tribology of the MAX phases and their composites during dry sliding: A review. *Wear* **2011**, *271*, 1878–1894. [CrossRef]
9. Hibi, Y.; Miyake, K.; Murakami, T.; Sasaki, S. Tribological behavior of SiC-reinforced Ti_3SiC_2-based composites under dry condition and under lubricated condition with water and ethanol. *J. Am. Ceram. Soc.* **2006**, *89*, 2983–2985. [CrossRef]

Disclaimer/Publisher's Note: The statements, opinions and data contained in all publications are solely those of the individual author(s) and contributor(s) and not of MDPI and/or the editor(s). MDPI and/or the editor(s) disclaim responsibility for any injury to people or property resulting from any ideas, methods, instructions or products referred to in the content.

Article

Tribological Performance of Steel/W-DLC and W-DLC/W-DLC in a Solid–Liquid Lubrication System Additivated with Ultrathin MoS$_2$ Nanosheets

Meirong Yi [1,2], Taoping Wang [1,2], Zizheng Liu [1,2], Jin Lei [1,2], Jiaxun Qiu [1,2] and Wenhu Xu [1,2,*]

[1] School of Advanced Manufacture, Nanchang University, Nanchang 330031, China; yimr18@ncu.edu.cn (M.Y.); 18779475197@163.com (T.W.); 15079551992@163.com (Z.L.); 15946993985@163.com (J.L.); 17746676641@163.com (J.Q.)
[2] Key Laboratory of Tribology, Nanchang University, Nanchang 330031, China
* Correspondence: flipreverse@126.com

Abstract: In this paper, MoS$_2$ nanosheets with an ultrathin structure were fabricated using a solvothermal method and further added into PAO oil, which was further combined with W-DLC coating to constitute a solid–liquid lubricating state. The influences of MoS$_2$ concentration, applied load and counter surfaces on the lubricating of the solid–liquid hybrid lubricating system were explored through a ball-on-disk tribometer. The friction results indicated that the steel/W-DLC and W-DLC/W-DLC tribopairs lubricated with ultrathin MoS$_2$ possessed better friction reduction and wear resistance behaviors in comparison to pure PAO oil. However, compared to the steel/steel couple case, the prepared MoS$_2$ nanosheets exhibited a more efficient lubricating effect for the W-DLC/W-DLC couple. The beneficial boundary lubricating impact of MoS$_2$ nanosheets on self-mated W-DLC coated rubbing surfaces could be attributed to the tribochemical reaction between MoS$_2$ and doping W element in DLC, resulting in a formation of a thin tribofilm at both counterparts. Meanwhile, the extent of graphitization of W-DLC film induced by friction was alleviated because of the lubrication and protection from the formation of MoS$_2$-based tribofilm at both counterparts.

Keywords: W-DLC coating; ultrathin MoS$_2$ nanosheets; lubricating additive; solid–liquid composite lubrication; friction; wear

1. Introduction

Solid–liquid composite lubricating systems have always been an important research focus, because they can largely offer excellent tribological property and undertake serious operating environments [1–3]. Diamond-like carbon (DLC) coating is a solid lubrication material with outstanding worn resistance, low friction coefficient and high hardness, which makes it suitable for protective films for multifarious mechanical parts [4,5]. Since traditional lubricant additives have been developed and customized for lubricating metallic surfaces, there was no consistent opinion concerning the lubrication efficacy of lubricating additives on the DLC coated surfaces. Tannous et al. [6] have compared the tribological properties of IF-MoS$_2$ in different tribopairs of steel/steel, alumina/alumina and DLC/DLC, respectively. They have suggested that the friction reduction behavior of MoS$_2$ particles was only achieved for steel surfaces, with no effect for other friction pairs of alumina or DLC. Meanwhile, they have also pointed out that a MoS$_2$-based tribofilm was merely found on the rubbed ferrous surfaces due to the occurrence of a tribochemical reaction between MoS$_2$ and the iron-based surfaces. However, Kalin et al. [7] have obtained a positive lubrication effect for the self-mated DLC-coated surfaces through utilizing MoS$_2$ nanotubes as oil additives. They have attributed this to the fact that MoS$_2$ flakes physically attached to the DLC-coated surfaces, which contributed to a formation of thin tribofilm at both counterparts. It can be seen that some conflicting opinions on the lubrication efficacy of DLC-coated friction surfaces lubricated with nanoparticle additivated oil. Accordingly, more work is required for the successful application of DLC coating under the liquid-lubricated case.

In recent decades, a huge amount of research has been directed to employing nanoparticles as lubricant additives, including carbon [8], metals [9], sulfides [10], hydroxides [11], borates [12], etc. Especially, MoS_2 nanoparticles have obtained impressing attention arising from their particular layered structure [13]. The tribological response of MoS_2 nanoparticles as lubricant additives for metal friction pairs has been extensively studied in previous research. As is widely acknowledged, MoS_2 nanosheets are excellent solid lubricant additives for steel-based friction surfaces [14,15]. A latest tendency in the preparation of MoS_2 lubricant additives is to synthesize MoS_2 nanosheets with ultrathin structure to facilitate their entrance into the rubbing interfaces and accordingly optimizing the lubrication performance of oils. For example, Yi et al. [16] recently reported that ultrathin MoS_2 nanosheets could better enhance the tribological behavior of oils, as they were penetrated into the rubbing surfaces easily. This stimulates to exploit MoS_2 nanosheets with an ultrathin structure as lubricating additives for DLC-based solid–liquid composite lubrication. As stated above, as compared to steel, DLC coating presents lower surface energy and higher chemical inertness [17–19]. Whether DLC coatings are able to react with MoS_2 nanosheets or play an inactive role remains mostly unknown. It is of importance to comprehend this so as to engineering valid DLC-based solid–liquid compound lubrication system.

In this work, the friction property of solid–liquid compound lubricating system with W-DLC coating was studied in the cases of lubricating by pure PAO oil and ultrathin MoS_2 nanosheets additivated oils. W-DLC was selected as the coating material because the W-DLC coating possess outstanding adhesion to substrate, low internal stress and excellent mechanical properties, and is one of the most promising films for industrial applications. In addition, the corresponding friction mechanism is discussed at the end.

2. Materials and Methods

2.1. Lubricants and Additives

Polyolefin oil (PAO 6, Shenzhen Huashengyuan Petroleum Technology Co., Ltd., Shenzhen, China) with the kinematic viscosity of 31.70 mm^2/s at 40 °C and 5.95 mm^2/s at 100 °C was used as the base oil. The MoS_2 nanosheets with an ultrathin structure applied in this paper were synthesized through a solvothermal approach, as explained in our previously published research [20].

2.2. Preparation and Characterization of W-DLC Coating

W-DLC film was deposited on the AISI 52,100 through the magnetron sputtering method. The W-DLC film was fabricated according to the published literature [21]. Samples in the shapes of balls and discs were applied as the substrates to deposit W-DLC coatings, respectively. The steel ball had a diameter of 10 mm, a surface roughness of 50 nm and a hardness of 61–63 HRC. The steel disc had a size of 24 mm in diameter and of 7.9 mm in thickness. Before deposition, the discs were ground and burnished to get a surface roughness of 50 nm. The thickness and cross-section image of the W-DLC film on the disc was detected through scanning electron microscopy (SEM, Thermo Fisher Scientific, Prague, Czech Republic). Prior to SEM observation, a small 5 × 5 × 7.9 piece was cut from the middle of the W-DLC coated disc by wire electrical discharge machining method. Its microhardness and elasticity modulus were tested by a nanoindentation tester (500-NHT3, Anton Paar, Boudry, Switzerland) with a diamond Berkovich (trihedral pyramid) cusp, and the test parameters were a load of 10 mN and a hold of 30 s. The surface roughness of the coating was examined by an atomic force microscopy (AFM, Bruker, New York, NY, USA). The chemical component of the as-deposited film was gauged through X-ray photoelectron spectroscopy (XPS, Thermo Fisher Scientific, Prague, Czech Republic).

2.3. Friction Test

The friction tests were performed on a ball-on-disk tribotester (UMT-3, Bruker, New York, NY, USA). An oscillating frequency of 4 Hz and a stroke of 5 mm were utilized, giving a sliding velocity of 0.04 m/s. Each friction test lasted for 30 min and was iterated at least three times. During the friction, the ball was applied to slide against the W-DLC coated disc. Prior to experiments, fabricated MoS$_2$ nanosheets were dispersed into PAO 6 oil by ultrasonic agitation of 30 min. Then, the oil sample was dripped on the surface of the W-DLC coated disc with an amount of about 0.05 mL by a pipette, completely covering the rubbing surfaces. After friction tests, the disc was rinsed with acetone and the worn scar depth, and wear volume was observed through a white light interferometer (GT-X, Bruker, New York, NY, USA). The worn surface was also observed by SEM with energy-dispersive X-ray spectroscopy (EDS). XPS and Raman analyses were also used to characterize the worn surfaces.

3. Results and Discussion

3.1. Structure and Morphology

Figure 1a,b present the TEM morphology of fabricated MoS$_2$ nanosheets, from which the ultrathin structure of MoS$_2$ nanosheets are observed. Meanwhile, the lateral size of the fabricated MoS$_2$ is around 30 nm. The FTIR spectrum of the prepared MoS$_2$ is illustrated in Figure 1c, in which the peaks at 2917, 2846, 1652, 1403 and 718 cm^{-1} could be assigned to the bands belonging to oleylamine [16]. In addition, the peak at 470 cm^{-1} can be assigned to the Mo-S bond [22,23]. This illustrates that the surfaces of synthesized MoS$_2$ nanosheets are attached by oleylamine molecules. Figure 1d is the Raman spectrum of the fabricated MoS$_2$, in which the peaks at 381.5 and 405.3 cm^{-1} belong to the E$_{12g}$ and A$_{1g}$ modes of MoS$_2$, respectively [24,25]. Figure 1e–h shows the XPS analysis results of the prepared MoS$_2$. As seen from Figure 1e, five elements of C, O, N, Mo, S are discovered, in which C and N are derived from oleylamine molecules, and O elements are attributed to pollutants adsorbed on the surface of MoS$_2$. The two peaks in the Mo 3d spectrum at 227.52 and 230.68 eV could be distributed to Mo 3d$_{5/2}$ and Mo 3d$_{3/2}$ of Mo^{4+} (Figure 1f). The S 2p spectrum has two peaks at 160.28 and 161.42 eV, which can point to S 2p$_{3/2}$ and S 2p$_{1/2}$ of S^{2-} (Figure 1g). Meanwhile, the peak positions of Mo^{4+} and S^{2-} are very close to those reported values for MoS$_2$ [26,27]. In addition, the peak located at the position of 398.83 eV in the N 1s XPS spectrum can be attributed to N 1s (Figure 1h), which again confirms the decoration of oleylamine on the surfaces of MoS$_2$.

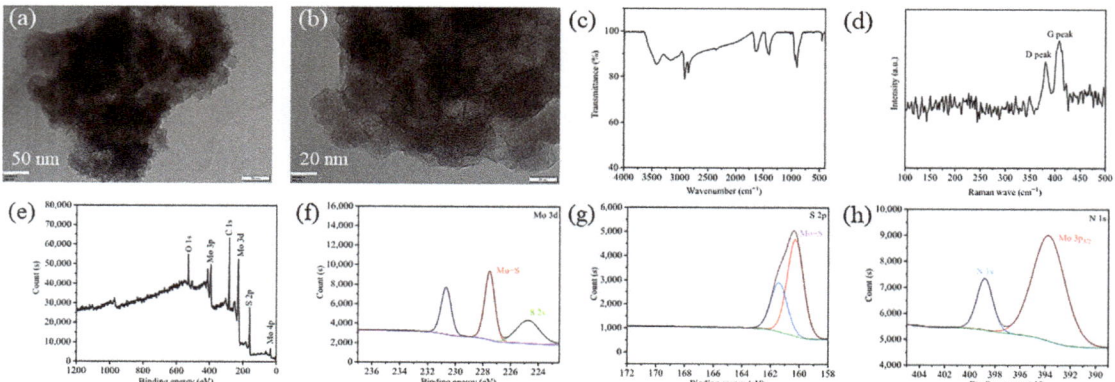

Figure 1. (**a**,**b**) TEM image, (**c**) FTIR spectra, (**d**) Raman spectra, (**e**) XPS survey spectrum and high-resolution XPS spectra of (**f**) Mo 3d, (**g**) S 2p, (**h**) N 1s for the fabricated MoS$_2$.

The cross-section of the W-DLC film on the steel disc is shown in Figure 2a,b. Obviously, the substrate surface is covered by a continuous layer of thin film. The thickness of the film is controlled at about 2.2 μm (Figure 2b). The Raman spectrum of the deposited W-DLC film is shown in Figure 2c. Two strong peaks corresponding to the typical Raman peaks of DLC films are observed at 1360 cm^{-1} (D band) and 1560 cm^{-1} (G band). Figure 2d exhibits the AFM topography of the W-DLC coating. No sharp rough peaks are observed, and the surface roughness of the W-DLC film is approximately 8.15 nm. Figure 2e shows the nanoindentation curve of the W-DLC coating. Its hardness and elastic modulus were gauged at 12.48 GPa and 125.8 GPa, respectively. As shown in Figure 2e, when the test load is 10 mN, the maximum indentation depth is 216 nm, which is close to 10% of the W-DLC coating thickness (2.2 μm), and hence, the influence of the steel matrix on the mechanical characters of W-DLC coating is avoided.

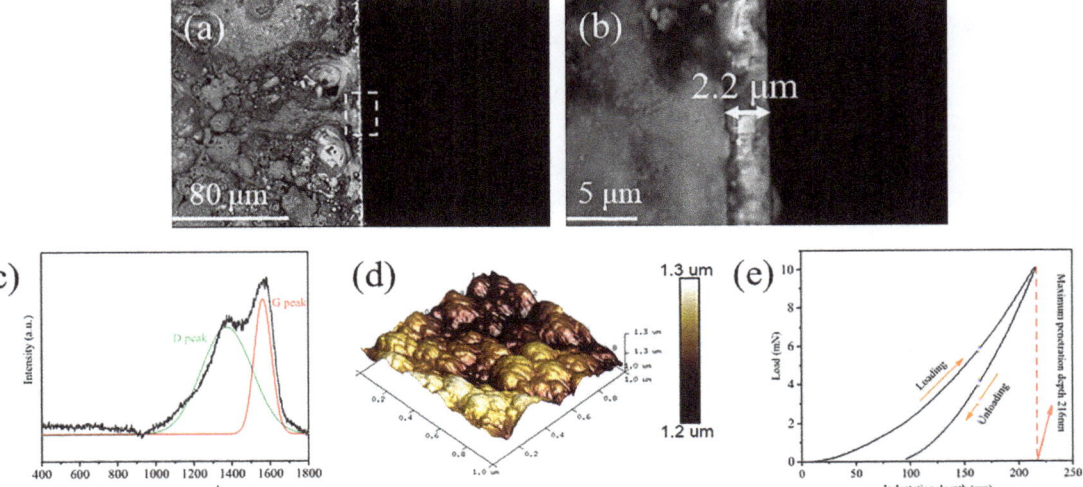

Figure 2. (a,b) TEM images, (c) Raman spectrum, (d) AFM image, (e) nanoindentation load-displacement curve of W-DLC coating.

Furthermore, the chemical component of the W-DLC coating was tested through XPS. Figure 3a shows the full XPS spectrum, in which the C, O and W elements are detected, with the corresponding atomic percentage present in Table 1. It shows that the W-DLC film is composed of 84.34 at% of C, 15.56 at% of O and 0.1 at% of W, respectively. The O element may have been introduced during the preparation of the film, or it is possible that the W-DLC film is placed in the air before the experiment. The XPS spectra of C 1s and W 4f are revealed in Figure 3b,c. The C 1s spectrum can be decomposed into five peaks, namely, WC (283.31 eV), W$_2$C (283.81 eV), sp^2-C (284.6 eV), sp^3-C (285.41 eV) and C-O (286.57eV) [28,29] (Figure 3b). The W 4f spectrum is shown in Figure 3c, and two states of W-C (at 33.61 and 35.69 eV) and W-O (at 34.64 and 37.15 eV) are found. As stated above, the oxygen element exists in the W-DLC coating as contaminants arising from external factors, and therefore, the W-O bond is observed.

Table 1. The XPS atomic percentage of W-DLC coatings.

Elements	C	O	W
Content, %	83.34	15.56	0.1

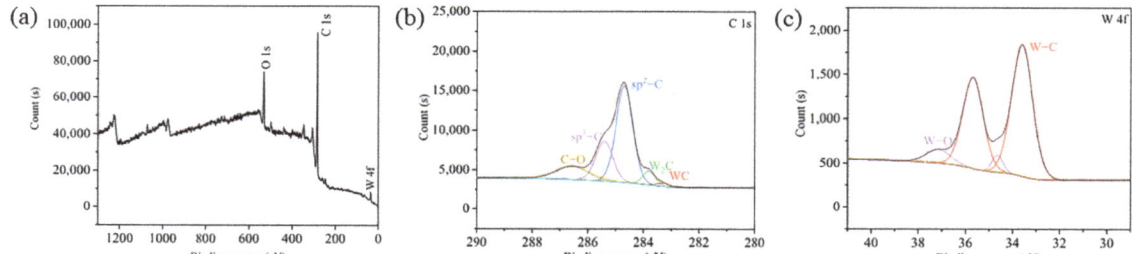

Figure 3. (a) XPS survey spectrum and high-resolution XPS spectra of (b) C 1s, (c) W 4f for W-DLC coating.

3.2. Tribological Property

Before friction tests, the dispersible stability of the lubricating oils additivated with MoS_2 nanosheets was examined by the precipitation observation method. The prepared MoS_2 were blended with the PAO oil at contents within a range of 1–5 wt%. Figure 4a shows the photographs of the oil samples after sitting for 3 h. Obviously, the appearance of the MoS_2-containing oils stays black after standing 3 h. Furthermore, no significant precipitation is found at the sample bottoms (Figure 4b). This illustrates the good dispersion stability of synthesized MoS_2 nanosheets in the oils during the tribological tests.

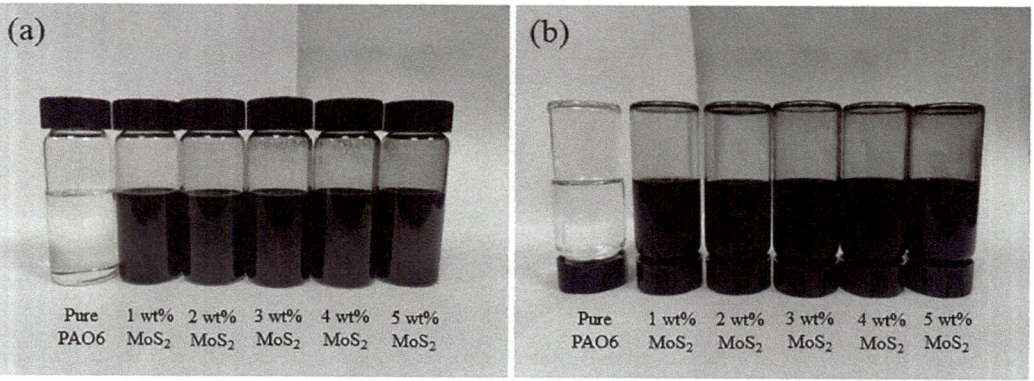

Figure 4. The photographs of oil samples after standing for 3 h with (a) sample bottles stood upright, (b) sample bottles upside down.

Figure 5 presents the tribological properties of steel/W-DLC and W-DLC/W-DLC tribopairs lubricated by the oils with different concentrations of fabricated MoS_2 (at the load of 20 N). As Figure 5a illustrated, the friction coefficient (COF) first increases and then decreases with increasing MoS_2 content in oils. A minimum COF value of 0.079 is obtained when the oil additivated with 4 wt% MoS_2. At most, the COFs are reduced by 12.79% (steel/W-DLC contact) and 14.78% (W-DLC/W-DLC contact) in comparison with that lubricated by pure PAO oil. A promotion of the MoS_2 content from 4 wt% to 5 wt% could lead to an increase in the COF. This is attributed to the unavoidable agglomeration and accumulation at a high concentration of MoS_2 nanosheets, resulting in a damage of lubricant film. Figure 5b exhibits the worn scar diameters (WSDs) on the upper tribopairs with the changes of the MoS_2 nanoparticles' concentration. Obviously, the WSDs tend to be parabolic with the MoS_2 additives' concentration increasing, irrespective of pairs of friction materials. This could be explained by the fact that the oils with a low content of MoS_2 nanosheets could not offer a sufficient supply of lubricant additives in the late-stage of the friction tests. On the contrary, the irreversible agglomeration and abrasive effects rising from the agglomeration of MoS_2 nanosheets at a high concentration greatly reduce the valid volume of lubrication constituents. For the steel/W-DLC couples, the WSDs are reduced by 3.08% and 0.84% after adding 3 wt% and 4 wt% into PAO oil, respectively, if compared to the pure PAO 6 oil. After adding 4 wt% MoS_2 nanosheets to PAO oil, the WSDs for the W-DLC/W-DLC tribopairs is reduced by 22.37%. As Figure 5 shows, the PAO oil with 4 wt% MoS_2 nanosheets exhibits a relatively better friction reduction and wear resistance properties, irrespective of tribopairs. Therefore, the MoS_2 nanosheets concentration in PAO 6 oil is fixed at 4 wt% for subsequent tribological tests.

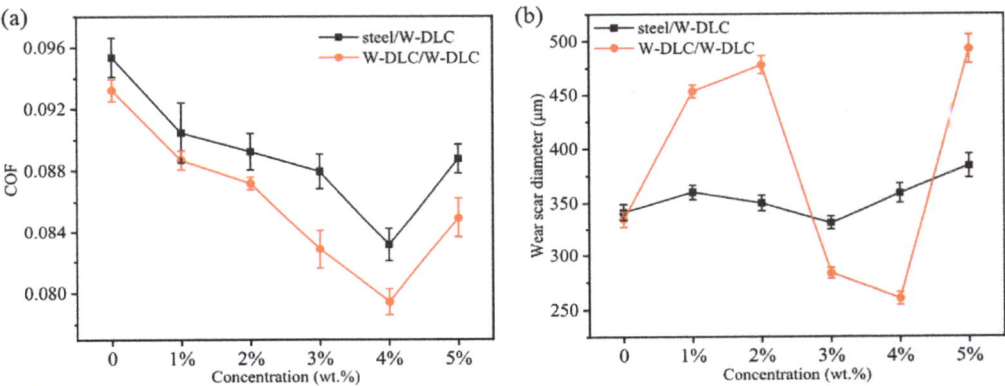

Figure 5. (a) COF and (b) wear scar diameter as functions of MoS_2 content in oils.

Figure 6 illustrates the influence of loads on the lubrication property of the steel/W-DLC and W-DLC/W-DLC contacts under different lubrication conditions. The experimental loads were set to 5N, 10 N, 20 N and 30 N, respectively. As Figure 6a shows, there is not significant change in the COF as the function of the load. However, the MoS_2 additivated oil exhibits a relatively smaller COF than that of the pure PAO oil. In the steel/W-DLC couple, the average COF is calculated to be decreased by 10.83% after adding fabricated MoS_2 into the PAO oil. With regard to the W-DLC/W-DLC couple, the introduction of MoS_2 nanosheets into the PAO oil reduces the average COF by 15.44% compared to the pure PAO oil. Figure 6b illustrates the maximum wear scar depth of the lower W-DLC discs with the changes of applied load. Obviously, the maximum wear scar depth increases gradually with the applied load, and there is a reduction in the maximum wear depth after adding fabricated MoS_2 into PAO oil. Moreover, it is clear from Figure 6b that the worn scar depth in the W-DLC/W-DLC couple is much shallower than that in the steel/W-DLC couple, irrespective of lubricating oils. For example, when tested with pure PAO oil, the average wear scar depth in the steel/W-DLC couple is 84.38% higher than that of W-DLC/W-DLC couple. As the MoS_2-containing oils used for lubrication, the average wear scar depth is about 15.61% lower for the steel/W-DLC contact and this for the W-DLC/W-DLC contact is about 41.09%, compared with that lubricated by pure PAO oil. From Figure 6, it is evident that the lubricating effect of the MoS_2 in the W-DLC/W-DLC couple is more benefitting in comparison with the steel/W-DLC couple.

The worn scars of the rubbed W-DLC film were analyzed using the white light interferometer, which also allows for the wear volume of scars to be measured. Figure 7 shows a comparison between wear scars generated from the steel/W-DLC and W-DLC/W-DLC tribopairs. As expected, the wear volume of scratches deepens with increased applied load. As Figure $7a_1$–a_4 show, deep furrows are found on the worn surface generated from the steel/W-DLC contact lubricated by base PAO oil. For the W-DLC/W-DLC lubricated by PAO oil, the wear scar is relatively flat, and very slight scratches occur on the rubbing surfaces (Figure $7b_1$–b_4). As the MoS_2-added oil used for lubrication, the wear scars in the steel/W-DLC couple turn much shallower in contrast to that lubricated by base PAO oil, and corresponding wear volumes are reduced significantly (Figure 7c$_1$–c$_4$). This result confirms the contribution of fabricated MoS$_2$ into improving the wear resistance property of PAO oil. However, no scratches are visible on the friction surface derived from W-DLC/W-DLC contact with an expectation at a high load of 30 N (Figure 7d$_1$–d$_4$).

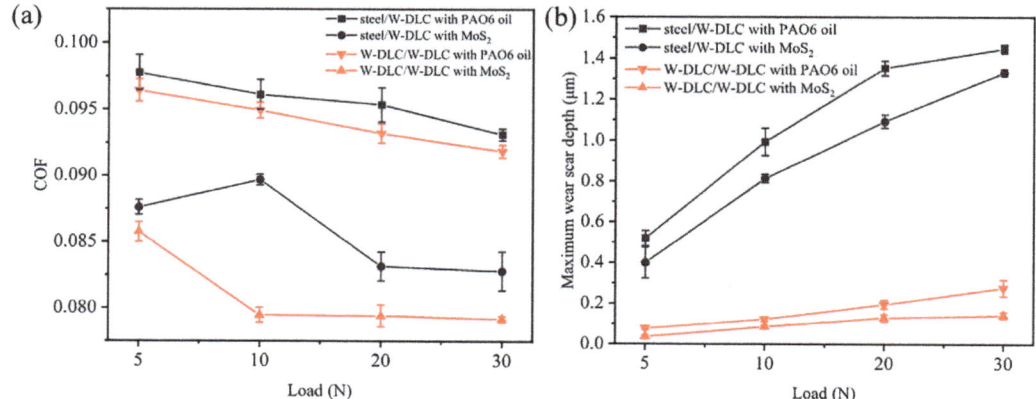

Figure 6. (**a**) COF and (**b**) maximum wear scare depth under different loads.

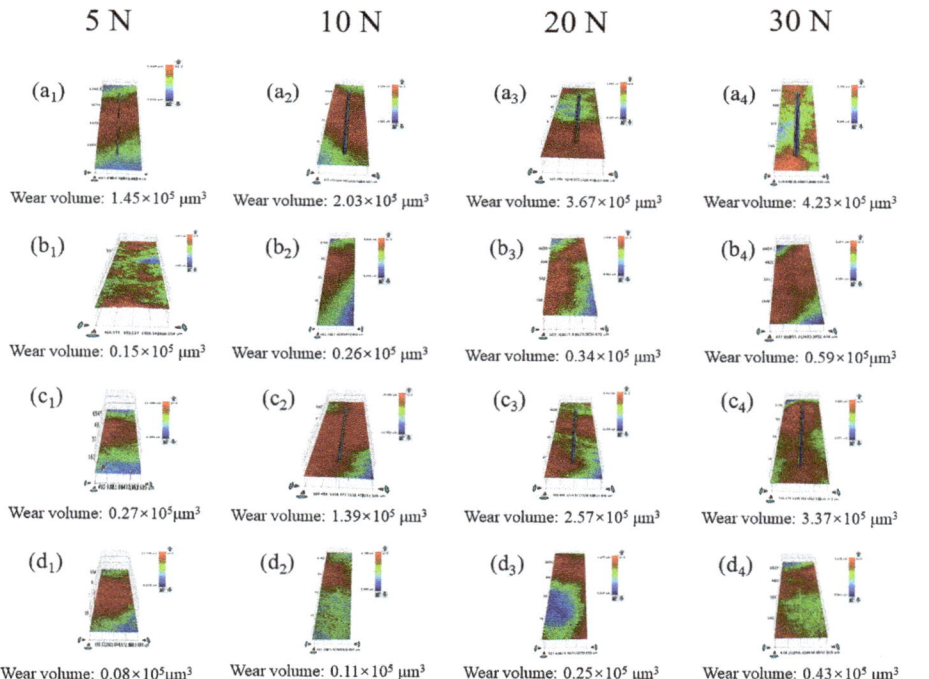

Figure 7. Three-dimensional images of the wear tracks derived from steel/W-DLC contact lubricated with (**a$_1$–a$_4$**) PAO oil or (**c$_1$–c$_4$**) MoS$_2$-added oil and W-DLC/W-DLC contact lubricated with (**b$_1$–b$_4$**) PAO oil or (**d$_1$–d$_4$**) MoS$_2$-added oil.

3.3. Wear Surface Analysis

The topographies of the above wear surfaces were investigated using SEM, with the results present in Figure 8. For the steel/W-DLC tribopairs, a lot of deep grooves are observed on the wear surface lubrication with pure PAO oil (Figure 8a_1–a_4), consistent with their 3D images, shown in Figure 7a_1–a_4. SEM images of worn surfaces for the W-DLC/W-DLC couple lubricated with pure PAO oil show relatively smooth wear surfaces only with some slight friction marks (Figure 8b_1–b_4). In the cases of the MoS_2-containing oil for lubrication, the wear damage in the steel/W-DLC contact is significantly alleviated compared to that tested with pure PAO oil (Figure 8c_1–c_4). At the load of 5 and 10 N, the wear surfaces are covered a layer of tribofilm (Figure 8c_1,c_2). However, at a higher load of 20 N, it is found that a certain quantity of W-DLC coating is peeled off and delaminated from the steel substrate, as evident in Figure 8c_3. When the imposed load increases to 30 N, the sign of grooves and plastic deformations are visible on the worn surface (Figure 8c_4). As for the W-DLC/W-DLC tribopairs, the worn regions tested with prepared MoS_2 nanosheets exhibit quite smooth characteristics at 5 and 10 N, with the presence of a small amount of pits (Figure 8d_1,d_2). Comparatively, in the tests at 20 N and 30 N, the wear surfaces show quite smooth appearances without any evidence no sign of scratches (Figure 8d_3,d_4).

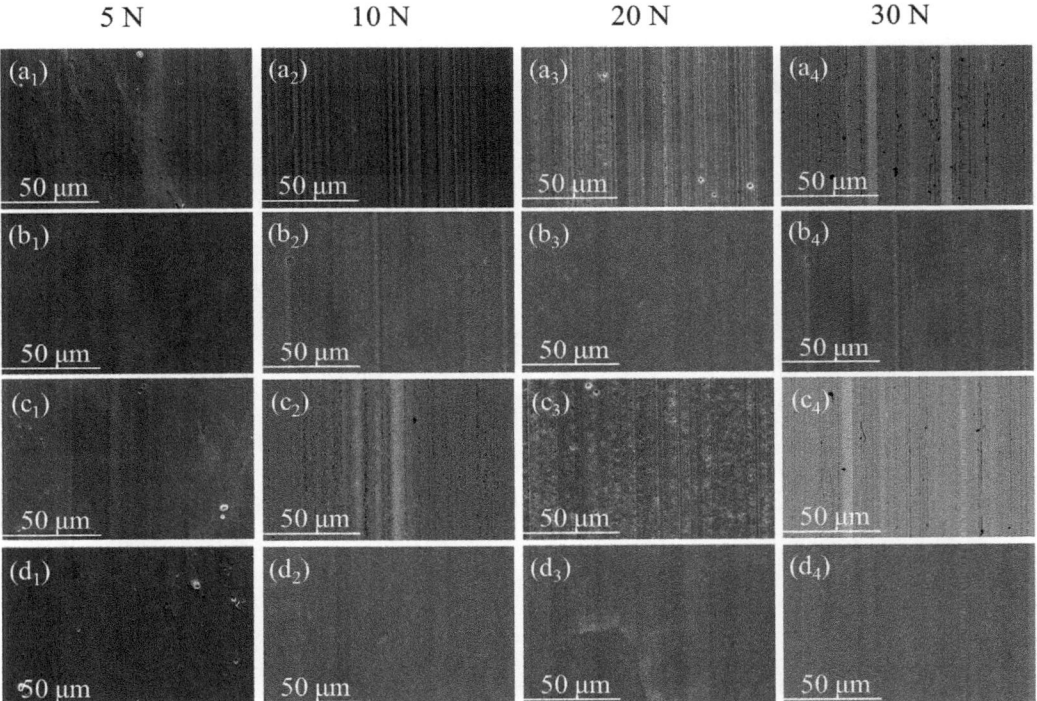

Figure 8. SEM images of the wear tracks from steel/W-DLC contact lubricated with (a_1–a_4) PAO oil or (c_1–c_4) MoS_2-added oil and W-DLC/W-DLC contact lubricated with (b_1–b_4) PAO oil or (d_1–d_4) MoS_2-added oil.

To analyze the elemental distribution of the worn marks on the W-DLC film coated discs tested in the presence of MoS_2 additives, EDS mapping was performed in the worn surfaces (Figure 8c_3,d_3), with the results shown in Figure 9. In both cases, the Mo and S elements appear on the worn scars. Prior to SEM analysis, the surfaces were ultrasonically cleaned with acetone, which guarantees that physically absorbed MoS_2 into rubbing surfaces can be removed. The remain of elements of Mo and S on worn surface, in spite of the ultrasonic cleaning, suggests a tribochemical reaction between the MoS_2 nanosheets and the W-DLC coating. Moreover, the worn scar derived from the W-DCL/W-DLC couple has more transferred Mo and S elements, indicates clearly the formation of a strong protective layer of MoS_2-based tribofilm.

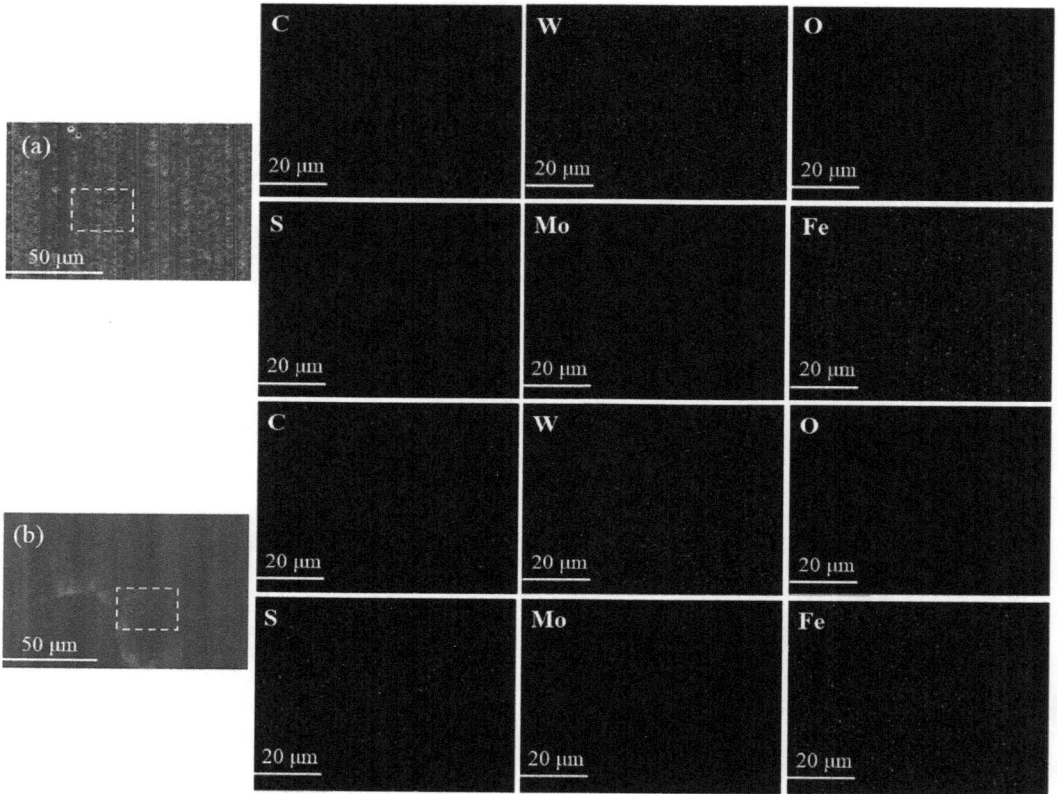

Figure 9. SEM images and the distribution of C, W, O, S, Mo and Fe elements on the wear tracks lubricated with MoS_2 from (**a**) steel/W-DLC contact and (**b**)W-DLC/W-DLC contact.

In order to further confirm whether the occurrence of interaction between the MoS_2 and the W-DLC coating, the worn scars on lower W-DLC specimens (Figure 8c_2,d_2) were examined by XPS analysis (Figure 10). Table 2 illustrates the quantification of XPS collected from the worn surface. The elements of C, W, O, Fe and Mo and S are discovered. The C, W and O elements could originate from the expected composition of their deposited coating. The minor Fe element could be derived from the steel substrate. Regardless of the type of tribopairs, Mo and S elements are found in the worn scars, which could be derived from the MoS_2 additives. In addition, a large concentration of O element, about 17–20 at%, is detected at the surfaces of W-DLC film, indicating that the W-DLC film was oxidized in the tribotest.

The XPS spectra of W 4f, C1s, Mo 3d and S 2p were collected from two mentioned tribopairs (Figure 9a,b) and are displayed in Figure 11. As Figure 11 shows, the Mo 3d, C 1s, W 4f and S 2p XPS spectra for the two tribopairs tested with MoS_2 nano-additives are similar. The C1s spectrum can be assigned into WC_2, WC, sp^2-C, sp^3-C and C-O bonding states, respectively (Figure 11a_1,b_1). The W 4f spectrum exhibits six peaks, which correspond to three tungsten-bonding states, i.e., W-C, WS$_2$ and WO$_3$, respectively (Figure 11a$_2$,b$_2$). In contrast to the as-received W-DLC coating (Figure 3c), the W present in the rubbed W-DLC coating is transferred from WO$_2$ into the oxidized form of WO$_3$. The Mo 3d$_{5/2}$ spectrum consists of the major peak belonging to MoS$_2$ and the minor peak arising from Mo-O (Figure 11a$_3$,b$_3$). As to the S 2p$_{3/2}$, it consists of the S-W bond and MoS$_2$ (Figure 11a$_4$,b$_4$). The discovery of the S-W species in the Figure 11a$_4$,b$_4$ indicates that the MoS$_2$ adhered onto the W-DLC coating via the bonding of W-S. Presumably, the generation of a tribofilm based on MoS$_2$ nanosheets is responsible for the worn resistance effect.

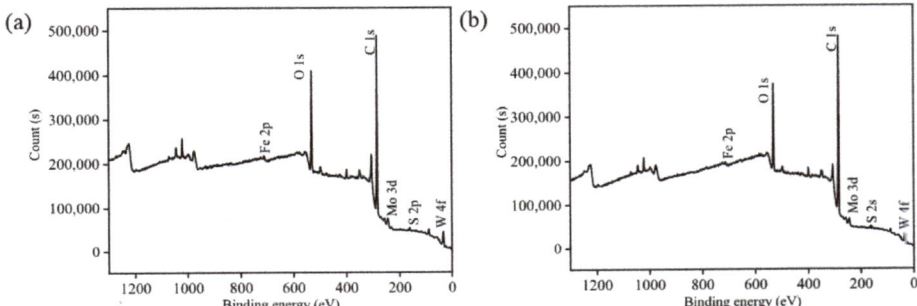

Figure 10. XPS survey spectra for the wear tracks lubricated with MoS$_2$ from (**a**) steel/W-DLC contact and (**b**) W-DLC/W-DLC contact.

Table 2. The XPS quantification for the steel/W-DLCW-DLC/W-DLC contacts.

Elements	C, %	O, %	W, %	S, %	Mo, %	Fe, %
steel/W-DLC	79.09	19.52	0.03	0.3	0.05	1.01
W-DLC/W-DLC	81.01	17.81	0.02	0.23	0.03	0.9

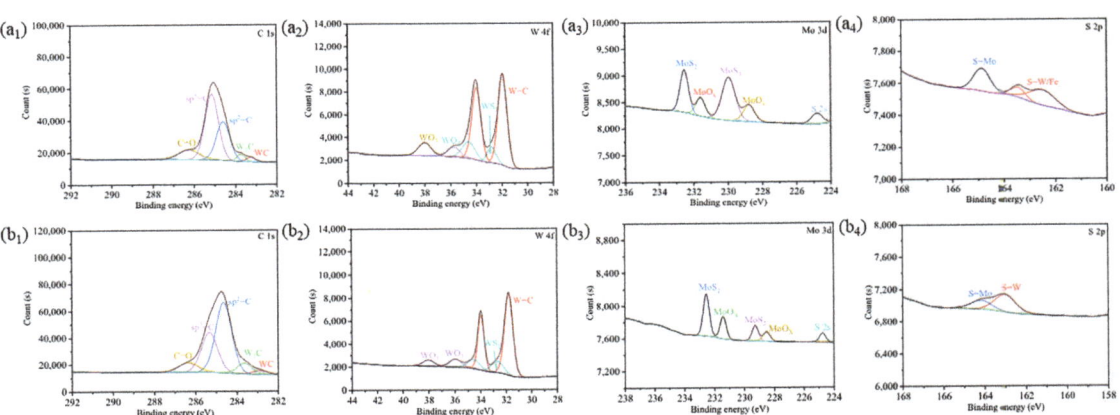

Figure 11. XPS spectra of C 1s, W 4f, Mo 3d and S 2p for wear tracks lubricated with MoS$_2$ from (**a$_1$–a$_4$**) steel/W-DLC and (**b$_1$–b$_4$**) W-DLC/W-DLC contact.

Furthermore, the wear marks on the W-DLC-coated discs (Figure 9a,b) were analyzed using Raman spectroscopy. As Figure 12 shows, the Raman spectra of the rubbed W-DLC coatings show a peak around 1360 cm^{-1}, corresponding to the D band and a peak around 1560 cm^{-1} attributed to the G band. However, the two rubbed surfaces exhibit a significant difference in the aspect of the peak ratio (I_D/I_G) of the D and G bands. It is well known that the peak ratio (I_D/I_G) value indicates the extent of graphitization of DLC coatings [30,31]. Moreover, the higher extent of graphitization is generally confirmed from the increase in the peak ratio (I_D/I_G) of the D and G bonds. Through curve fitting, the rubbed W-DLC coating generated from the steel/W-DLC contact present an I_D/I_G value of about 6.6, and this for the W-DLC/W-DLC contact is 0.82. Obviously, the peak ratio (I_D/I_G) in the rubbed W-DLC coating from the steel/W-DLC couple is significantly higher than that for the W-DLC/W-DLC couple. This suggests that higher extent of graphitization occurred at the rubbed W-DLC surface generated from the steel/W-DLC couple as compared to the W-DLC/W-DLC couple. It is speculated that the microhardness of the W-DLC coating decrease arising from the increased extent of graphitization under applied loads, and therefore, induced the increase of wear rates of the W-DLC coating.

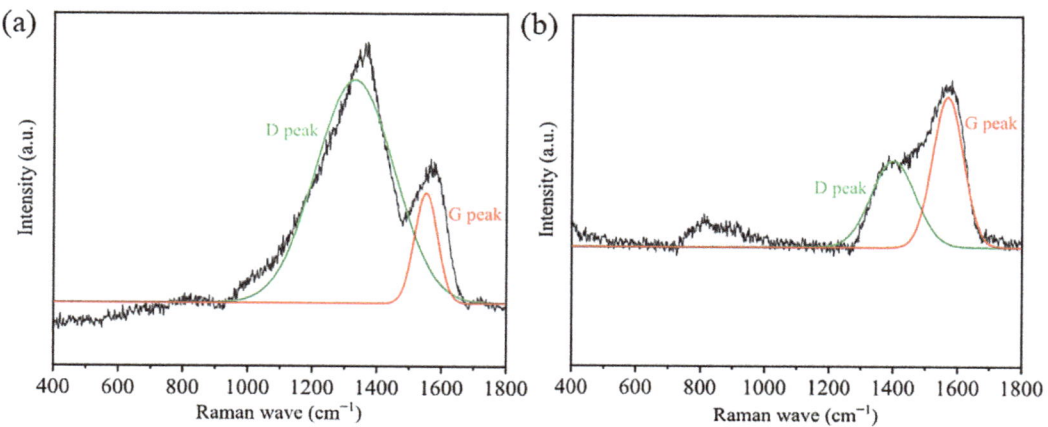

Figure 12. Raman spectra for the wear tracks lubricated with MoS_2 from (**a**) steel/W-DLC contact and (**b**) W-DLC/W-DLC contact.

Figure 13 illustrates the SEM images of the wear ball region derived from the steel/W-DLC and W-DLC/W-DLC tribopairs experimented with MoS_2 (at 20 N). It is obvious that compared to steel/W-DLC contact, the WSD of W-DLC/W-DLC is significantly reduced. For the steel/W-DLC contact (Figure 13a), there are a few debris particles scattered at the edge of the wear mark. Meanwhile, the Mo and S elements are mainly discovered at the margin, which may result from the broken MoS_2 nanosheets or their hard oxides. C and W elements are also observed on the worn ball surface. This can be interpreted as the W-DLC film being transferred to the ball when the naked ball slides on the W-DLC coated surface. With regard to the W-DLC/W-DLC couple (Figure 13b), some particle-like debris observed on the wear mark. Elements including C, W, O, Fe, Mo and S exist on the worn surface. In addition, there is an obvious distribution of Mo and S elements detected, indicating a continuous MoS_2-based tribofilm on the worn surface.

3.4. Discussion

In this paper, the tribological behaviors of steel/W-DLC and W-DLC/W-DLC contacts were tested in the presence of base PAO 6 oil and ultrathin MoS_2-containing oil. As seen from Figures 6 and 7, the friction is generally lower when two contact regions were coated with W-DLC than when only one counter-body was coated, which is observed in pure PAO oil and MoS_2-additivated PAO oil.

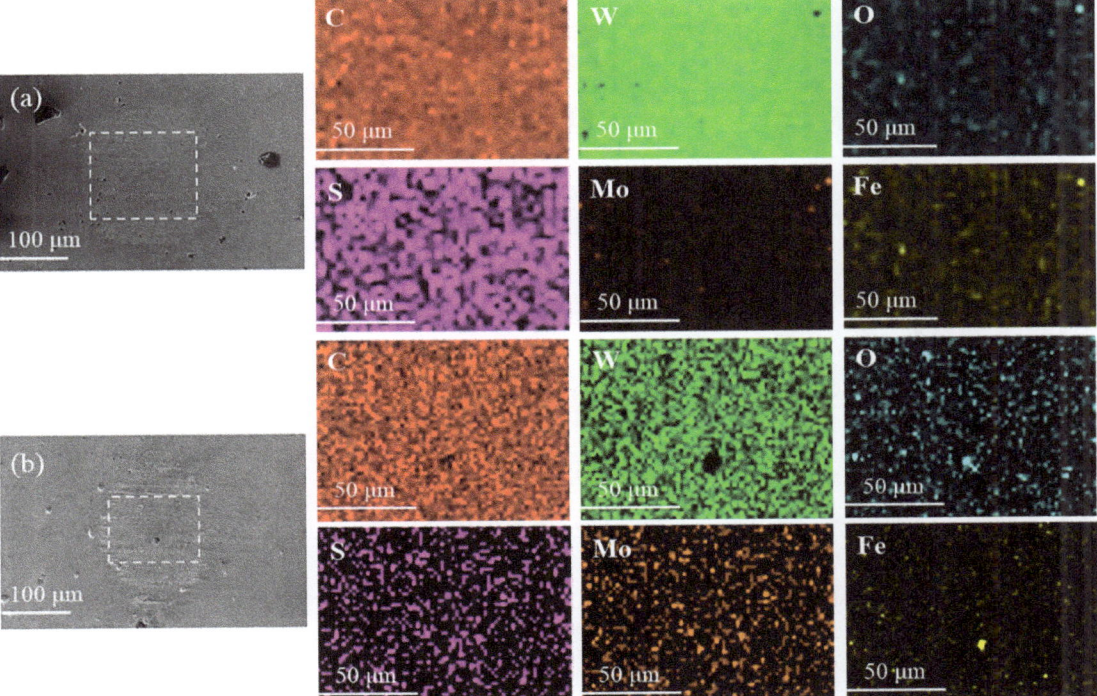

Figure 13. SEM images of the wear ball surfaces and the distribution of C, W, O, S, Mo and Fe elements, where tracks were lubricated with MoS$_2$ from (**a**) steel/W-DLC contact and (**b**) W-DLC/W-DLC contact.

As well known, the lubrication region of a ball–disc point contact pair can be determined by the film thickness ratio λ illustrated in Equation (1), which depends not only on the oil film ply, but also on the roughness of the frictional surfaces [32]:

$$\lambda = \frac{h}{\sqrt{R_{a1}^2 + R_{a2}^2}} \quad (1)$$

where R_{a1} and R_{a2} are the roughness of the two frictional surfaces, respectively, and h is the oil film thickness. When $\lambda \leq 1$, two frictional surfaces are lubricated at the boundary status; when $1 < \lambda \leq 3$, the mixed lubrication state occurred on two frictional pairs; when $\lambda > 3$, two frictional surfaces are absolutely separated with a lubricating film and stay in the fluid lubricant state.

The relevant oil film thickness (h_c) in the central area can be calculated according to the Hamrock–Dowson formula [32]:

$$H_c^* = \frac{2.69 G^{*0.53} U^{*0.67}(1 - 0.61 e^{-0.73k})}{W^{*0.067}} \quad (2)$$

$$h_c = H_c^* R' \quad (3)$$

$$G^* = \alpha E', \quad U^* = \frac{\eta_0 U}{E' R'}, \quad W^* = \frac{W}{E' R'^2} \quad (4)$$

where h_c is the oil film plying in the central area, α is the viscosity coefficient of the oil (3.5×10^{-8} m^2/N), E' is the equivalent elasticity modulus (172.28 Gpa), U is the entrainment speed (0.02 m/s), η_0 is the viscosity of the oil, R' is the equivalent radius, W is the applied load, and k is the ovality (1.03). According to the Equations (1)–(4), the results are present in Table 3, in which the λ value exhibits a value lower than 1, indicating that a boundary lubrication regime occurs during the friction process.

Table 3. Thickness of oil film under different loads.

Lubrication Conditions	Steel/W-DLC			W-DLC/W-DLC		
Load, N	10	20	30	10	20	30
Oil film thickness, nm	45.7	43.6	42.4	46.6	44.5	43.1
Composite roughness, nm	55.90	55.90	55.90	50.64	50.64	50.64
$\lambda_{initial}$	0.82	0.78	0.77	0.92	0.88	0.85

It is highly appreciated that an effective tribofilm exercise important influences on the friction behavior of oils in the boundary lubricating regime. Generally, it has been documented that the doping metal elements in DLC coating could provide reactive sites for the coating–lubricant interactions of forming a tribofilm [33]. Such behavior has been observed for the W-DLC coating in our paper, as confirmed by the XPS results shown in Figure 11. However, the friction could also induce the graphitization of W-DLC film, which resulted in local lamination of W-DLC films [34]. It is shown in Figure 13 that in contrast to the steel/W-DLC couple, the graphitization extent of W-DLC film in the self-mated W-DLC/W-DLC couple was significantly decreased. Two possible friction mechanisms are schematically presented in Figure 14. For the steel/W-DLC contact (Figure 14a), the pressure and temperature in the contact provoked the tribochemical reaction between MoS$_2$ and the bare steel ball or the disc coated with W-DLC film forming a tribofilm separating two reciprocating surfaces. However, the W-DLC film was largely graphitized and oxidized under the friction conditions, resulting in the drop of its microhardness, and thus causing local delamination of W-DLC films. In addition, the carbon atoms in the DLC coating could be diffused into the ferrous surface because of the thermo-chemical interaction between the DLC coating and steel surface, which accelerated W-DLC coatings' wear [35]. Thus, the shift of W-DLC stuff from the lower W-DLC coated disc into the upper steel ball was observed in Figure 12a. In addition, the MoS$_2$ nanosheets adhere onto the bare steel surface, and later, those nanoparticles slid against the W-DLC film, which could accelerate the wear of coated surface. As to the self-mated W-DLC/W-DLC couple, as shown in Figure 14b, a rather thick layers of tribofilm adheres to the DLC-coated surface through the tribochemical reaction between MoS$_2$ and the W element doped in W-DLC, which has been demonstrated by an SEM observation in Figure 10. Furthermore, the tribofilm is formed at both counterparts, and they may keep the contacting asperities separated. In this case, the extent of graphitization of W-DLC film was alleviated because of the lubrication and protection from the MoS$_2$-based tribofilm on two friction surfaces, and a better lubrication effect was obtained.

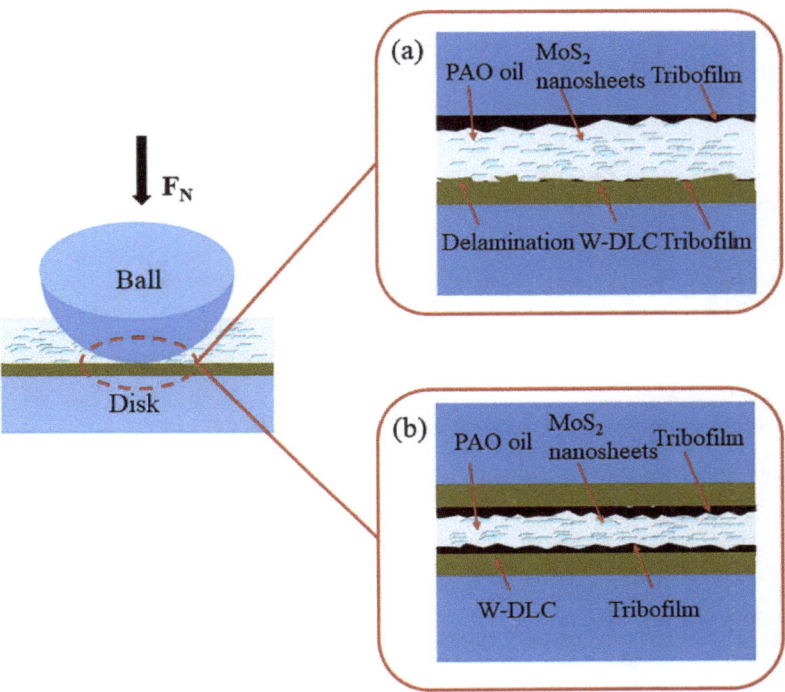

Figure 14. Schematic diagram of friction mechanisms for (**a**) steel/W-DLC and (**b**) W-DLC/W-DLC contact lubricated with MoS_2.

4. Conclusions

The friction property of the steel/W-DLC and W-DLC/W-DLC tribopairs were investigated under the boundary lubrication regime with MoS_2-containing oil for lubrication. The introduction of ultrathin MoS_2 nanosheets into the PAO oil could importantly improve the friction behavior for the steel/W-DLC and W-DLC/W-DLC couples. And best results were gained for the W-DLC/W-DLC couple lubricated with ultrathin MoS_2 nanosheets, in which the friction coefficient and wear were decreased by 13.98% and 41.09% respectively, in comparison with the case of the W-DLC/W-DLC couple lubricated by pure PAO oil. Meanwhile, the reductions in the friction coefficient and wear were 4.98% and 89.10%, respectively, in contrast to the steel/W-DLC contact tested with ultrathin MoS_2 nanosheet additivated oils. The reason of friction reduction is the formation of MoS_2-based friction film on rubbing regions. The tribofilm was formed by a tribochemical reaction between MoS_2 and the doping W elements in the DLC coating. Meanwhile, the extent of graphitization of W-DLC film was alleviated because of the lubrication and protection from the MoS_2-based tribofilm. Wear marks were not obvious on the W-DLC/W-DLC couple, which suggested that combined with the good friction reduction and wear resistance behaviors of ultrathin MoS_2 additivated oil, the application of W-DLC coating prevented serious wear damage taking place. It can be deduced that an incorporation of W-DLC/W-DLC contact and MoS_2 nanosheets as oil additives provided an efficient solid–liquid composite lubricating effect.

Author Contributions: Conceptualization, Methodology, Investigation, Writing—review & editing, M.Y.; Writing—original draft, Investigation, T.W.; Data curation, Z.L.; Valition, J.L.; Valiation, J.Q.; Validation, Resources, W.X. All authors have read and agreed to the published version of the manuscript.

Funding: This research was financially supported by the National Natural Science Foundation of China (52065042), Natural Science Foundation of Jiangxi Province (20212BAB214059 and 20232BAB204040) and Tribology Science Fund of State Key Laboratory of Tribology (SKLTKF19B06).

Data Availability Statement: Data will made available on request.

Acknowledgments: Specific thanks to School of Advanced Manufacture and Key Laboratory of Tribology, Nanchang University for empowering in the accomplished of this study.

Conflicts of Interest: The authors declare no conflict of interest.

References

1. Qi, J.; Wang, L.; Yan, F.; Xue, Q. The tribological performance of DLC-based coating under the solid-liquid lubrication system with sand-dust particles. *Wear* **2013**, *297*, 972–985.
2. Fan, X.; Xue, Q.; Wang, L. Carbon-based solid-liquid lubricating coatings for space applications: A review. *Friction* **2015**, *3*, 191–207.
3. Zhang, T.; Wan, Z.; Ding, J.; Zhang, S.; Wang, Q.; Kim, K. Microstructure and high-temperature tribological properties of Si-doped hydrogenated diamond-like carbon films. *Appl. Surf. Sci.* **2018**, *435*, 963–973.
4. Liu, Y.; Zhang, H. The synergistic mechanism between graphitized tribofilm and graphitized surface of diamond-like carbon film under different temperature environments. *Diam. Relat. Mater.* **2022**, *123*, 108875.
5. Liu, Y.; Zhang, H. Roles of transfer layer and surface adhesion on superlubricity behaviors of diamond-like carbon film depending on rotating and reciprocating motion. *Appl. Surf. Sci.* **2022**, *604*, 154538.
6. Tannous, J.; Dassenoy, F.; Lahouij, I.; Le Mogne, T.; Vacher, B.; Bruhacs, A.; Tremel, W. Understanding the tribochemical mechanisms of IF-MoS_2 nanoparticles under boundary lubrication. *Tribol. Lett.* **2011**, *41*, 55–64.
7. Kalin, M.; Kogovsek, J.; Remskar, M. Nanoparticles as novel lubricating additives in a green, physically based lubrication technology for DLC coatings. *Wear* **2013**, *303*, 480–485.
8. Patel, J.; Kiani, A. Effects of reduced graphene oxide (rGO) at different concentrations on tribological properties of liquid base lubricants. *Lubricants* **2019**, *7*, 11.
9. Wang, J.; Zhang, H.; Hu, W.; Li, J. Modified Ni nanoparticles as additives in various greases: Assessment of comparative performance potential. *Lubricants* **2022**, *10*, 367.
10. Freschi, M.; Di Virgilio, M.; Zanardi, G.; Mariani, M.; Lecis, N.; Dotelli, G. Employment of micro- and nano-WS_2 Structures to enhance the tribological properties of copper matrix composites. *Lubricants* **2021**, *9*, 53.
11. Li, Z.; Chang, Q.; Meng, Y.; Yang, H.; Hao, L. In situ visualization study of tribofilm growth process from magnesium silicate hydroxide nanoparticles. *Tribol. Int.* **2023**, *187*, 108725.
12. Saffari, H.; Soltani, R.; Alaei, M.; Soleymani, M. Tribological properties of water-based drilling fluids with borate nanoparticles as lubricant additives. *J. Pet. Sci. Eng.* **2018**, *171*, 253–259. [CrossRef]
13. Yi, M.; Zhang, C. The synthesis of MoS_2 particles with different morphologies for tribological applications. *Tribol. Int.* **2017**, *116*, 285–294.
14. Chen, J.; Xu, Z.; Hu, Y.; Yi, M. PEG-assisted solvothermal synthesis of MoS_2 nanosheets with enhanced tribological property. *Lubr. Sci.* **2020**, *32*, 273–282. [CrossRef]
15. Chen, Z.; Liu, X.; Liu, Y.; Gunsel, S.; Luo, J. Ultrathin MoS_2 Nanosheets with Superior Extreme Pressure Property as Boundary Lubricants. *Sci. Rep.* **2015**, *5*, 12869.
16. Yi, M.; Qiu, J.; Xu, W. Tribological performance of ultrathin MoS_2 nanosheets in formulated engine oil and possible friction mechanism at elevated temperatures. *Tribol. Int.* **2022**, *167*, 107426. [CrossRef]
17. Zahid, R.; Masjuki, H.; Varman, M.; Mufti, R.; Kalam, M.; Gulzar, M. Effect of lubricant formulations on the tribological performance of self-mated doped DLC contacts: A review. *Tribol Lett.* **2015**, *58*, 32.
18. Shang, L.; Gou, C.; Li, W.; He, D.; Wang, S. Effect of microstructure and mechanical properties on the tribological and electrochemical performances of Si/DLC films under HCl corrosive environment. *Diam. Relat. Mater.* **2021**, *116*, 108385. [CrossRef]
19. Zhang, M.; Wu, G.; Lu, Z.; Shang, L.; Zhang, G. Corrosion and wear behaviors of Si-DLC films coated on inner surface of SS304 pipes by hollow cathode PECVD. *Surf. Topogr. Metrol. Prop.* **2018**, *6*, 034010. [CrossRef]
20. Yi, M.; Zhang, C. The synthesis of two-dimensional MoS_2 nanosheets with enhanced tribological properties as oil additives. *RSC Adv.* **2018**, *8*, 9564–9573. [CrossRef]
21. Li, D.; Kong, N.; Li, R.; Zhang, B.; Zhang, Y.; Wu, Z.; Zhang, Q. The tribological performance of W-DLC in solid–liquid lubricationsystem addivated with Cu nanoparticles. *Surf. Topogr. Metrol. Prop.* **2021**, *9*, 045043.
22. Rawat, S.; Harsha, A.; Agarwal, P.; Kumari, S.; Khatri, P. Pristine and alkylated MoS_2 nanosheets for enhancement of tribological performance of paraffin grease under boundary lubrication regime. *J. Tribol.* **2019**, *141*, 072102.
23. Chouhan, A.; Sarkar, K.; Kumari, S.; Vemuluri, S.; Khatri, P. Synergistic lubrication performance by incommensurately stacked ZnO-decorated reduced graphene oxide/MoS_2 heterostructure. *J. Colloid Inter. Sci.* **2020**, *580*, 730–739.
24. Kumari, S.; Chouhan, A.; Sharma, P.; Kuriakose, S.; Tawfik, A.; Spencer, S.; Walia, S.; Sugimura, H.; Khatri, P. Structural defects-mediated grafting of alkylamine on few-layer MoS_2 and its potential for enhancement of tribological properties. *ACS Appl. Mater. Inter* **2020**, *12*, 30720–30730. [CrossRef] [PubMed]

25. Li, H.; Zhang, Q.; Yap, C.; Tay, B.; Edwin, T.; OlivIer, A.; Baillargeat, D. From bulk to monolayer MoS2: Evolution of Raman scattering. *Adv. Funct. Mater.* **2012**, *22*, 1385–1390. [CrossRef]
26. Ma, C.; Qi, X.; Chen, B.; Bao, S.; Yin, Z.; Wu, X.; Luo, Z.; Wei, J.; Zhang, H.; Zhang, H. MoS2 nanoflower-decorated reduced graphene oxide paper for high-performance hydrogen evolution reaction. *Nanoscale* **2014**, *6*, 5624–5629.
27. Wang, H.; Wang, B.; Wang, D.; Lu, L.; Wang, J.; Jiang, Q. Facile synthesis of hierarchical worm-like MoS2 structures assembled with nanosheets as anode for lithium ion batteries. *RSC Adv.* **2015**, *5*, 58084–58090.
28. Gao, K.; Zhang, L.; Wang, J.; Zhang, B.; Zhang, J. Further improving the mechanical and tribological properties of low content Ti-doped DLC film by W incorporating. *Appl. Surf. Sci.* **2015**, *353*, 522–529.
29. Takeno, T.; Komiyama, T.; Miki, H.; Takagi, T.; Aoyama, T. XPS and TEM study of W-DLC/DLC double-layered film. *Thin Solid Film* **2009**, *517*, 5010–5013. [CrossRef]
30. Habibi, A.; Khoie, S.; Mahboubi, F.; Urgen, M. Raman spectroscopy of thin DLC film deposited by plasma electrolysis process. *Surf. Coat. Technol.* **2017**, *309*, 945–950.
31. Wong, P.; He, F.; Zhou, X. Interpretation of the hardness of worn DLC particles using micro-raman spectroscopy. *Tribol. Int.* **2010**, *43*, 1806–1810. [CrossRef]
32. Wen, S.; Huang, P. *Principles of Tribology*; Tsinghua University Press: Beijing, China, 2012.
33. Miyake, S.; Saito, T.; Yasuda, Y.; Okamoto, Y.; Kano, M. Improvement of boundary lubrication properties of diamond-like carbon (DLC) films due to metal addition. *Tribol. Int.* **2004**, *37*, 751–761. [CrossRef]
34. Zhang, Y.; Zhang, S.; Sun, D.; Yang, G.; Gao, C.; Zhou, C.; Zhang, C.; Zhang, P. Wide adaptability of Cu nano-additives to the hardness and composition of DLC coatings in DLC/PAO solid-liquid composite lubricating system. *Tribol. Int.* **2019**, *138*, 184–195. [CrossRef]
35. Mannan, A.; Sabri, M.; Kalam, M.; Hassan, M. Tribological performance of DLC/DLC and steel/DLC contacts in the presence of additivated oil. *Int. J. Surf. Sci. Eng.* **2018**, *12*, 60–75. [CrossRef]

Disclaimer/Publisher's Note: The statements, opinions and data contained in all publications are solely those of the individual author(s) and contributor(s) and not of MDPI and/or the editor(s). MDPI and/or the editor(s) disclaim responsibility for any injury to people or property resulting from any ideas, methods, instructions or products referred to in the content.

Article

Graphite Fluoride as a Novel Solider Lubricant Additive for Ultra-High-Molecular-Weight Polyethylene Composites with Excellent Tribological Properties

Guodong Huang [1,*], Tao Zhang [1], Yi Chen [1], Fei Yang [1], Huadong Huang [2] and Yongwu Zhao [3]

[1] School of Mechanical Technology, Wuxi Institute of Technology, Wuxi 214121, China; zhangt@wxit.edu.cn (T.Z.)
[2] Department of Precision Manufacturing Engineering, Suzhou Institute of Industrial Technology, Suzhou 215104, China
[3] School of Mechanical Engineering, Jiangnan University, Wuxi 214122, China
* Correspondence: huanggd@wxit.edu.cn

Abstract: The tribological properties of ultra-high-molecular-weight polyethylene (UHMW-PE) play a significant role in artificial joint materials. Graphite fluoride (GrF), a novel solid lubricant, was incorporated into ultra-high-molecular-weight polyethylene (UHMW-PE) at different concentrations via ball milling and heat pressing to prepare the GrF-UHMW-PE composites. The structure, hardness, and tribological behavior of the composites were investigated using X-ray diffraction (XRD), Fourier-transform infrared (FT-IR) spectrometry, ball indentation hardness, and a reciprocating ball-on-plane friction tester, respectively. The results of FT-IR showed that hydrogen bonds (C-F···H-C) could be formed between GrF and UHMW-PE. The hardness of the composites was significantly enhanced by increasing the GrF concentrations. GrF in the composites displayed superior lubricant properties and the coefficient of friction (COF) of the composites was significantly decreased at lower concentrations of GrF viz. 0.1 and 0.5 wt%. The addition of GrF also significantly enhanced the anti-wear properties of the composites, which was a combined effect of lubrication as well as hardness provided by GrF. At 0.5 wt% GrF concentration, the COF and the wear rate were reduced by 34.76% and 47.72%, respectively, when compared to UHMW-PE. As the concentration of GrF increased, the wear modes of the composites transitioned from fatigue wear to abrasive wear. Our current work suggested that GrF-UHMW-PE composites could be a suitable candidate for artificial joint materials.

Keywords: polymer composites; ultra-high-molecular-weight polyethylene (UHMW-PE); graphite fluoride (GrF); hardness; wear mechanism

Citation: Huang, G.; Zhang, T.; Chen, Y.; Yang, F.; Huang, H.; Zhao, Y. Graphite Fluoride as a Novel Solider Lubricant Additive for Ultra-High-Molecular-Weight Polyethylene Composites with Excellent Tribological Properties. *Lubricants* **2023**, *11*, 403. https://doi.org/10.3390/lubricants11090403

Received: 25 August 2023
Revised: 12 September 2023
Accepted: 13 September 2023
Published: 15 September 2023

Copyright: © 2023 by the authors. Licensee MDPI, Basel, Switzerland. This article is an open access article distributed under the terms and conditions of the Creative Commons Attribution (CC BY) license (https://creativecommons.org/licenses/by/4.0/).

1. Introduction

Ultra-high-molecular-weight polyethylene (UHMW-PE) is considered a unique engineering plastic with excellent performance [1]. Owing to its chemical inertness [2], low friction coefficient [3], wear resistance [4], and good biocompatibility [5], UHMW-PE has already been applied as a bearing surface for total hip replacements since the 1960s [6] and is currently regarded as the gold standard material [7] used in this domain. However, due to the prolonged use of hip prostheses, wear debris of UHMW-PE is generated and can induce periprosthetic osteolysis [8], which may eventually lead to the failure of the total hip arthroplasty. This causes patient suffering and increased financial burden. In order to improve the longevity of total hip and alleviate the patient's pain, it is important to improve the anti-wear properties of UHMW-PE materials. Therefore, researchers have developed numerous approaches [7,9,10] to enhance the wear resistance of UHMW-PE, and one effective method has been the incorporation of reinforcing fillers.

Carbon-based materials, such as graphite [11], graphene nanoplatelets (GNPs) [12,13], graphene oxide (GO) [14], carbon nanotubes (CNTs) [15], and carbon fiber (CFs) [16], have not only exceptional mechanical and tribological properties, but also excellent biocompatibility, which have been utilized in reinforcing fillers for polymers. Currently, much study has already been conducted on this subject. Adding GO to UHMW-PE can significantly improve the hardness [17],yield strength [18], and anti-wear resistance [19]. UHMW-PE-filled CNTs [7] can efficiently enhance the tribological properties of UHMW-PE. As a superior reinforcement material for UHMW-PE, graphene can improve the tensile and creep-resistance properties of UHMW-PE [20]. And graphene-UHMW-PE composites have also shown higher hardness and lower friction [21] than pure UHMW-PE. Furthermore, Reddy K.S.N. et al. [22] also pointed out that UHMW-PE-filled graphite could significantly improve its tribological behavior. According to the aforementioned research findings, it can be inferred that UHMW-PE-filled carbon materials have the potential to extend the service life of artificial joints.

Graphite fluoride (GrF) is an important derivative of graphite with the molecular formula $(CF_x)_n$ ($0 < x < 1.25$), which is obtained by the direct reaction of graphite with fluoride gas at a controlled pressure and temperature. During the chemical reaction process, fluorine atoms bind with carbon atoms to form covalent C-F bonds, which result in the conversion of some C-C bonds from sp2 to sp3 hybridization [23]. The presence of F atoms in graphite endows GrF with unique lubrication properties, excellent chemical stability, extremely low surface energy, outstanding thermal conductivity, and stability. As a result, GrF has gained substantial attention in recent years, and it has already been employed in a variety of disciplines such as solar cells [24], batteries [25], sensors [26], supercapacitors [27], hydrophobic coatings [28], and lubricants [29]. Among the various applications, GrF, a well-known solid lubricant, has gained increasing attention. In comparison to graphite and molybdenum disulfide (MoS2), GrF has demonstrated a superior lubrication performance relative to conventional solid lubricants due to its expanded interlamellar space [30] and its low surface energy [31]. Burnished GrF on stainless-steel disks showed lower friction coefficients than MoS2 and graphite, and the wear lives of GrF film were also longer than them at 400 °C [32]. At high temperature, polyimide-bonded GrF [33] showed a lower friction and longer lifetime than polyimide-bonded MoS2. Comparing and analyzing the tribological properties of polytetrafluoroethylene (PTFE)-GrF and PTFE-MoS$_2$ composites, Yan et al. [34] found that GrF could considerably reduce the friction coefficient and enhance the anti-wear performance of PTFE in comparison to MoS2. These experiments focused primarily on the lubricant properties of GrF at a high temperature. However, the lubricative property of GrF in polymer composites has been little-reported at room temperature, and the anti-wear mechanism of GrF-polymer composites is ambiguous. In addition, the effect of GrF addition content on their tribological properties remains unclear.

According to our knowledge, there have been few reports on GrF-UHMW-PE composites. In this study, GrF is used as the solid lubricant additive and incorporated into a matrix of UHMW-PE. GrF-UHMW-PE composites are prepared. The effect of the GrF content on the mechanical and tribological properties of a UHMW-PE matrix was systematically investigated.

2. Materials and Methods

2.1. Materials

The Commercial UHMW-PE (GRU1050) powder was offered by Ticona/Celanese (Dallas, TX, USA). GrF powder with a 52–60 wt% fluoride content was purchased from Nanjing XFNANO Materials Technology Co., Ltd. (Nanjing, China). Other experimental reagents were analytically pure.

2.2. Preparation of GrF-UHMW-PE Composites

The percentage of GrF added plays an important role in the properties of the polymer. Sun et al. [35] reported that the incorporation of 0.5 wt% GrF into polyamide 6 made an optimal enhancement to its tensile strength and elastic modulus. Moreover, they [36] added 0.5 wt% and 1.0 wt% of GrF to polyamide 66 and observed a significant improvement in its tribological properties. In accordance with the aforementioned literature, the current study involved the incorporation of GrF at varying percentages: 0 wt%, 0.1 wt%, 0.5 wt%, and 1.0 wt%.

GrF-UHMW-PE composites were prepared as follows. Various amounts, i.e., 0 wt%, 0.1 wt%, 0.5 wt%, 1.0 wt% of GrF, were mixed with the UHMW-PE powder. The powder mixture was firstly mixed evenly using mechanical ball milling, which was operated for 5 h at a rotation speed of 400 rpm at room temperature. Then, the uniformly dispersed mixture was placed into a prefabricated metal mold and cold-pressed under 5 MPa for 15 min at room temperature. Immediately after, the mixture was transferred into a hot-air oven and heated for 2 h at 200 °C. After the heating process, the sample was taken out into air and pressed at 10 MPa until it cooled down to room temperature. According to the process described above, GrF-UHMW-PE composites were finally obtained, and are shown in Figure 1.

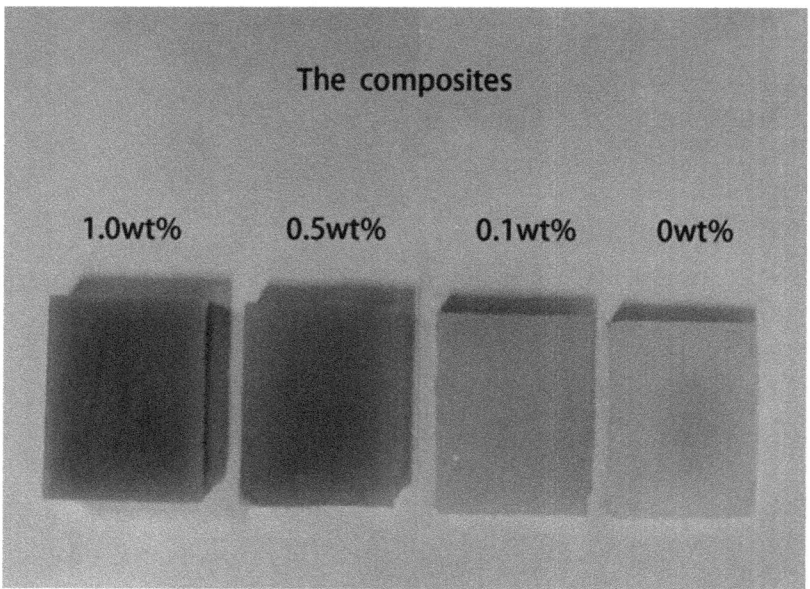

Figure 1. Photograph of GrF-UHMW-PE composites.

2.3. X-ray Diffraction (XRD) and Fourier-Transform Infrared Spectroscopy (FT-IR)

The XRD data of the samples were recorded with a Bruker D8 diffraction using a Cu K-alpha radiation. The FT-IR data of the samples were measured in Attenuated Total Reflectance (ATR) mode, by using an Alpha FT-IR spectrometer (Bruker Corporation, Ettlingen, Germany). The spectra were recorded at room temperature in a wave number range of 4000–600 cm^{-1}.

2.4. Ball Indentation Hardness

The ball indentation hardness (H) of the composites was determined based on the ISO 2039-73 standard [37]. The hardness values were measured as follows. In this experiment, a Si_3N_4 ball with a diameter of 5 mm was used as the indenter. At the beginning of testing, an initial preload of 9.8 N was carried out for 30 s. Immediately after, the sample was loaded to a peak force of 132 N and the force remained constant for 30 s. Finally, the sample was unloaded. The hardness (H) for each sample was calculated according to Equation (1):

$$H = \frac{P \propto}{h_r \pi D(h - h_0)}; \quad h_0 = h_r - \alpha \tag{1}$$

where P was the test load in N/mm, h was the maximum indentation depth in mm, D (=5) was the ball diameter in mm, α (=0.21) was a constant and h_r (=0.25) was the reduced depth of impression. Each sample's hardness test was repeated five times and the final mean value obtained from the five test results was calculated.

2.5. Tribological Testing

The contact model for the friction process is shown in Figure 2. When external force (F_n) was applied to the Si_3N_4 ceramic ball, the force and the average pressure ($\overline{p}$) between the sample and the Si_3N_4 ball could be calculated based on Hertz contact theory [38]. The calculated equations were as follows:

$$r = \sqrt[3]{\frac{3 \cdot F_n \cdot R}{4 \cdot E^*}} \tag{2}$$

$$\frac{1}{E^*} = \frac{1-v_1^2}{E_1} + \frac{1-v_2^2}{E_2} \tag{3}$$

$$\frac{1}{R} = \frac{1}{R_1} + \frac{1}{R_2} \tag{4}$$

$$\overline{P} = \frac{F}{\pi \cdot r^2} \tag{5}$$

where $\overline{p}$ was the average pressure in MPa, r was the contact radius in mm, F_n was the external force in N, E^* was the Effective Young's modulus in GPa, E_1 = (1 GPa) was Young's modulus of UHMW-PE in GPa, E_2 = (300 GPa) was Young's modulus of the Si_3N_4 ball in GPa, R_1 (=4.75) was the radius of the Si_3N_4 ball in mm, R_2 (=+∞) was the radius of the sample in mm, v_1 (=0.24) was Poisson's ratio of the Si_3N_4 ball and v_1 (=0.46) was Poisson's ratio of UHMW-PE. According to the research of Radovan et al. [39], the maximum contact pressure range between the ceramic ball and UHMW-PE in artificial joints during human rapid motion is 45.1 to 58.2 MPa. Therefore, based on the aforementioned equation, the predicted normal force applied was 30 N during tribological testing.

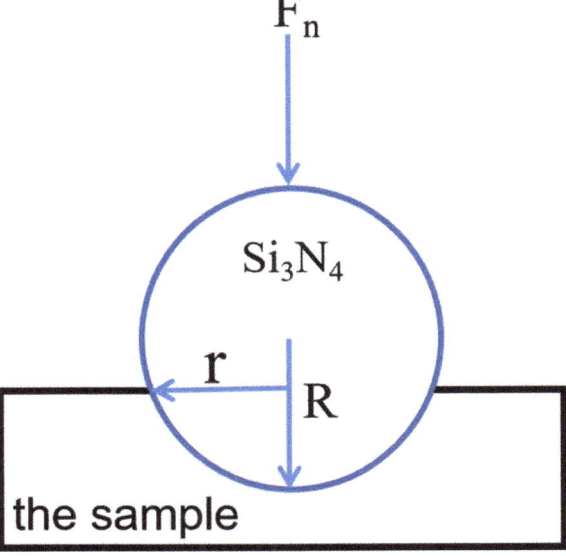

Figure 2. The contact model.

The friction and wear tests were performed on an Rtec MFT-5000 multi-function tribometer (Rtec Instruments, San Jose, CA, USA) in a linear reciprocating ball-on-flat mode. Figure 3 shows a schematic diagram of reciprocating testing. A Si_3N_4 ceramic ball with a diameter of 9.5 mm was used as the composite block counterpart. The tribological test parameters were as shown below: normal load (F_n) of 30 N, sliding frequency of 1 Hz [40], stroke length of 13 mm, and test time of 60 min [40]. During the tribological process, friction force (F_f) was recorded automatically by the MFT-500 tribometer, and the coefficient of friction (COF) was calculated simultaneously according to Equation (6):

$$\mu = \frac{F_f}{F_n} \qquad (6)$$

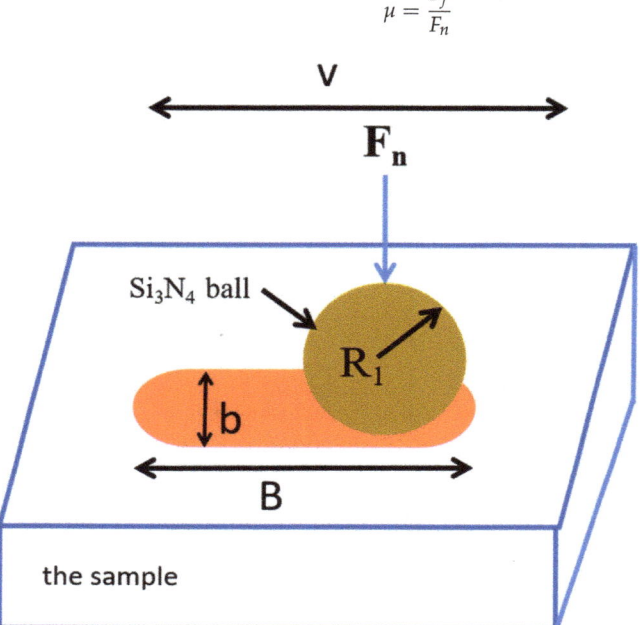

Figure 3. Schematic diagram of reciprocating testing.

After tribological tests, the wear volume (V) loss was calculated using Equation (7):

$$V = B\left[\frac{\pi R_1^2}{180}\arcsin\left(\frac{b}{2R_1}\right) - \frac{b}{2}\sqrt{R_1^2 - \frac{b^2}{4}}\right] \qquad (7)$$

where B, b, and R_1 indicated the wear length, wear width, and the radius of a Si_3N_4 ball, respectively. The 3D and 2D profiles of the wear track were obtained using an UP-3000 3D Optical Profilometer (Rtec Instruments, USA). B and b can be calculated according to these profiles. The average wear rate (K_0) was calculated according to the following Equation (8):

$$K_0 = \frac{V}{F_n \times L} \qquad (8)$$

where L represented the total sliding distance.

2.6. Scanning Electron Microscope (SEM)

The surface morphology of the tensile cross-section and worn surface was observed using a Scanning Electron Microscope (SEM, ZEISS Sigma HD, Oberkochen, Germany).

The samples were placed a disc, sprayed with gold and then transferred to the SEM sample platform.

3. Results and Discussion

3.1. SEM of Tensile Sections

In order to investigate the dispersion of GrF in the composites, the surface morphology of tensile cross sections was observed using SEM, as depicted in Figure 4. According to Figure 4, it found that GrF had good dispersion. It was noted that when GrF was a high filler (1.0 wt%), GrF particles were near together and might have formed the aggregation.

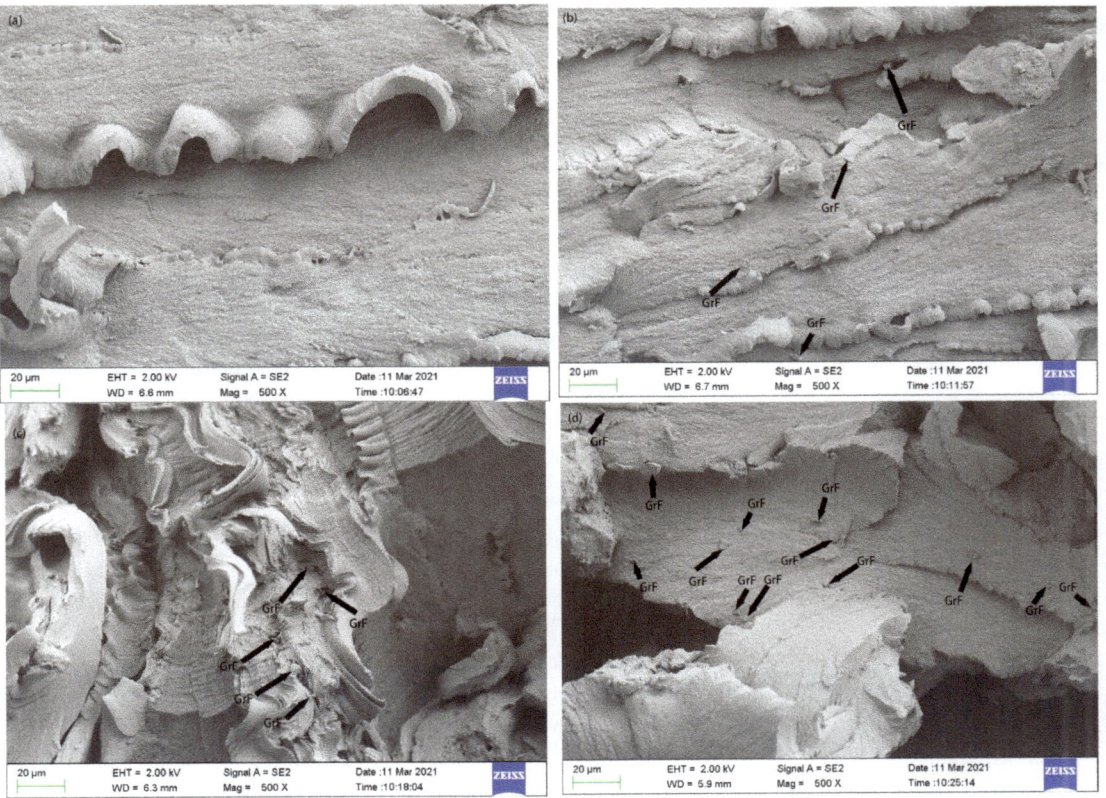

Figure 4. SEM of tensile sections: (**a**) Pure UHMW-PE, (**b**) filled 0.1 wt% GrF, (**c**) filled 0.5 wt% GrF, and (**d**) filled 1.0 wt% GrF.

3.2. XRD Patterns of the Composites

XRD is a powerful tool that we used to investigate the phase structure of materials and the dispersion of the filler in the polymer matrix. Figure 5 displays the XRD patterns of GrF and polymer composites. The laminated structure of GrF showed two characteristic diffraction peaks at $2\theta = \sim 12.85°$ and $\sim 40.70°$, which were assigned to the (001) and (100) reflections, respectively. The strongest (001) diffraction peak indicated a hexagonal system compound with a much higher fluorine content. Furthermore, a broad (002) diffraction peak suggests that GrF exhibited poor ordering along the stacking direction. Moreover, a very weak peak appeared at $2\theta = \sim 27.46°$, indicating the presence of non-fluorinated graphite. Figure 5 also demonstrates that the main XRD peaks of UHMW-PE were located at 2θ = ~21.39°, 24.05°, and 36.20°, which correspond to the (110), (200) and (020) crystal planes of UHMW-PE, respectively. According to XRD peak analysis, it was observed that the addition of GrF to UHMW-PE showed no noticeable change, suggesting that GrF was uniformly dispersed in the UHMW-PE matrix and a partially intercalated and exfoliated structure was formed in the composites. Similar XRD results were also reported for GO-UHMW-PE [41], graphene-UHMW-PE [42], and CNTs-UHMW-PE [43].

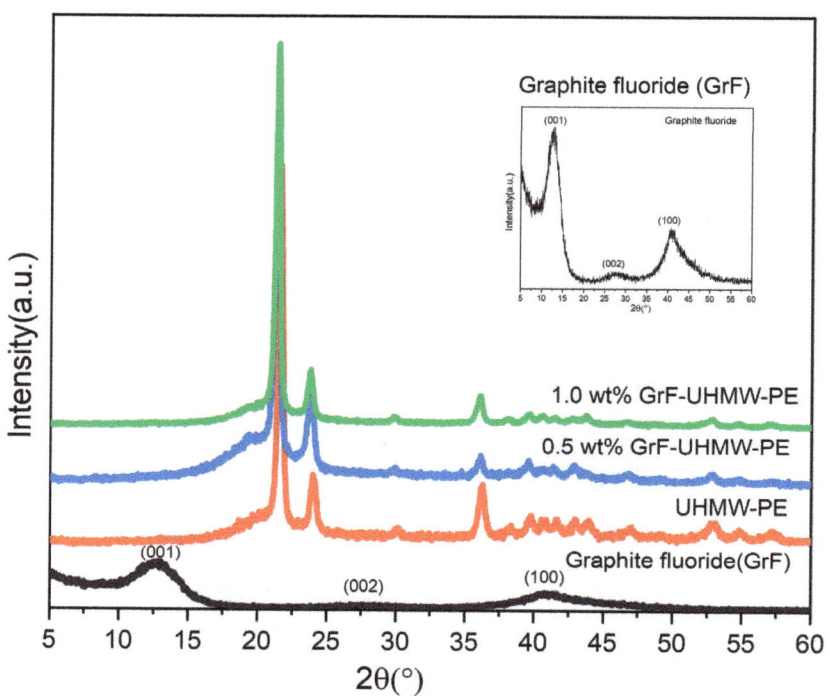

Figure 5. The XRD patterns of GrF powder and the composites.

3.3. FT-IR Spectrometry of the Composites

Figure 6 depicts the results of the FT-IR analysis of GrF and the composites, which aimed to investigate the impact of GrF on the structure of UHMW-PE. GrF exhibited two characteristic FT-IR peaks. The strongest peak at 1206 cm^{-1} was attributed to the F-C stretching vibration, while the overlapping peak at 1322 cm^{-1} represented the vibration of –CF$_2$ groups. Pure UHMW-PE has four main peaks at 720 cm^{-1}, 1462 cm^{-1}, 2840 cm^{-1}, and 2918 cm^{-1}, assigned to –CH$_2$ rock, C-H bending, C-H stretching and C-H stretching, respectively. However, when GrF (1.0 wt.%) was added to UHMW-PE, two new peaks at 1205 cm^{-1} and 1217 cm^{-1} emerged in the FT-IR spectrometry of UHMW-PE. The signal at 1205 cm^{-1} was obviously the typical absorption peak of GrF, showing that GrF was uniformly dispersed in the UHMW-PE matrix.

The emergence of a new peak at 1217 cm^{-1} in close proximity to 1205 cm^{-1} may be attributed to the formation of hydrogen bonds between GrF platelets and UHMW-PE molecular chains. A proposed mechanism for the formation of hydrogen bonds is illustrated in Figure 7. During the manufacturing process, F atoms on the GrF platelets could react with H atoms on UHMW-PE molecules to form C-F···H-C hydrogen bonds, which resulted in a new peak at 1217 cm^{-1} appearing. Similar experimental results have also been observed in GrF-PV6 composites [35]. The interaction between GrF and PV6 led to the formation of the hydrogen bonds, which contributed to the good adhesion and dispersion of GrF in the PV6 matrix. In addition, Thalladi et al. [44] noted that certain fluorinated compounds contained hydrogen bonds that could impact the properties of the compound. Thus, it was suggested that hydrogen bonding in the GrF-UHMW-PE composites could enhance the interaction between the fillers and the matrix and improve their mechanical and tribological properties.

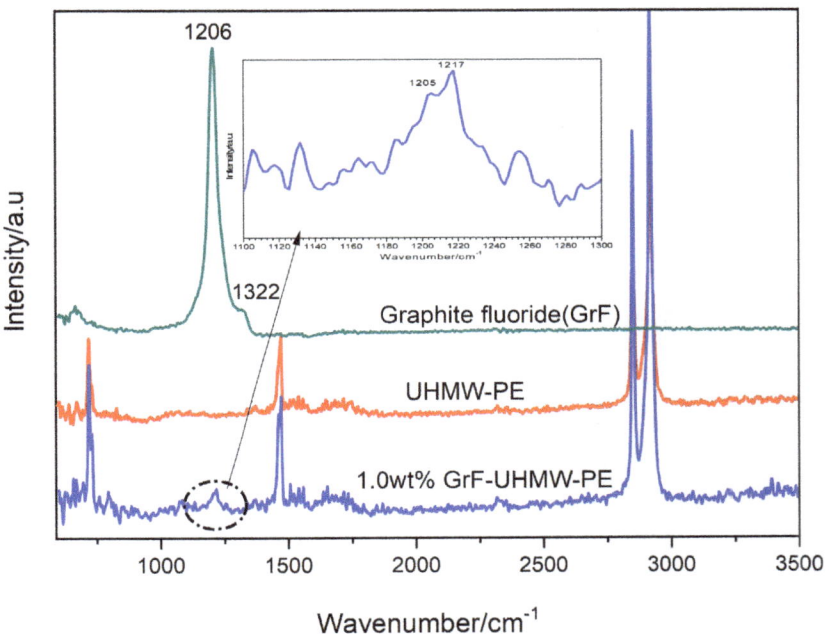

Figure 6. FT-IR spectrometry of GrF powder and the composites.

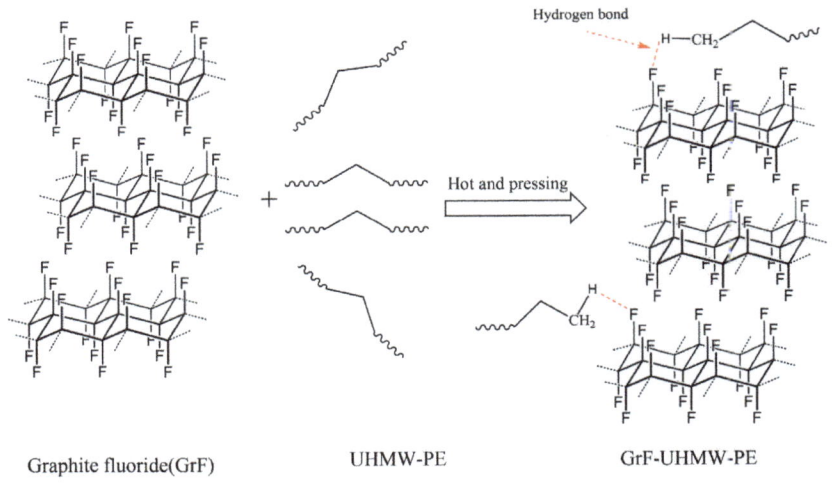

Figure 7. Proposed hydrogen bond formation mechanism.

3.4. The Hardness of the Composites

In order to investigate the effect of GrF filler on the mechanical properties of the UHMW-PE matrix, the experimental results of hardness are shown in Figure 8. As can be noted, the hardness of the GrF-UHMW-PE composites gradually increased with increasing GrF filler concentrations. The incorporation of 0.1 wt% GrF significantly increased the hardness of pure UHMW-PE and the hardness increased from an initial value of 23.97 MPa to 29.77 MPa, an increment of 24%. With the further addition of GrF to a concentration of 1.0 wt%, the hardness showed a significant increase to 32.95 MPa with an increment of 37%. These experimental results indicated that adding a lower amount of GrF could remarkably enhance the hardness of UHMW-PE.

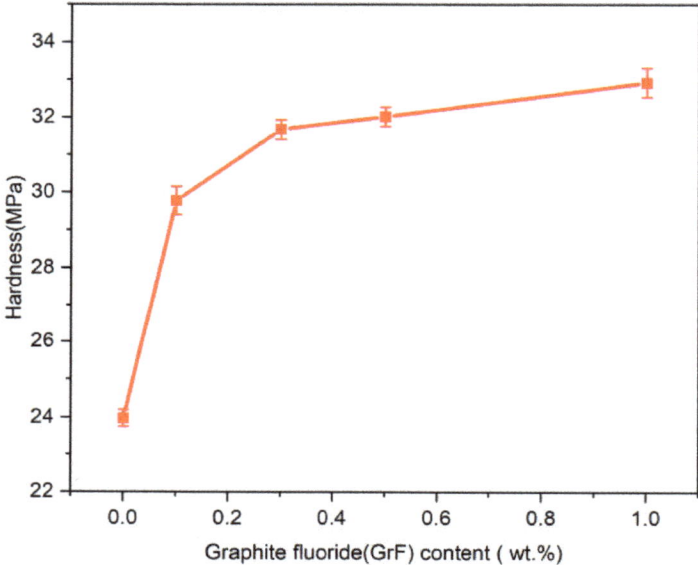

Figure 8. The hardness of the composites.

The hardness results of GrF-UHMW-PE composites showed a similar trend to those of the GO-UHMW-PE [45] and graphite-UHMW-PE [46] composites. According to the findings of Chen et al., [45] the addition of 1.0 wt.% GO resulted in an approximately 15% increase in the hardness of UHMW-PE. Similarly, Lorenzo-Bonet et al. [46] reported that the addition of 1.0 wt.% graphite enhanced the hardness of UHMW-PE by approximately 35%. In general, carbon fillers including GO, graphite, and GrF, exhibited superior mechanical properties compared to pure UHMW-PE. The presence of these enhanced particles in the composites allowed them to bear partial loads and facilitate the transmission of loads within the UHMW-PE matrix. As a result, incorporating these enhanced particles significantly enhanced the mechanical properties of UHMW-PE. Furthermore, in the case of the GrF-UHMW-PE composites, the presence of hydrogen bonds enhanced interface interactions that restricted the molecular-chain mobility of UHMW-PE around GrF platelets. This interaction was beneficial for transmitting stress from the UHMW-PE matrix to GrF, thereby enhancing the mechanical performance of the UHMW-PE matrix.

3.5. Tribological Properties of the Composites

3.5.1. The Coefficient of Friction (COF)

The addition of GrF had a significant impact on the COF of UHMW-PE, as depicted in Figure 9. Figure 9 illustrates the variation curve of the COF curve as a function of time and the average COF values at different filler contents. Three samples with filler contents lower than 0.5 wt% initially experienced a rapid decrease in COF for a brief period, followed by a gradual increase. After a running-in period, the COF curve stabilized. However, when 1.0 wt% GrF was added, the COF initially increased for 1100 s and then remained stable. Additionally, it was observed that the curves for pure UHMW-PE, 0.1 wt% GrF-UHMW-PE, and 0.5 wt% GrF-UHMW-PE were relatively flatter compared to the curve of 1.0 wt% GrF-UHMW-PE.

The average COF of pure UHMW-PE was approximately 0.1020. The addition of a small amount of GrF (<0.5 wt%) led to a significant reduction in the COF of UHMW-PE. Notably, when 0.5 wt% of GrF was added, the lowest COF of 0.0666 was achieved. This represented a 34.76% decrease compared to pure UHMW-PE. However, further increasing the content of GrF to 1.0 wt% resulted in a significant increase in the COF, raising it from 0.0666 to 0.1402. This corresponded to a 37.31% increase compared to pure UHMW-PE.

It was well-known that UHMW-PE has a lower COF due to its self-lubricating properties [47]. Although carbon fillers including graphite, graphene, GO, and CNTs, have been shown to have excellent lubricating performances, their impact on the lubricity of UHMW-PE was generally limited and might have a negative effect in some circumstances. For example, when layered graphite [46] (0.1 wt%) was added to UHMW-PE, only a 13% reduction in COF was observed. However, based on the results of the above experiment, the addition of 0.5 wt% GrF resulted in a reduction of the COF of UHMW-PE to 35%. GrF, as an excellent solid lubricant, can reduce COF even in small amounts. The reason can be attributed to the characteristic structure of GrF. The fluoride atom bonded to the carbon atom in GrF resulted in an interlayer spacing of (001) planes of 6.88 Å (the above experimental results), which was larger than the d-spacing of graphite (3.40 Å) [48]. As a result, the expansion of the carbon layer planes facilitates sliding between the carbon layers under reciprocal frictional forces. In addition, during the friction process, GrF on the worn surface of the GrF-UHMW-PE composites was exposed and released, acting as a solid lubricant to reduce COF. The transfer films formed on the surface of the Si_2N_3 counterpart ball can also help reduce COF. On the contrary, a high filler content (1.0 wt%) in UHMW-PE could produce larger aggregated particles and increased surface roughness during the wear process (see SEM of worn surfaces), resulting in a higher COF than pure UHMW-PE.

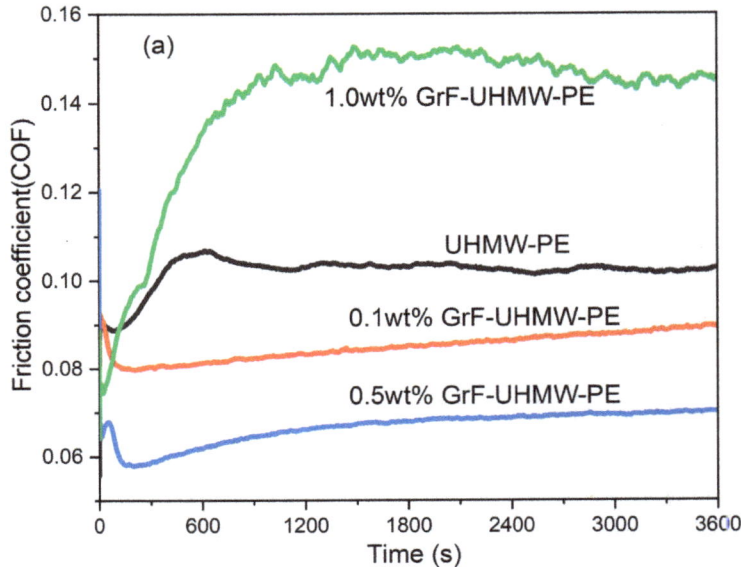

Figure 9. *Cont.*

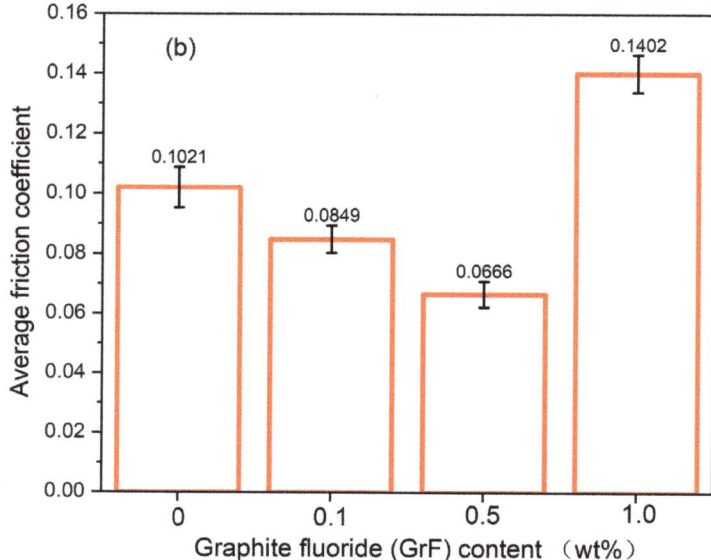

Figure 9. (a) Variation in COF with time, and (b) average COF.

3.5.2. The Wear of the Composites

In the reciprocating sliding tests, wear scars were produced by the Si_3N_4 counterparts. Figure 10 depicted the wear scars' three-dimensional and two-dimensional topography. The influence of the addition of GrF on the depth of wear scars can be observed in Figure 10. Adding 0.1 wt% GrF can significantly reduce the depth of wear. The further addition of GrF resulted in a further reduction in the depth of the wear scars, and the depth became the shallowest when the addition reached 0.5 wt%. However, when the addition amount was 1.0 wt% GrF, the depth of the composites was deeper than that of the composites with 0.5 wt% GrF.

Figure 11 depicts the composites' wear rates based on the results of Equation (8). The wear rates were determined by the amount of GrF. The wear rate of pure UHMW-PE was measured to be 8.13×10^{-5} mm^3 N^{-1} m^{-1}. With the addition of 0.1 wt% GrF in the UHMW-PE matrix, the wear rate decreased by 16.48%. Increasing the addition to 0.5 wt% GrF resulted in the wear rate reaching a minimum value of 4.25×10^{-5} mm^3 N^{-1} m^{-1}, reducing it by 47.72% compared to pure UHMW-PE. In contrast, raising the GrF content form 0.5 wt% to 1.0 wt% increased the wear rate. However, the wear rate was still lower in comparison to that of UHMW-PE.

These results of wear rates demonstrated that a lower addition of GrF had the potential to enhance the anti-wear performance of UHMW-PE. This enhancement can be attributed to the exceptional lubricant properties and mechanical characteristics of GrF. By acting as a solid lubricant, GrF effectively decreased the COF of UHMW-PE, resulting in a significant reduction in wear rate. Furthermore, the addition of GrF improved the hardness of the composites, thereby enhancing their tribological properties. In general, the wear resistance of the composites strongly influenced by their hardness [49]. Therefore, increasing the hardness further improved their anti-wear resistance. However, it is important to note that the high addition of GrF (1.0 wt%) led to accelerated wear, which can be attributed to its aggregation. The addition of 1.0 wt% GrF caused the formation of larger aggregated particles, leading to a high COF. This effect was stronger than that of solider lubrication.

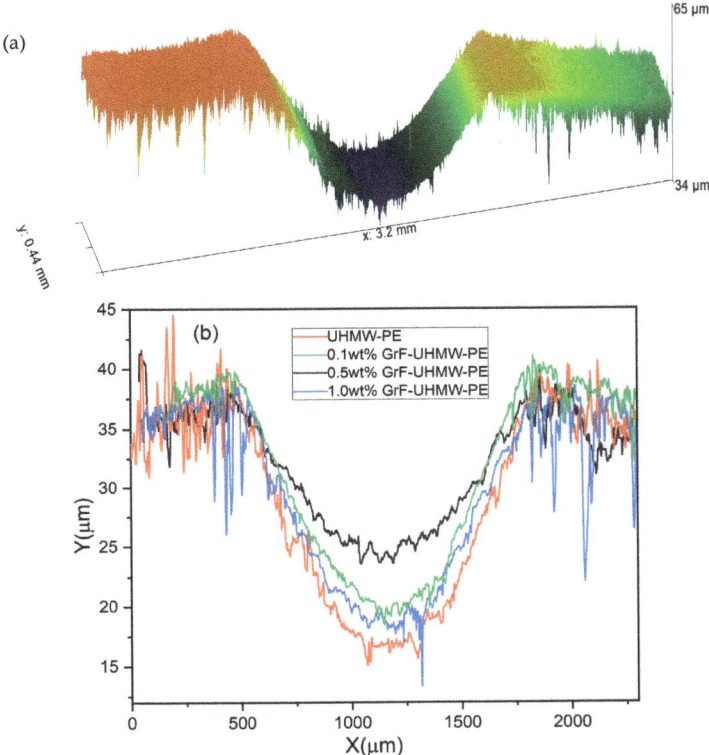

Figure 10. Wear profiles of the composites: (**a**) 3D profile of wear track, (**b**) 2D profile of wear track.

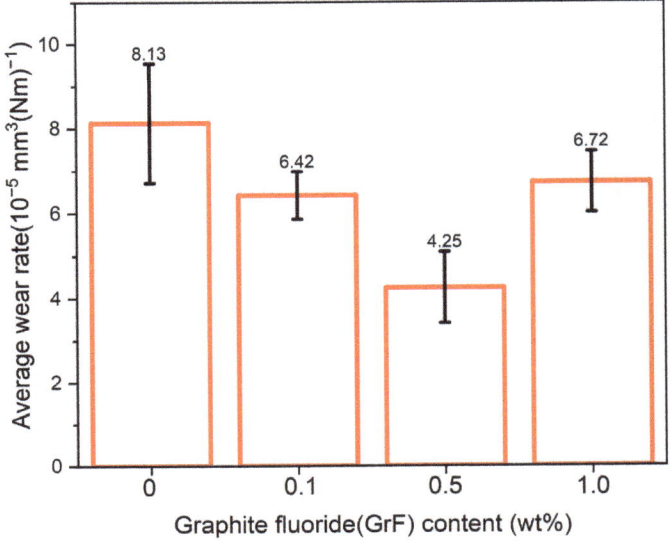

Figure 11. Wear rates of the composites.

3.5.3. SEM of Worn Surfaces and Wear Mechanism of the Composites

In order to investigate the wear mechanism of the composites, Figure 12 displays SEM images of the worn surface morphology of pure UHMW-PE and its composites with various GrF contents. As can be easily observed, the worn surface of pure UHMW-PE (Figure 12a) exhibited a number of cracks and delamination, which indicates that fatigue wear was the dominant mechanism for UHMW-PE. However, the incorporation of 0.1 wt% GrF into the UHMW-PE matrix significantly improved the worn surface. No visible cracks or delamination were found (Figure 12b), while some deeper scratches and extensive small pitting appeared, which suggests that the main wear mechanisms for this composite were fatigue wear and abrasive wear. As the GrF content increased to 0.5 wt%, the worn surface became relatively smooth (Figure 12c). As a result, it had a lower COF, which was consistent with the experimental results. Some pitting was also observed on the worn surface. Fatigue wear remained the predominant wear mechanism. Further increasing the GrF content to 1.0 wt% resulted in a significant deterioration of the worn surface (Figure 12d). Deeper abrasive grooves and larger wear debris particles characterized the severe wear scar surface, leading to increased surface roughness. Consequently, the COF also increased, which was consistent with the experimental results. The primary wear mechanism observed under these conditions was abrasive wear.

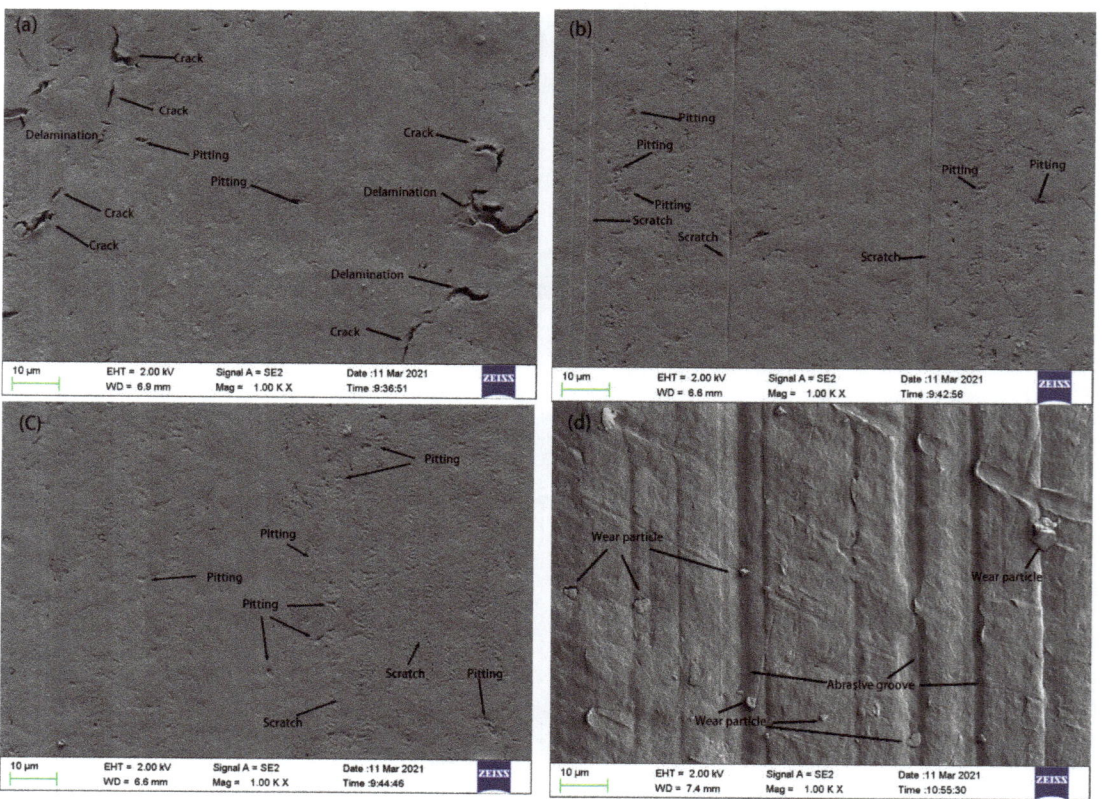

Figure 12. (**a**) Pure UHMW-PE, (**b**) filled 0.1 wt% GrF, (**c**) filled 0.5 wt% GrF, and (**d**) filled 1.0 wt% GrF.

In general, the main wear mechanism in polymer composites was abrasive wear [50] and fatigue wear [51]. These types of wear were characterized by the presence of cracks, scratches, furrows, and plastic deformation along the worn surface, which were caused by point contact loads. During the friction between the polymer and Si_2O_3 ceramic ball, heat generation occurred, and as a result, thermal softening took place along the friction surface, which led to plastic deformation, scratches, and furrows on the wear surface. Figure 13 displayed a diagrammatic representation of the wear mechanisms. For the pure UHMW-PE friction process, the transfer membranes were formed on a ceramic ball and the friction process belonged to homogeneous material interaction. Fatigue cracks and delamination occurred under reciprocal cyclic stress due to the low mechanical properties of pure UHMW-PE. However, once heterogeneous GrF (0.1 wt%) was added and the mechanical properties of UHMW-PE were enhanced, abrasive scratches occurred because tougher wear particles, transfer films, and samples formed a three-body wear that was more likely to form scratches under cyclic stress. As the cycling process continued, micro-cracks were generated and propagated deeply into the polymer matrix, until one crack became large enough to break from the polymer, resulting in the occurrence of pitting. The addition of 0.5 wt% GrF to UHMW-PE matrix resulted in improved mechanical properties that effectively mitigated the detrimental effects of fatigue wear and abrasive wear, which led to a significant reduction in the occurrence of scratches and pitting. As a result, the worn surface was improved and exhibited a smoother surface. However, with the addition of GrF at higher concentrations, much larger wear particles were easily generated during the friction process due to the aggregation, which exacerbated three-body abrasion wear. Consequently, deeper abrasive grooves developed and caused an increase in the COF.

According to the above results, it can be concluded that the composites containing a lower content of GrF could efficiently enhance anti-wear properties and demonstrate resistance against fatigue-induced wear. The wear mechanisms were strongly dependent on GrF content. With the addition of GrF, the wear mechanism of the composites transformed from fatigue wear to abrasive wear.

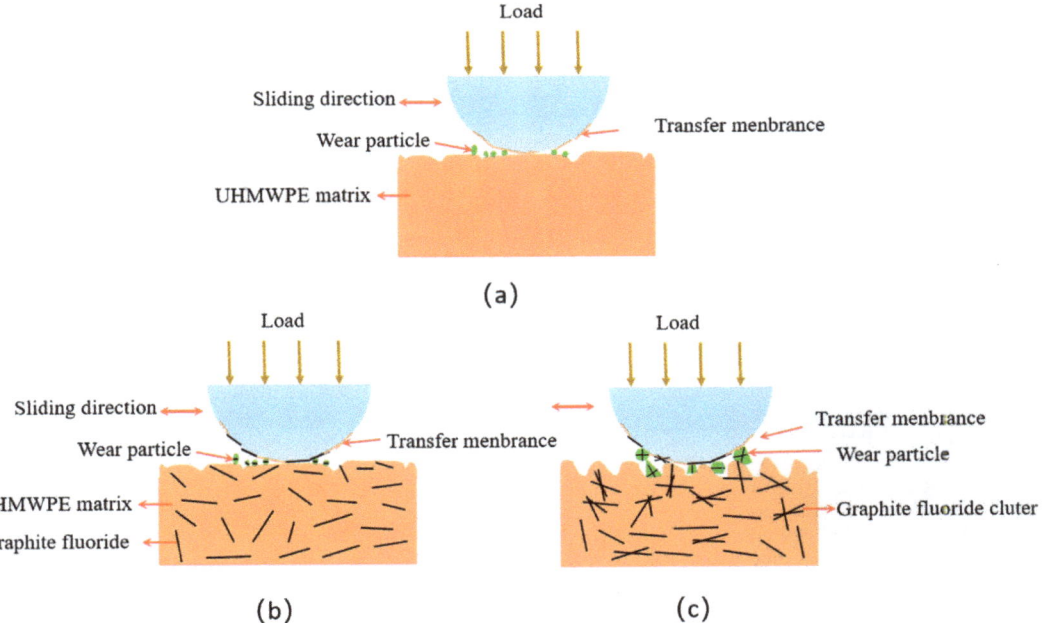

Figure 13. Schematic illustration of the wear mechanisms. (**a**) No addition, (**b**) lower addition (<=0.5 wt%), and (**c**) higher addition (1.0 wt%).

4. Conclusions

GrF, a novel solid lubricant, was incorporated into UHMW-PE and the GrF-UHMW-PE composites were successfully prepared. The structure, hardness, and tribological properties of the composites were investigated. The results were as follows:

1. The hydrogen bonds (C-F···H-C) were formed and enhanced interfacial interaction between GrF and UHMW-PE, according to the results of FT-IR spectrometry.
2. Adding GrF can significantly enhance the hardness of the composites. At 1.0 wt% GrF, the hardness increased by 37%, compared to pure UHMW-PE.
3. At a lower GrF concentration, GrF exhibited remarkable lubricant properties and anti-wear properties. At 0.5 wt% GrF concentration, the COF and wear rate were reduced by 34.76% and 47.72%, respectively, compared to the UHMW-PE. At a high GrF concentration, the COF significantly increased while the anti-wear properties decreased.
4. As the concentration of GrF increased, the wear modes of the composites transitioned from fatigue wear to abrasive wear.

Author Contributions: Conceptualization, G.H. and Y.C.; methodology, G.H.; software, H.H.; validation, G.H., F.Y. and Y.Z.; formal analysis, G.H.; investigation, G.H.; resources, G.H.; data curation, F.Y.; writing—original draft preparation, G.H.; writing—review and editing, G.H.; visualization, H.H.; supervision, Y.Z.; project administration, T.Z.; funding acquisition, T.Z. All authors have read and agreed to the published version of the manuscript.

Funding: This research was funded by the Jiangsu Natural Science Foundation (No. BK20201142), Topnotch Academic Programs Project of Jiangsu Higher Education Institutions (Grant no. PPZY2015B186), Suzhou Key Laboratory Support Project (Grant no. SZS201815) and the National Natural Science Foundation of China (Grant no. 51675232).

Data Availability Statement: Not applicable.

Conflicts of Interest: The authors declare no conflict of interest.

References

1. Kurtz, S.M. *UHMWPE Biomaterials Handbook: Ultra High Molecular Weight Polyethylene in Total Joint Replacement and Medical Devices*; Academic Press: Cambridge, MA, USA, 2009.
2. Anaya-Garza, K.; Domínguez-Crespo, M.; Torres-Huerta, A.; Brachetti-Sibaja, S.; Moreno-Palmerin, J. UHMWPE/OPA Composite Coatings on Ti6Al4V Alloy as Protective Barriers in a Biological-Like Medium. In *Environmental Concerns and Remediation: Proceedings of F-EIR Conference 2021*; Springer: Cham, Switzerland, 2022; pp. 1–12.
3. Liu, Z.; Du, Y.; Ma, H.; Li, J.; Zhang, X.; Zhu, E.; Shi, C.; Zhu, Z.; Zhao, S. Mechanism of boron carbide particles improving the wear resistance of UHMWPE: Structure-property relationship. *Polymer* **2022**, *245*, 124733. [CrossRef]
4. Rahman, M.M.; Biswas, M.A.S.; Hoque, K.N. Recent development on micro-texturing of UHMWPE surfaces for orthopedic bearings: A review. *Biotribology* **2022**, *31*, 100216. [CrossRef]
5. Patil, N.A.; Njuguna, J.; Kandasubramanian, B. UHMWPE for biomedical applications: Performance and functionalization. *Eur. Polym. J.* **2020**, *125*, 109529. [CrossRef]
6. Bistolfi, A.; Giustra, F.; Bosco, F.; Sabatini, L.; Aprato, A.; Bracco, P.; Bellare, A. Ultra-high molecular weight polyethylene (UHMWPE) for hip and knee arthroplasty: The present and the future. *J. Orthop.* **2021**, *25*, 98–106. [CrossRef]
7. Baena, J.C.; Wu, J.; Peng, Z. Wear performance of UHMWPE and reinforced UHMWPE composites in arthroplasty applications: A review. *Lubricants* **2015**, *3*, 413–436. [CrossRef]
8. Kandahari, A.M.; Yang, X.; Laroche, K.A.; Dighe, A.S.; Pan, D.; Cui, Q. A review of UHMWPE wear-induced osteolysis: The role for early detection of the immune response. *Bone Res.* **2016**, *4*, 16014. [CrossRef]
9. Dwivedi, Y.; Laurent, M.P.; Sarvepalli, S.; Schmid, T.M.; Wimmer, M.A. Albumin protein cleavage affects the wear and friction of ultra-high molecular weight polyethylene. *Lubricants* **2017**, *5*, 33. [CrossRef]
10. Dangsheng, X. Friction and wear properties of UHMWPE composites reinforced with carbon fiber. *Mater. Lett.* **2005**, *59*, 175–179. [CrossRef]
11. Wang, L.L.; Zhang, L.Q.; Tian, M. Mechanical and tribological properties of acrylonitrile–butadiene rubber filled with graphite and carbon black. *Mater. Des.* **2012**, *39*, 450–457. [CrossRef]
12. Moghadam, A.D.; Omrani, E.; Menezes, P.L.; Rohatgi, P.K. Mechanical and tribological properties of self-lubricating metal matrix nanocomposites reinforced by carbon nanotubes (CNTs) and graphene—A review. *Compos. Part B Eng.* **2015**, *77*, 402–420. [CrossRef]

13. Khun, N.W.; Zhang, H.; Lim, L.H.; Yang, J. Mechanical and tribological properties of graphene modified epoxy composites. *Appl. Sci. Eng. Prog.* **2015**, *8*, 101–109.
14. Miao, X.; Li, Z.; Liu, S.; Hou, K.; Wang, J.; Yang, S. Ionic bridging strengthened MXene/GO nanocomposite films with extraordinary mechanical and tribological properties. *Appl. Surf. Sci.* **2023**, *625*, 157181. [CrossRef]
15. Huang, Z.; Zheng, Z.; Zhao, S.; Dong, S.; Luo, P.; Chen, L. Copper matrix composites reinforced by aligned carbon nanotubes: Mechanical and tribological properties. *Mater. Des.* **2017**, *133*, 570–578. [CrossRef]
16. Zhou, S.; Zhang, Q.; Wu, C.; Huang, J. Effect of carbon fiber reinforcement on the mechanical and tribological properties of polyamide6/polyphenylene sulfide composites. *Mater. Des.* **2013**, *44*, 493–499. [CrossRef]
17. Pang, W.; Wu, J.; Zhang, Q.; Li, G. Graphene oxide enhanced, radiation cross-linked, vitamin E stabilized oxidation resistant UHMWPE with high hardness and tensile properties. *RSC Adv.* **2017**, *7*, 55536–55546. [CrossRef]
18. Ni, Z.; Pang, W.; Chen, G.; Lu, P.; Qian, S. The influence of irradiation on thermal and mechanical properties of UHMWPE/GO nanocomposites. *Russ. J. Appl. Chem.* **2017**, *90*, 1876–1882. [CrossRef]
19. Tai, Z.; Chen, Y.; An, Y.; Yan, X.; Xue, Q. Tribological behavior of UHMWPE reinforced with graphene oxide nanosheets. *Tribol. Lett.* **2012**, *46*, 55–63. [CrossRef]
20. Bhattacharyya, A.; Chen, S.; Zhu, M. Graphene reinforced ultra high molecular weight polyethylene with improved tensile strength and creep resistance properties. *Express Polym. Lett.* **2014**, *8*, 74–84. [CrossRef]
21. Chih, A.; Ansón-Casaos, A.; Puértolas, J. Frictional and mechanical behaviour of graphene/UHMWPE composite coatings. *Tribol. Int.* **2017**, *116*, 295–302. [CrossRef]
22. Reddy, K.S.N.; Unnikrishnan, D.; Balachandran, M. Investigation and optimization of mechanical, thermal and tribological properties of UHMWPE–graphite nanocomposites. *Mater. Today Proc.* **2018**, *5*, 25139–25148. [CrossRef]
23. Di Vittorio, S.; Dresselhaus, M.; Dresselhaus, G. A model for disorder in fluorine-intercalated graphite. *J. Mater. Res.* **1993**, *8*, 1578–1585. [CrossRef]
24. Yang, L.; Li, Y.; Wang, L.; Pei, Y.; Wang, Z.; Zhang, Y.; Lin, H.; Li, X. Exfoliated fluorographene quantum dots as outstanding passivants for improved flexible perovskite solar cells. *ACS Appl. Mater. Interfaces* **2020**, *12*, 22992–23001. [CrossRef] [PubMed]
25. Giraudet, J.; Claves, D.; Guérin, K.; Dubois, M.; Houdayer, A.; Masin, F.; Hamwi, A. Magnesium batteries: Towards a first use of graphite fluorides. *J. Power Sources* **2007**, *173*, 592–598. [CrossRef]
26. Kumaran, A.A.; Chithrambattu, A.; Vedhanarayanan, B.; Rajukrishnan, S.B.A.; Praveen, V.K.; Kizhakayil, R.N. Fluoride-philic reduced graphene oxide–fluorophore anion sensors. *Mater. Adv.* **2022**, *3*, 6809–6817. [CrossRef]
27. Zhao, F.-G.; Kong, Y.-T.; Pan, B.; Hu, C.-M.; Zuo, B.; Dong, X.; Li, B.; Li, W.-S. In situ tunable pillaring of compact and high-density graphite fluoride with pseudocapacitive diamines for supercapacitors with combined predominance in gravimetric and volumetric performances. *J. Mater. Chem. A* **2019**, *7*, 3353–3365. [CrossRef]
28. Lei, F.; Wu, B.; Sun, H.; Jiang, F.; Yang, J.; Sun, D. Simultaneously improving the anticorrosion and antiscratch performance of epoxy coatings with graphite fluoride via large-scale preparation. *Ind. Eng. Chem. Res.* **2018**, *57*, 16709–16717. [CrossRef]
29. Mittal, D.; Singh, D.; Sharma, S.K. Thermal characteristics and tribological performances of solid lubricants: A mini review. *Adv. Rheol. Mater.* **2023**. [CrossRef]
30. Fusaro, R.L.; Sliney, H.E. Graphite fluoride (CFx) n—A new solid lubricant. *Asle Trans.* **1970**, *13*, 56–65. [CrossRef]
31. Thomas, P.; Bilas, P.; Molza, A.; Legras, L.; Mansot, J.L.; Guérin, K.; Dubois, M. 14—Fluorinated Nanocarbons for Lubrication. In *New Fluorinated Carbons: Fundamentals and Applications*; Boltalina, O.V., Nakajima, T., Eds.; Elsevier: Boston, MA, USA, 2017; pp. 325–360.
32. Fusaro, R.L.; Sliney, H.E. *Preliminary Investigation of Graphite Fluoride (CF X) N as a Solid Lubricant*; National Aeronautics and Space Administration: Washington, DC, USA, 1969.
33. Fusaro, R.L.; Sliney, H.E. Lubricating characteristics of polyimide bonded graphite fluoride and polyimide thin films. *ASLE Trans.* **1973**, *16*, 189–196. [CrossRef]
34. Yan, Y.T.; Wang, R.; Song, W.L. The preparation and Tribological Properties of graphite fluoride. *Adv. Mater. Res.* **2014**, *941*, 1544–1547. [CrossRef]
35. Sun, H.; Jiang, F.; Lei, F.; Chen, L.; Zhang, H.; Leng, J.; Sun, D. Graphite fluoride reinforced PA6 composites: Crystallization and mechanical properties. *Mater. Today Commun.* **2018**, *16*, 217–225. [CrossRef]
36. Sun, H.; Li, T.; Lei, F.; Yang, M.; Li, D.; Huang, X.; Sun, D. Graphite fluoride and fluorographene as a new class of solid lubricant additives for high-performance polyamide 66 composites with excellent mechanical and tribological properties. *Polym. Int.* **2020**, *69*, 457–466. [CrossRef]
37. Lu, P.; Wu, M.; Ni, Z.; Huang, G. Oxidative degradation behavior of irradiated GO/UHMWPE nanocomposites immersed in simulated body fluid. *Polym. Bull.* **2021**, *78*, 5153–5164. [CrossRef]
38. Hertz, H. The contact of elastic solids. *J. Reine. Angew. Math.* **1881**, *92*, 156–171.
39. Zdero, R.; Bagheri, Z.S.; Rezaey, M.; Schemitsch, E.H.; Bougherara, H. The biomechanical effect of loading speed on Metal-on-UHMWPE contact mechanics. *Open Biomed. Eng. J.* **2014**, *8*, 28. [CrossRef]
40. Kapps, V.; Maru, M.M.; Kuznetsov, O.; Achete, C.A. Identifying differences in the tribological performance of GUR 1020 and GUR 1050 UHMWPE resins associated to pressure × velocity conditions in linear reciprocating sliding tests. *J. Mech. Behav. Biomed. Mater.* **2023**, *145*, 106038. [CrossRef] [PubMed]

41. Pang, W.; Ni, Z.; Chen, G.; Huang, G.; Huang, H.; Zhao, Y. Mechanical and thermal properties of graphene oxide/ultrahigh molecular weight polyethylene nanocomposites. *Rsc Adv.* **2015**, *5*, 63063–63072. [CrossRef]
42. Shafiee, M.; Aamazani SA, A. Optimization of UHMWPE/graphene nanocomposite processing using ziegler-natta catalytic system via response surface methodology. *Polym.-Plast. Technol. Eng.* **2014**, *53*, 969–974. [CrossRef]
43. Zhao, Y.; Wang, M.; Tang, Z.; Wu, G. Radiation effects of UHMW-PE fibre on gel fraction and mechanical properties. *Radiat. Phys. Chem.* **2011**, *80*, 274–277. [CrossRef]
44. Thalladi, V.R.; Weiss, H.-C.; Bläser, D.; Boese, R.; Nangia, A.; Desiraju, G.R. C−H···F Interactions in the Crystal Structures of Some Fluorobenzenes. *J. Am. Chem. Soc.* **1998**, *120*, 8702–8710. [CrossRef]
45. Chen, Y.; Qi, Y.; Tai, Z.; Yan, X.; Zhu, F.; Xue, Q. Preparation, mechanical properties and biocompatibility of graphene oxide/ultrahigh molecular weight polyethylene composites. *Eur. Polym. J.* **2012**, *48*, 1026–1033. [CrossRef]
46. Lorenzo-Bonet, E.; Hernandez-Rodriguez, M.A.L.; Perez-Acosta, O.; De la Garza-Ramos, M.A.; Contreras-Hernandez, G.; Juarez-Hernandez, A. Characterization and tribological analysis of graphite/ultra high molecular weight polyethylene nanocomposite films. *Wear* **2019**, *426–427*, 195–203. [CrossRef]
47. Aliyu, I.K.; Mohammed, A.S.; Al-Qutub, A. Tribological Performance of UHMWPE/GNPs Nanocomposite Coatings for Solid Lubrication in Bearing Applications. *Tribol. Lett.* **2018**, *66*, 144. [CrossRef]
48. Tsai, J.-L.; Tu, J.-F. Characterizing mechanical properties of graphite using molecular dynamics simulation. *Mater. Des.* **2010**, *31*, 194–199. [CrossRef]
49. Momber, A.W.; Irmer, M.; Marquardt, T. Effects of polymer hardness on the abrasive wear resistance of thick organic offshore coatings. *Prog. Org. Coat.* **2020**, *146*, 105720. [CrossRef]
50. Unal, H.; Sen, U.; Mimaroglu, A. Abrasive wear behaviour of polymeric materials. *Mater. Des.* **2005**, *26*, 705–710. [CrossRef]
51. Zhao, G.; Wang, T.; Wang, Q. Friction and wear behavior of the polyurethane composites reinforced with potassium titanate whiskers under dry sliding and water lubrication. *J. Mater. Sci.* **2011**, *46*, 6673–6681. [CrossRef]

Disclaimer/Publisher's Note: The statements, opinions and data contained in all publications are solely those of the individual author(s) and contributor(s) and not of MDPI and/or the editor(s). MDPI and/or the editor(s) disclaim responsibility for any injury to people or property resulting from any ideas, methods, instructions or products referred to in the content.

MDPI AG
Grosspeteranlage 5
4052 Basel
Switzerland
Tel.: +41 61 683 77 34

Lubricants Editorial Office
E-mail: lubricants@mdpi.com
www.mdpi.com/journal/lubricants

Disclaimer/Publisher's Note: The statements, opinions and data contained in all publications are solely those of the individual author(s) and contributor(s) and not of MDPI and/or the editor(s). MDPI and/or the editor(s) disclaim responsibility for any injury to people or property resulting from any ideas, methods, instructions or products referred to in the content.

www.ingramcontent.com/pod-product-compliance
Lightning Source LLC
LaVergne TN
LVHW070645100526
838202LV00013B/887